PLANT SCIENCE RESEARCH AND PRACTICES

THE BOOK OF FLAVONOIDS

PLANT SCIENCE RESEARCH AND PRACTICES

Additional books and e-books in this series can be found on Nova's website under the Series tab.

PLANT SCIENCE RESEARCH AND PRACTICES

THE BOOK OF FLAVONOIDS

CHAO-HUI FENG
AND
JUAN FRANCISCO GARCÍA MARTÍN
EDITORS

nova
science publishers
New York

DOI: https://doi.org/10.52305/WGCH6543

NOTICE TO THE READER

The Publisher has taken reasonable care in the preparation of this book, but makes no expressed or implied warranty of any kind and assumes no responsibility for any errors or omissions. No liability is assumed for incidental or consequential damages in connection with or arising out of information contained in this book. The Publisher shall not be liable for any special, consequential, or exemplary damages resulting, in whole or in part, from the readers' use of, or reliance upon, this material. Any parts of this book based on government reports are so indicated and copyright is claimed for those parts to the extent applicable to compilations of such works.

Independent verification should be sought for any data, advice or recommendations contained in this book. In addition, no responsibility is assumed by the Publisher for any injury and/or damage to persons or property arising from any methods, products, instructions, ideas or otherwise contained in this publication.

This publication is designed to provide accurate and authoritative information with regard to the subject matter covered herein. It is sold with the clear understanding that the Publisher is not engaged in rendering legal or any other professional services. If legal or any other expert assistance is required, the services of a competent person should be sought. FROM A DECLARATION OF PARTICIPANTS JOINTLY ADOPTED BY A COMMITTEE OF THE AMERICAN BAR ASSOCIATION AND A COMMITTEE OF PUBLISHERS.

Additional color graphics may be available in the e-book version of this book.

Library of Congress Cataloging-in-Publication Data

ISBN: 978-1-68507-376-3

Published by Nova Science Publishers, Inc. † New York

Contents

PREFACE

Flavonoids are one of the most important groups of dietary phenolics. They are widely present in vegetables, fruits, cereals and crops, but are not synthesized by the human body. Flavonoids are compounds with low molecular weight. Their structure is based on two benzene rings linked to a heterocyclic pyran ring containing oxygen.

The loss of the effectiveness of antimicrobials, mainly due to inadequate prescription methods and misuse of antibiotics, represents a threat to human health. Hence, the search for new antimicrobials is mandatory. Flavonoids have been demonstrated to have various bioactivity properties in humans: antioxidant, anticancer, antiviral, anti inflammatory, anti-allergic, antimicrobial, antiproliferative and so on. The variety of flavonoids in nature along with their numerous bioactivities in humans have made researchers to put attention to the determination, extraction and characterization of flavonoids.

Many analytical techniques can be found in the literature for the determination of flavonoids. With regard to their extraction, the chemical groups linked to the main flavonoids' structure make them more complex to be extracted from their natural matrices, as well as to differentiate among them. Sammani and Cerdà provide in Chapter 1 a general overview of the main traditional and modern analytical techniques that have been used for flavonoids determination in their natural sources along with their manufacturing. A detailed description of flow techniques and their evolution, together with their role in the automation of the sample pretreatment and its combination with separation techniques, is included. Finally, these authors highlight the advantages and disadvantages of the different techniques.

With regard to the sources of flavonoids, plants are the primary source of flavonoids because of the ease of access and low cost. Medicinal plants are worldwide used for the treatment of various diseases. This is due to the presence of chemical constituents in plant extracts such as flavonoids. Since medicinal plants are natural products, the population believes they have fewer side effects and no toxicity, which is not entirely true.

In the following six chapters, Dr. Lima and her research group characterize several natural plants in terms of phenolic and flavonoid content and assess their antioxidant activity and toxicity. Thus, Emediato et al. study in Chapter 2 the species *Bidens subalternans* DC, which leaves are used in traditional medicine to treat hepatitis, jaundice, fever, throat disorders, and coughs. Notwithstanding, the biological activity of this plant is still poorly known. Therefore, Emediato et al. performed the phytochemical screening, the determination of phenol and total flavonoid content, and to evaluate the larvicidal and antimicrobial activities of the ethanol extract from the leaves of *B. subalternans* DC, showing their antibacterial potential and that the extract is not toxic.

In the following chapter (Chapter 3), Mano-Sousa et al. evaluate the extraction efficiency of phenolic compounds and flavonoids obtained in the flowers of *Matricaria recutita* L. (German chamomile) by Soxhlet extraction and ultrasound-assisted extraction as well as to evaluate the antioxidant and larvicidal potential of the extracts and their fractions. This plant is widely used by the population due to its therapeutic potential, which is attributed to the various phytochemicals present in this species, including flavonoids such as chlorogenic acid, apigenin, catechin gallate, and galangin. In spite of showing antioxidant and larvicidal activity, the authors point out that further studies are needed to assess the toxic effect and antioxidant activity for safe therapeutic use of the plant.

Similarly to Chapter 3, Azevedo et al. evaluate in Chapter 4 the extraction efficiency of phenolic compounds and flavonoids obtained from *Psidium guajava* L. fresh leaves by Soxhlet and ultrasound-assisted extraction as well as to the antioxidant and larvicidal potential of the extracts. These authors found phenolic compounds and flavonoids, such as gallic acid isomers and quercetin derivatives, in the leaves extracts, but the antioxidant activity of extracts could not be correlated with the total phenolic compounds and total flavonoids content.

In the following chapter, Gonçalves et al. are focused on the species *Tecoma stans* (L.) Juss. ex Kunth (Bignoniaceae), originally from Mexico and the southern United States and that can be found in North, Central and South America. Data in the literature relates its biological activities with the phenolic compounds identified in the phytochemical analysis, highlighting the presence of flavonoids in different parts of the plant. Hence, the authors review in Chapter 5 the anti-hyperglycemic, anti-inflammatory and antinociceptive potential along with the antimicrobial, antioxidant effect, anticancer, wound healing and protective hepatorenal activities of *T. stans*. Further information about other plants of the family Bignoniaceae is provided in Chapter 10.

Within this genus *Smilax*, the plants *S. brasiliensis* Sprengel and *S. fluminensis* Steud are endemic in Brazil, native to the Cerrado, and poorly studied. The same goes for the species *Solanum lycocarpum*, also found in the Brazilian Cerrado, In Chapters 6 and 7, Silva et al and Amado et al. review the available literature on the phenolic and flavonoid compounds found in *S. brasiliensis* and *S. fluminensis*, and in *S. lycocarpum*, respectively, along the biological activities associated with them. As a result, it can be stated that *S. brasiliensis* Sprengel shows antioxidant, allelopathic, antimicrobial, anticancer, antihyperglycemic and antihyperlipidemic activities, *S. fluminensis* Steud exhibits toxic, antineoplastic, and antioxidant activities, and *S. lycocarpum* has antioxidant, antibacterial, anti-inflammatory, anti-tumor, anti-nociceptive, allelopathic and larvicidal activities, which are correlated with the phenolic compounds and flavonoids present in these species.

In the following two chapters, Dr Castro and her research team extract and identify flavonoids from plants of the *Bauhinia* genus. Thus, in Chapter 8, Teixeira et al. evaluate and identify the flavonoids present in the leaves of *B. holophylla*, a woody species from the Brazilian Cerrado, finding the presence of flavonoid-*O*-glycoside derivatives, to subsequently attempt to relate them to the medicinal properties attributed to this species. In the following chapter (Chapter 9), Cardoso et al. establish calli cultures from leaf explants of *B. variegata*, an ornamental species originally from India, and produce phenolic compounds, especially flavonoids, in *in vitro* cultures. The authors conclude that 2,4-dichlorophenoxyacetic acid and 6-benzylaminopurine, as well as the presence or absence of light, strongly influence the induction, growth, consistency, color and production of phenolic compounds in *B. variegata* calli.

To finish this part of the book focused on plants as source of flavonoids, Carvalho et al. summarizes in Chapter 10 the occurrence of flavonoids in the tribe Bignonieae, the most diverse and abundant clade of lianas in the Neotropics comprising around 393 species and 20 genera. The authors study the biological activity of these compounds highlighting their pharmacological activity and the medicinal potential of members of tribe Bignonieae, being the antiprotozoal, antiradical, and antitumoral properties the main activities found in Bignonieae flavonoids.

Finally, fruit are also another important source of flavonoids. Among them, oranges can be highlighted. In Chapter 11, Alés and Álvarez describe the history of the bitter oranges of Seville: i.e., when were introduced, their cultivation and myths. Subsequently, the chapter deals with the essential oils and naringin (a bioflavonoid) than can be extracted from bitter oranges, to finally focuses on the action of bioflavonoids against COVID-19. Thus, a certain intake of citrus fruits or their extracts could prevent or minimize COVID-19 infection. In the last chapter of this book (Chapter 12), Feng et al. overviews the recent findings in this field, highlighting the flavonoid hesperidin, which structure has been demonstrated to be able to bind to key proteins for the functioning of SARS-CoV-2 virus.

In: The Book of Flavonoids
Editors: Chao-Hui Feng et al.
ISBN: 978-1-68507-376-3

Chapter 1

Sample Pre-Treatment and Flavonoids Analytical Methodologies for the Quality Control of Foods and Pharmaceuticals Matrices

Mohamad Subhi Sammani and Víctor Cerdà***
Department of Chemistry, University of the Balearic Islands,
Palma de Mallorca, Balearic Islands, Spain

Abstract

The variety of flavonoids in nature, besides their numerous bioactivities in human and animal bodies, where they cannot be synthesized biologically, have increased the attention of scientists to include them in a suitable form to facilitate their consumption by humans and animals.

* Corresponding Author's E-mail: m.s.sammani@hotmail.com.

Many products have been continuously found to serve these substances, securing the daily recommended need of flavonoids.

Moreover, the notable prevention and treatment behaviour of these natural substances on non-communicable diseases raises the interest in the importance of knowing which are the influence compounds and their impact concentration on the daily consumption of foods.

This variety of nature and the manufacturing productions of flavonoids requires strict control of the original source, the production stages as well as the final products.

Therefore, the development of suitable analytical methodologies is the target of scientists to fulfil the requirements of quality control, achieving the requested accuracy and sensitivity.

On the other hand, most of the flavonoids' matrices required a pretreatment process before the analysis regarding their intricacy. This stage is considered tedious and the most consuming of time and solvents compared to the entire analysis process. In this context, the development of quicker pretreatment methods that can consume less solvents and time and produce less generated waste is interesting.

In this chapter, the analytical techniques that have been developed and employed for the determination of flavonoids in nature, commercial products, pharmaceuticals, and dietary supplements are discussed focusing on the trend of analytical methodologies in this field.

The required sample pretreatment procedures prior to the analysis stage are included, concentrating on the state-of-the-art systems that have been recently employed for the development of fully automated sample pretreatment methodologies for this purpose, such as flow analysis systems and microfluidic systems that have not yet been widely discussed in the literature concerning flavonoids' determination.

The advantages and disadvantages of each analytical and pretreatment technique are mentioned. The superiority of the developed flow analytical systems for the automation of the flavonoids' sample pretreatment over the existing manual ones are also explained.

Keywords: flavonoids, natural products, pharmaceutical formulations, food supplements, non-communicable diseases, analytical techniques, adsorption spectroscopic techniques, ultra-violet spectroscopy, visible spectroscopy, fluorescence spectroscopy, nuclear magnetic resonance spectroscopy, X-ray crystallography, circular dichroism spectroscopy, vibrational spectroscopic techniques, near-infrared spectroscopy, Raman spectroscopy, terahertz spectroscopy, hyperspectral imaging spectroscopy, mass spectroscopy, chromatographic techniques, paper chromatography, thin-layer chromatography, gas chromatography, supercritical fluid chromatography, high-performance liquid chromatography, sample pretreatment techniques, solid-phase extraction, liquid-liquid extraction, liquid-liquid micro-extraction, dispersive liquid-liquid micro-extraction, hollow fibre liquid phase micro-extraction, solid-phase extraction, solid-phase micro-extraction, dispersive solid-phase micro-extraction, matrix solid-phase dispersive extraction, maceration extraction, soxhlet extraction, hot water extraction, heat reflux extraction, microwave-assisted extraction, ultrasound-assisted extraction, pressurized liquid extraction, shockwave-assisted extraction, supercritical fluid extraction, high-speed counter-current chromatography, flow analysis systems, lab-on-valve, multisyringe flow-injection system, microfluidic systems, lab-off chip, lab-on-chip, paper-based microfluidic system

INTRODUCTION

The first flavonoid was discovered by Szent-Gyorgyi in 1937 and was classified as vitamin P [1]. This compound was later named flavonoid rutin and was known as an individual biochemical group, "flavonoids." Then, they discovered its bioactivity decreased cardiovascular disease in the French population, which was associated with red wine consumption despite their popular diet that consisted of a high amount of saturated fat, as also known the French paradox [2], led to an attractive interest in the study of this bio-group.

Flavonoids are larger widespread secondary metabolism compounds occurring in the plant kingdom [3]. The main skeleton of these compounds consists of two benzene rings (A and B) linked through a heterocyclic pyran ring (C) (Figure 1) and they share 15 carbon atoms in the phenylbenzopyran (C_6-C_3-C_6) framework [4]. Regarding the linkage of the aromatic ring to the benzopyrano moiety, they are divided into three main groups: flavonoids (2-phenylbenzopyrans), isoflavonoids (3-benzopyrans) and neoflavonoids (4-benzopyrans) (Figure 1).

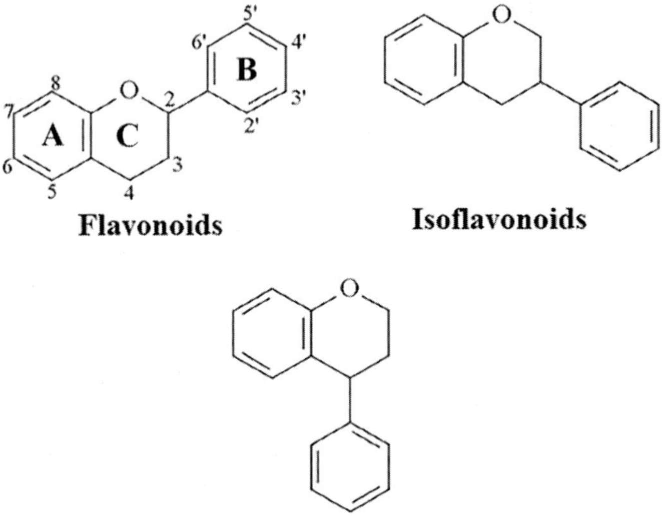

Figure 1. The main skeleton of flavonoids, isoflavonoids and neoflavonoids.

The classification of flavonoids is based on the oxidation and saturation degree on position 3 and/or position 4 in the C-ring and may be distinguished from many major subclasses: flavanes, flavones, isoflavones, flavanones, flavanonols, flavanols, flavonols, anthocyanidins and chalcones (Figure 2).

The variation of flavonoid structures mainly regarding their hydroxylated positions 3, 5, 7, 3′, 4′ and 5′ which are frequently found methylated, prenylated, acetylated, sulphated, or glycosylated [5].

Naturally, flavonoids can reside in plants as aglycone, or in a more stable form "glycoside" by linking one sugar molecule or more, fundamentally, L-rhamose, D-glucose, glucorhamnose, galactose or arabinose [6].

In nature, flavonoids are present in vegetables, fruits, grains and especially in the green leaves and located in their cell vacuoles in all plant parts (flowers, leaves, roots, seeds and fruits) [7]. They are directly responsible for the colour and aroma of fruits and flowers, and play a role in the interaction between the plant and the environment [8]. In addition, they play the role of the first line of defense in plants against foreign factors and pollution such as UV-filters, temperature acclimation, frost hardiness and antibacterial functions [8].

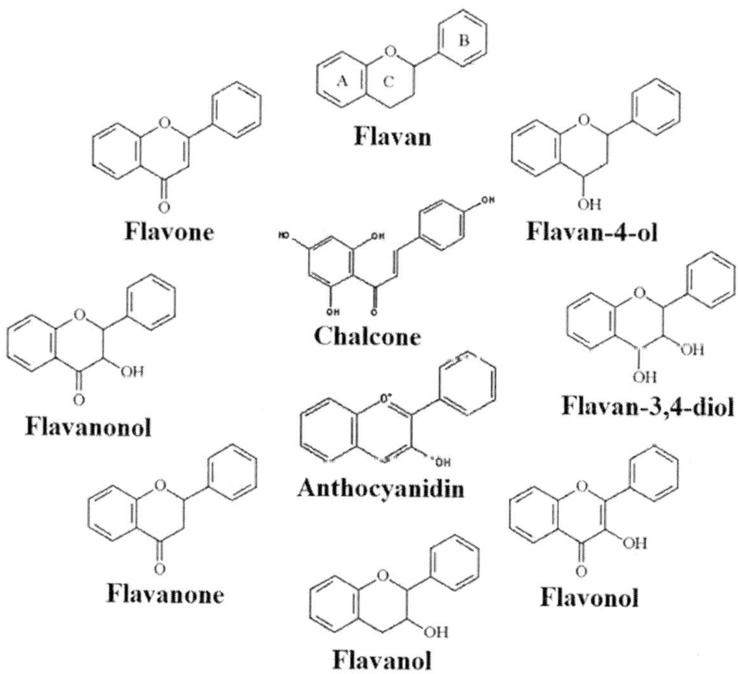

Figure 2. The main subclasses of flavonoids.

Flavonoids are not biologically synthesized in human bodies but they enter by the consumption of adequate rich-flavonoids foods. However, they have countless biological and pharmacological activities ranging between prevention, protection, treatment and reduce the risk of numerous chronic diseases e.g., communicable and non-communicable diseases [9]. The great property of almost all flavonoids is an antioxidant capacity that protects the body's cells against reactive oxygen species (ROS) and free radicals [10]. Against the communicable diseases (CDs), they show anti-chronic inflammatory disorders activity [11], antimicrobial activity [12], antifungal activity [13] and antiviral activity, for instance, Herpes simplex virus type 1 and 2 [14] and HIV-1 and HIV-

2 [15]. Recently, some works showed notable affection for Severe Acute Respiratory Syndrome (SARS) and Middle-East Respiratory Syndrome-Coronavirus (COV) [16-18]. For non-communicable diseases (NCDs), the numerous actions of flavonoids can prevent cancer [19], diabetes [20], cardiovascular diseases [21], ageing, and neurodegenerative diseases [22]. Regarding all these considerable actions, these phytochemical compounds should be included in the human and animal daily diet. Whereas, the estimated mean daily total flavonoid intake in the USA is 189.7 mg per day and is much lower in adults than other generations, this directly depends on the diet, meaning consumption of vegetables, fruits, grains, coffee, tea, and much more besides personal income [23]. Increasing the daily intake level of flavonoids might cause opposite behaviours. The overdose level of flavonoids cannot be observed in the food because of their low toxicity (LD_{50} 2-10 g) [10], while in the manufacturing of products, they usually exist in a high concentration and can reach potentially toxic levels [24].

On the other hand, it is difficult to reach the recommended daily intake dose of flavonoids (70 and 170 mg per day per human adult) [25] because of their low concentration in vegetables and fruits, and their low absorption in the human body. These are causing us to rely on pharmaceutical and food supplements that contain concentrated natural flavonoids (which can reach up to a 500 times higher concentration than food or plants), therefore, they offer easier consumption.

To date, numerous dietary supplement products of flavonoids located in the market, besides, many pharmaceutical drugs have been found. Regarding the Drug Bank database [26], there are 11 flavonoids approved, 33 flavonoids and another 50 compounds that have been investigated and are in the experimental stage [26].

This variety of the natural sources of flavonoids and their commercial products lead us to continuously develop new analytical methods or even improve the existing ones to follow the development in the flavonoids products and to control the safety, quality, and efficacy of these products. On the other hand, research carried out over two decades has provided data suggesting that diet is an essential factor influencing the risk of development of NCDs. The increasing knowledge on chemopreventive properties of certain food ingredients, in particular, flavonoid compounds, opened the door to extensive epidemiological and clinical studies of the role of these substances in the prevention of chronic diseases [27]. Different flavonoid compounds may have different effects on NCDs. The major flavonoids intake studies were carried out by the estimation of total flavonoids depending on the US or European phenol-Explorer database [28]. However, values based on estimations using databases are only approximate and are inaccurate, as the content of flavonoids may vary significantly according to place, harvest time or food processing. In this sense, the exact knowledge of the applicable flavonoids and their real concentrations in food and pharmaceutical studied samples play an important role in the accuracy of the results in this type of studies. Furthermore, the studies of the relation between the intake of flavonoids and their effectiveness on the prevention and/or treatment of NCDs could not be accurate without the exact knowledge of the kinds and the concentration of flavonoids located in the tested samples, as well as their bioavailability in the hosted bodies. Hence, the extraction, purification, and analysis of these samples are essential to determine which flavonoid compounds are associated with the treatment of NCDs.

Many analytical techniques have been applied for the determination of flavonoids in their different matrices. There is not an ideal technique to determine all flavonoids, the choice of a suitable one depends on the target analysts, the complexity of the matrix and the goal of this analysis. Sample pretreatment is mandatory prior to many analytical techniques and also depends on the complexity of studied matrices.

In this context, the development of faster, effectiveness, costless, and more environmentally friendly methodologies are the target of analytical chemists. Nevertheless, the most of existing developed techniques are based on manual approaches involving time, money, and solvents consumptions as well environmentally pollutants.

Flow analysis and microfluidic systems that are principally used for environmental pollution analysis have been recently implemented in the case of flavonoids determination. These systems notably reduce time and generated solvents while increased the analytical throughput, accuracy, and sensitivity. Regarding their ability to automate the sample pretreatment process with the capability to online transferring of the eluent to the coupled analytical instrument or detection systems [29].

In this chapter, a general overview of the main traditional and modern analytical techniques, and sample pretreatment techniques that have been used for flavonoids determination in their natural sources, and their manufacturing products are extensively discussed. A detailed description of flow techniques and their evolution, together with their role in the automation of the sample pretreatment and its combination with separation techniques are mentioned. The advantages and disadvantages of these applied techniques are also included.

SAMPLE PRETREATMENT TECHNIQUES OF FLAVONOIDS

Sample preparation is the most important stage that emphasizes the accuracy of the analytical results and can also enhance the sensitivity of the analysis. This stage causes a major delay over other analytical steps [30]. The type of preparation depends on the complexity of the sample, the concentration of the target analytes, their chemical structure, their chemical behaviour, the analytical technique used, and the goal of this analysis.

Sample preparation of flavonoids is quite diverse. In the case of pharmaceutical formations, direct dissolving and filtering helped by assistant tools such as shaking, vertexing or sonicating are usually used. These simple preparation approaches are generally employed for these types of formulations because the flavonoids are present in a pure form as an active pharmaceutical ingredient (API) with a relatively high concentration to be detectable, besides the non-significant effect of the existing excipients in the pharmaceutical drugs on the determination of these components [31].

On the other hand, more complex samples such as plants, natural extracts, foods and food supplements require grinding, drying or freezing. As they contain several derivatives of each flavonoid and the target flavonoid might be in trace concentration, more complex sample preparation work is needed, for instance, solid-liquid extraction (SLE) [32], liquid-liquid extraction (LLE) [33] and solid-phase extraction (SPE) [34].

The physicochemical properties of the interested flavonoids play an important role in the choosing of the appropriate extraction solvents as well as the used extraction technique. Generally, flavonoid aglycons are poorly soluble in water while the attaching of sugar moieties increase the water solubility of these flavonoids' glycoside forms, thus increasing the likelihood of using more extraction solutions [35]. The polarity also is focused here, the less polar flavonoids' subgroups i.e., isoflavones, flavanones, are normally extracted by chloroform, dichloromethane, and ethyl acetate, while the more polar flavonoids are extracted by alcohol or a water-alcohol mixture [36]. Moreover, the stability of flavonoids is also a substantial consideration, some flavonoids are relatively stable under ambient conditions, whereas others such as anthocyanins are labile. Mainly, these compounds are sensitive to elevated temperature, light, air, mechanical processes that limit the available extraction options, collection samples procedures, sample storage conditions, and the applied extraction techniques [37].

Analytical techniques and the goal of analysis, the applied analytical technique depends on the objective of the analysis. For instance, in case the objective is the determination of total flavonoids content, total antioxidant power, the biological action studies of total flavonoids content in the sample, non-selective analytical techniques are still useful and simple preparation sample is sufficient i.e., the preparation sample before spectrophotometric techniques [38] and the study of total flavonoids intake associated with the prevention of some chronic diseases [39]. On the other hand, the individual flavonoids' detection or the study of the biological behaviour of a specific flavonoid in the complex sample require separative and selective analytical techniques, such as chromatographic techniques, where the isolation, extraction, and purification procedures are mandatory. For instance, the study of the effective specific flavonoids' intakes from the Finnish local foods on the risk of the chronic diseases [40]. This type of study required extraction and selective analytical technique to exactly determine which flavonoids located in the test foods have an effective influence on the studied diseases.

In this context, sample pretreatment techniques can be divided into conventional extraction techniques (CETs), assisted extraction techniques (AETs), alternative modern extraction techniques (AMETs) and automated extraction systems (AES). Each one includes a number of implemented techniques in the case of flavonoids extraction. In conventional techniques, they can include maceration extraction (ME) [41], Soxhlet extraction (SE) [42], hot water bath extraction (HWBE) [43], and heat reflux extraction (HRE) [44]. The assisted techniques consist of microwave-assisted extraction (MAE) [45], ultrasound-assisted extraction (UAE) [46], pressurized liquid extraction (PLE) [47], and shockwave-assisted extraction (SWAE) [48]. While in the modern techniques they include LLE [49], SPE [50] and their derivatives, supercritical fluid extraction (SFE) [51] and high-speed counter-current chromatography system (HSCCC) [52]. The automated systems that have been developed recently for flavonoids' extraction are based on flow analysis [53] and microfluidic systems [54].

Conventional Extraction Techniques

Maceration extraction (ME) is based on the soaking of a ground dried sample in a stoppered container with a proper solvent. Then, the mixture is frequently shaken and stands for up to 7 days [55]. This technique is usually carried out for plant samples in which the solvent softens and breaks the wall's cells and releases the soluble flavonoids. Afterwards, it is followed by filtration of the mixture. Nowadays, this technique is not often used in this field due to the long time needed, solvent consumption and the low resulting extraction yield.

Reflux, HWBE, HRE, and SE are high-temperature extraction procedures. SE is the wider traditional method that has been carried out for flavonoids' extraction from plants, vegetables, and fruits because of its ease of operation, cheap equipment and low cost of maintenance. Methanol, acetonitrile, acetone, and ethyl acetate are the common extraction solvents used in these techniques [56]. The extraction process is varied and can range between 2 and 48 h [41, 44, 57-59]. The main constraints of these techniques for flavonoids' extraction are its long extraction time, elevated operational temperature, huge quantity of consuming solvents, and the required evaporation step to concentrate the extractant prior to its analysis in the case of reflux and SE techniques [42]. These inconveniences lead to a decrease in the extraction yield, increase detection errors, and environmental pollution. Regarding that, nowadays, CETs are less frequently applied for flavonoids applications.

The optimization of operational conditions has great importance. In general, CETs' main factors that should be optimized to avoid negatively affecting the extraction yield are the temperature, the type of extraction solvents, pH, and the extraction time.

Many works have compared the CETs techniques concerning flavonoids extraction. The comparison between SE, ME and other AETs for the extraction of total phenols from *Pinus radiata* Bark [60] has been done. In this work, SE showed the best extraction yield of phenolic compounds while ME was the worst using the same quantity of extraction solvent (200 mL) and the same extraction time (180 min). Another work also showed that the SE technique provides a better yield than ME for the extraction of total flavonoids from *Spirulina platensis* sample [41]. Analyzing data, the higher yield of SE was performed in less extraction time (4 h) than ME (2 days). Another work has compared the ambient temperature extraction between HRE, and SE for the extraction of total flavonoids from *Saussurea medusa* Maxim [61]. The work concluded that the required extraction time decreased from the ambient temperature extraction (24 h) to HRE (6 h), while SE was medial (20 h). On the other hand, SE showed the maximum yield (4.1%).

In general, SE provided better recoveries than other CETs, but was still much less effective than AETs, AMETs and AES [60, 61]. Table 1 shows the comparison between the recent CETs applied to flavonoids determination. The conveniences and inconveniences of these techniques are summarized in Table 6.

Table 1. The comparison between the recent CETs applied to flavonoids determination

Extract. Tech.	Matrix	Ext.	Extraction conditions			Yield (%)	Refer.
			Solvent (mL)	Temp. (°C)	Time		
HRE	*S. medusa*	TFC	10 (80% EtOH)	90	6 h	3.9	[61]
HRE	*T. chebula*	TFC	25 (EtOH)	40	2 h	0.47	[62]
SE and ME	*C. album*	TFC	(acet.)	-	7 days	0.73	[55]
SE	*S. medusa*	TFC	10 (EtOH)	90	20 h	4.1	[61]
SE	*P. radiata* Bark	TPC	200 (acet.:water, 7:3 v/v)	82	180 min	12	[60]
SE	*S. platensis*	TFC	100 (EtOH)	60-80	4 h	5.3	[41]
SE	*M. oleifera L.*	TFC	4 (pure EtOH)	90	3 h	23.2	[63]
ME	*P. radiata Bark*	TPC	200 (acet.:water, 7:3 v/v)	40	180 min	8.7	[60]
ME	*S. platensis*	TFC	100 (EtOH)	-	2 days	2.7	[41]
ME	*Cassia alata*	TFC, KA	1:20 ratio of 100% EtOH	60	2 h	7.0 1.2	[64]

Acet.: acetone; EtOH: ethanol; Ext. extractant; Extract.: extraction; HRE: heat reflux extraction; KA: kaempferol; ME: maceration extraction; SE: Soxhlet extraction; TFC: total flavonoids content; TPC: total phenolic content; Temp.: temperature; Tech. technique.

Assisted Extraction Techniques

Some assistant tools have been developed in order to reduce the extraction time and solvent consumption of the CETs and to increase the contacting of the sample with the extracting solvents so to improve the efficiency of the extraction. Those frequently used for flavonoids extraction are MAE, UAE, PLE, and recently SWAE.

MAE and UAE are the common techniques carried out for the extraction of flavonoids from food and plant samples.

MAE generates microwave energy in an electromagnetic spectrum of radiation between 300 MHz and 300 GHz (Figure 3) and is usually combined with HRE [62]. Microwaves directly heat the targeted materials based on their dielectric constants, followed by ionic conduction and dipole rotation of the molecules [65] causing higher yield, equilibrium concentration and effective diffusion coefficient of flavonoids [62]. Also, it has been reported that MAE effectively reduces the extraction time due to the increase of the mass transfer rate, resulting in less extraction solvents need, and therefore a less extraction cost than CETs [60, 61]. On the other hand, microwave energy generates heat in the solvents that lead directly to decomposing of the phytochemical compounds and a decrease in their biological properties thus reducing the extraction yield [66]. In this case, MAE is not suitable if the biological activity of flavonoids is the objective.

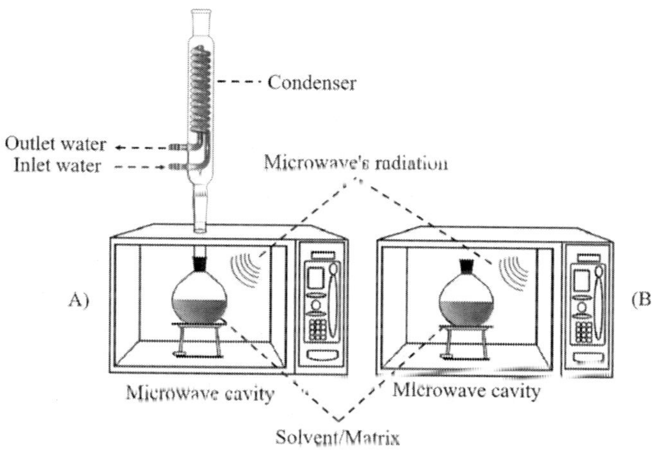

Figure 3. A): microwave-assisted extraction technique combined with a reflux condenser, B): the simple microwave-assisted extraction instrument.

UAE generates ultrasound radiation based on the phenomenon of acoustic cavitation, which consists of the formation of gas bubbles caused by the passage of ultrasonic waves through the liquid and subsequently ruptures the plant cells and releases bioactive compounds from the cells. In this way, the analysts can diffuse easily and dissolve the analytes in the extraction solvent [67] (Figure 4). The potential of the UAE is its application in the food and pharmaceutical industries [68].

This technique reduces the required extraction time, solvents and increases the extraction yield of flavonoids compared to the CETs [60, 61]. Nevertheless, MAE is still the preferable technique for the extraction of bioactive substances since it is much more effective, needs less time, and provides a higher extraction yield than UAE [60, 61, 64]. However, for the biological studies of flavonoids, UAE is the technique of choice since the ultrasonication is not using thermal generation power thus improving the biological properties of the extract [69]. Nonetheless, increasing the applied ultrasound frequency results in the free radical formation, producing unacceptable changes of extracting components, and increasing the temperature of the solvent that might lead to flavonoids' degradation [70]. For all these considerable limitations, the affected factors in MAE and UAE should be optimized carefully to increase the extraction yield of flavonoids and prevent their degradation. The main factors are time, temperature, solid-solvent ratio, solvent volume, solvent polarity, irradiation, and frequency of intensity [68, 71].

PLE was developed as an alternative technique to the ones mentioned previously (Figure 5). It has also been named pressurized fluid extraction (PFE), accelerated solvent extraction (ASE) and pressurized solvent extraction (PSE) [67]. This technique is working under high pressure and temperature to maintain the solvent in the liquid state under a higher temperature than its boiling point. Working under these conditions lead to an increase in the mass transfer from the solid sample into the extraction solvent, providing a faster extraction process, less needed amount of solvent, and higher extraction yield.

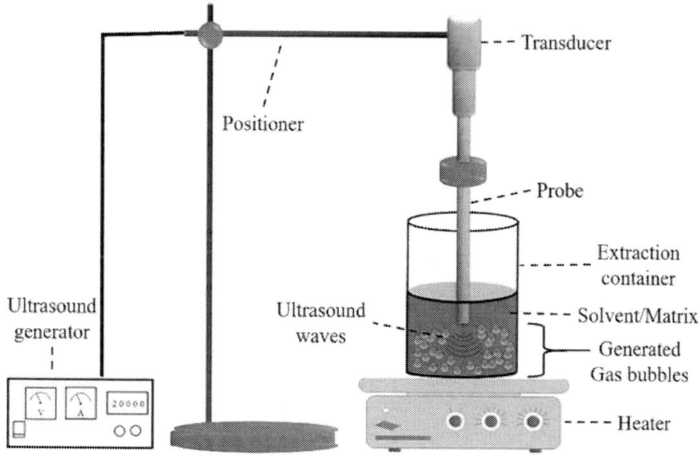

Figure 4. The illustration of ultrasound-assisted extraction technique.

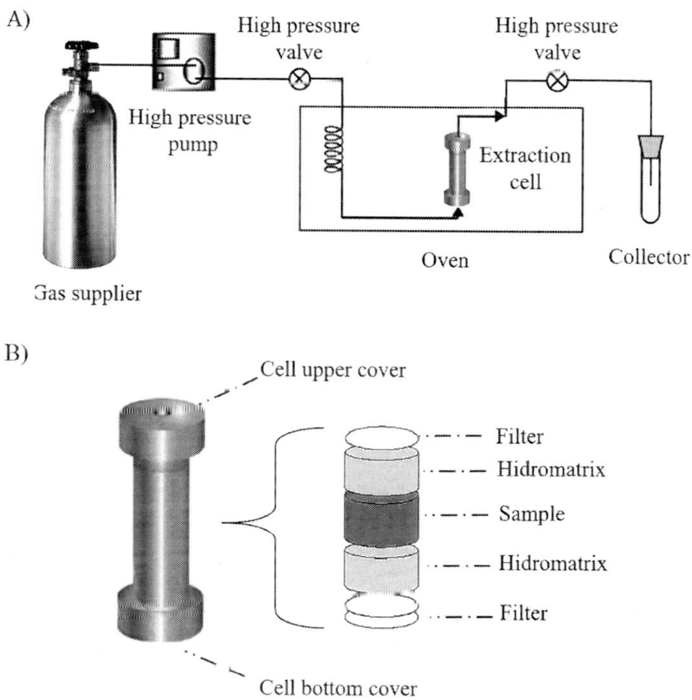

Figure 5. A): The basic scheme of the pressurized liquid extraction technique. B): The diagram of used extraction cell.

On the other hand, the equipment is costly and needs a careful optimization of the PLE factors to avoid applying an elevated pressure and temperature that directly affects the degradation of flavonoids [72]. These variables are mainly the sample size, solvent composition, pressure, temperature, pH, flow rate (in case of dynamic mode) and the extraction time [73].

SWAE is the widely attained technique in medical and biomedical applications to mediate drug delivery, needleless injection, thrombus ablation, pain release, enhancement of bone fusion to treat skull bone defects, and gene transfection [74]. This technique is based on causing instantaneous high pressure to propagate plant media at rates exceeding the speed of sound. The pressure divides into penetration and reflected waves upon a change in density. These generate the splitting open of cell structures instantly generating multiple cracks on a cell's wall. Thus, quickly reaching the biological compounds located in their storage [75] (Figure 6). As the result, it is a suitable technique for the extraction of biological components from plant tissues. This technique has been implemented recently to extract the flavonoids from plant matrices. It was reported for the first time, the extraction of total flavonoids and total phenolic acids from *Eysenhardtia polystachya heartwood* applying 1500 shock waves producing 34.54% and 31.95% higher contents than Soxhlet and UAE, respectively. Extraction times using shock waves (5 min) were much shorter than all other tested techniques [48]. The main advantage of SWAE is the prevention of the degradation of flavonoids that can be faced off in a high-pressure operation due to a long applying time such as in the case of PLE, as well as a much shorter extraction time and solvent consumption compared to PLE, MAE and UAE. The reason is the fact of applying high pressure only for some microseconds.

The recent applications of AETs for flavonoids extraction from food and plant materials are illustrated in Table 2. The benefits and disadvantages of these techniques are summarized in Table 6.

Table 2. The comparison between recent AETs techniques applied to flavonoids determination

Extract. Tech.	Matrix	Ext.	Extraction conditions			Yield ($mg\ g^{-1}$)	Refer.
			Solvent (mL)	Temp (°C)	Time (min)		
MAE	*T. chebula*	TFC	200 (water)	100	1	23.4	[62]
MAE	*H. sabdariffa*	Antho. flav.	(water)	-	8	1.3	[76]
MAE	*T. foenum-graecum*	TFC	1 (60% EtOH)	70	3	-	[68]
MAE	*C. alata*	TFC, KA	1:20 ratio of 100% EtOH	-	4	135.2 17.6	[64]
MAE	*M. oleifera*	TFC	4 (52% EtOH)	-	20	36.1	[63]
UAE	*C. alata*	TFC, KA	1:20 ratio of 100% EtOH	-	5	86.7 14.2	[64]
UAE	Guayule	TFC	25 (water or 25% EtOH)	25	60	14	[71]
UAE	Red grape skins	19 flav.	1:80 ratio of mobile phase	50	15	-	[77]
UAE	*C. tinctorius*	TFC	56.37 mL g	41.4	55.9	55.4	[78]
UAE	*M. oleifera*	TFC	4 (52% EtOH)	30	20	44.0	[63]
PLE	Goldenberry	QU, RU, MA	1-3 mL min^{-1} (70% EtOH)	25	10-60	0.6×10^{-3} 1.2×10^{-3} 3.6×10^{-3}	[79]
PLE	Orange peel	HES	2.37 g min^{-1} (75% EtOH)	65	5-40	19.3	[80]
PLE	Grape marc	15 Antho.	5 g min^{-1} (50% EtOH)	40	220	10.2	[73]
PLE	*L. citriodora*	7 flav.	(46% EtOH)	200	17	-	[81]
SWAE	*E. polystachya*	TFC	0.7 (water)	30	26	-	[48]

ACN: acetonitrile; Antho: anthocyanins; EtOH: ethanol; Ext. extractant; Extract.; extraction; Flav.; flavonoids; HES: hesperidin; KA: kaempferol; MA, mangiferin; QU: quercetin; RU: rutin TFC: total flavonoids content, Temp.: temperature; Tech. technique.

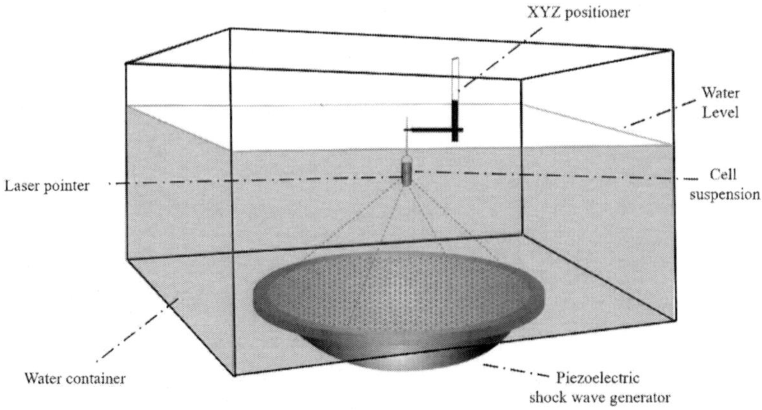

Figure 6. The illustration of underwater shock waves setup.

Alternative Modern Extraction Techniques

These techniques are the ones that have been developed as more effective alternative extraction techniques to the ones previously described. Generally, AMETs have enhanced the efficiency of the extraction of flavonoids, decreased the negative error that occurred through their degradation by applying elevated temperature and/or pressure along the processing time. These features not only allow for reducing the extraction time, cost, solvent consumptions, and waste, but they also increase the environmental sustainability, analytical method throughput, recovery, and sensitivity. Further, they can eliminate the interfering substances such as waxes, fats, terpenes and chlorophylls that are usually located in the natural samples. They clean-up and concentrate the analytes of interest, besides increasing the possibility to automate the extraction procedure.

Liquid-Liquid Extraction

Principally, LLE is based on the distribution of analyte according to its solubility in two different immiscible solvents [82] (Figure 7). It is one of the traditional extraction techniques of flavonoids from plants, vegetables, and fruits depending on its ease of use, its simple apparatus, and saving time, solvent, and sample consumptions compared to the CETs [83]. Pure organic solvents such as methanol, ethanol, ethyl acetate, and acetonitrile are the common extraction solvents for less polar flavonoids, while the mixing of these solvents with water is a suitable extraction mixture for the more polar compounds conducted at room temperature [33]. Nowadays, the traditional LLE procedure is not implemented widely for the extraction of flavonoids. That's because of its high consumption of toxic and expensive organic solvents, a relatively long processing time, low preconcentration factor, low selectivity, the need for an additional purification stage as the solvent evaporates to concentrate the analyst.

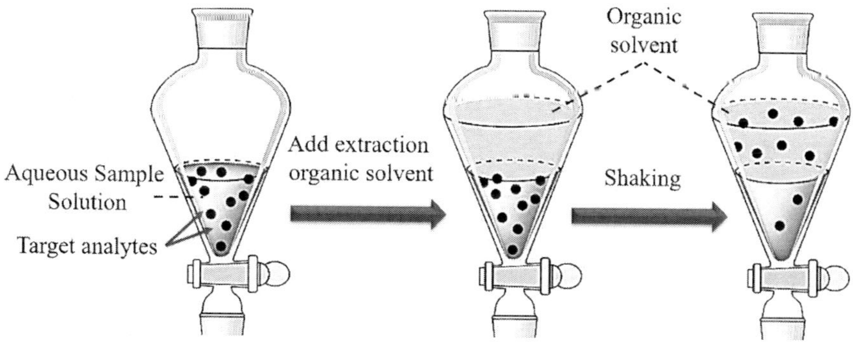

Figure 7. Liquid-liquid extraction procedure.

Moreover, it might be because of the non-suitability of the extraction solvent to the applicable analytical method, and also it's a difficult phase separation compared to other AMETs [84]. For instance, the extraction of four flavanones (hesperidin, naringenin, naringin, and poncirin) from industrial, hand-squeezed orange juices and fresh-in squeeze machines orange juices consumed 300 mL of ethyl acetate as an extraction solvent and 40 min were required [33]. After that, the evaporation task was carried out at 40°C before dissolving the analyte again in 10 mL of methanol. The extraction of total flavonoids from *Aurea Helianthus* flower required 75% ethanol, three times for a total time of 90 min followed by an evaporation step for 20 h. Then the re-solved extract was purified by AB-8 macroporous resin [85]. On the other hand, pH, temperature, sample-solvent ratio, and the number and time intervals of individual extraction steps played an important role in this extraction procedure.

Therefore, some improved LLE methods have been developed to simplify and miniaturize the sample preparation procedures. Liquid-liquid micro-extraction (LLME) improves the efficiency of the extraction as the use of organic solvents in the level of microliters. Thus, it provides a high preconcentration factor, reduces the solvents and time consumption [86]. This technique was reported for the green extraction of quercetin from food samples at a trace level using 200 μL of N,N-dimethyl-n-octylamine as the extracting solvent, needing 21 min including the vortex and the centrifugation of the solution [87].

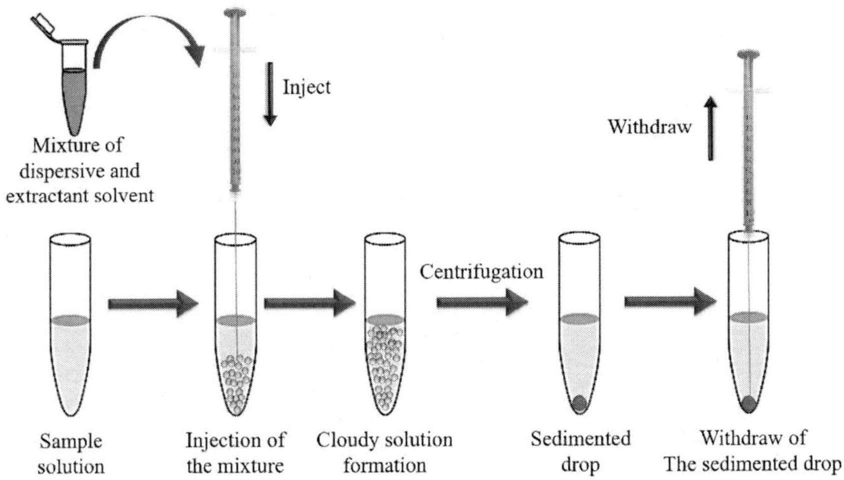

Figure 8. Dispersive liquid-liquid microextraction procedure.

Dispersive liquid-liquid micro-extraction (DLLME) and hollow fibre liquid phase micro-extraction (HF-LPME) have been developed within the last decade as a derivative to LLME and recently have been used for flavonoids extraction. DLLME depends on preparing the mixture of solvent and dispersive solutions, then the microliters of this mixture are rapidly injected into the aqueous sample solution as shown in Figure 8. These steps allowed the formation of a cloudy solution that can partition the analytes to the extraction phase instantaneously [88]. DLLME is an eco-friendly method regarding the small used organic solvents in the range of microliters that also lead to enhance the enrichment factor and high-recovery due to the concentrate of the analysts in these micro-drops of extraction solution. Thus the required extraction time is even less than LLME. For instance, the rapid DLLME method has been developed for the extraction of eight flavonoid aglycone compounds, baicalein, hesperitin, fisetin, naringenin, chrysin, myricetin, quercetin and kaempferol, from honey samples and related products. 1.5 mL of acetonitrile as disperser and 150 μL of chloroform as extractant were consumed and approximate 1 min for centrifugation was reported [89].

As observed above, the mentioned methods in almost all cases requested a centrifugation step to separate the extraction layer that showed difficulty to automate these procedures. Otherwise, the use of a lighter extraction solvent than water in DLLME shows the main limitation in the application of this technique manually. Solving this matter, Multisyringe flow injection analysis (MSFIA) was reported as an effective system for the automation of these extraction methods. It facilitates the collection of the extraction solvent's micro-drops that settle in the bottom (solvent heavier than water) or in the upper (solvent lighter than water) of the extraction syringe (the syringe can be changed its position depending on each matter [90]. Regrettably, LLME and DLLME based-MSFIA have not yet been implemented for the extraction of flavonoids. Basically, HF-LPME relies on the migration of the target analytes through a membrane, and from there into an acceptor phase by passive diffusion [91] (Figure 9). This method provides a rapid and simple operation, much reducing the consumption of solvent, and time, ease of its automation, excellent sample clean-up, and not requiring any additional purification processes [92]. The main drawback is the possibility of fibre pores getting blocked, limited availability of the commercial system, and the contamination of the sample solution due to the contamination of the used magnetic stirrer [93]. As an example, the efficient three-phase HF-LPME technique was developed for the extraction of 6 flavonoids (morin, naringenin, quercetin, luteolin, kaempferol and apigenin) from *Echinophora platyloba* DC. and *Mentha piperita* using 1-octanol as extraction solvent, and 30 μL of 100 mmol L^{-1} borates as an acceptor [94]. This procedure was carried out in 80 min and achieved an acceptable recovery, between 92% and 99%.

Overcoming the obstacles of harmful organic solvents and respecting the environmental concerns, alternative extraction solvents have been suggested, such as ionic liquids (ILs) and deep eutectic solvents (DESs) [95]. ILs consist of organic cations and organic/inorganic anions, consequently, the unique features are observed as negligible vapour pressure, high viscosity, thermal stability, high tunability, miscibility with water and organic solvents, good solubility for organic and inorganic substances, and the ability to efficiently transfer and absorb microwave energy [87, 92, 96, 97]. DESs have similar properties to ILs plus they have a comparatively lower melting point, are cheaper, and less toxic than ILs [98]. Recently, these alternative solvents have been widely used for flavonoids' extraction.

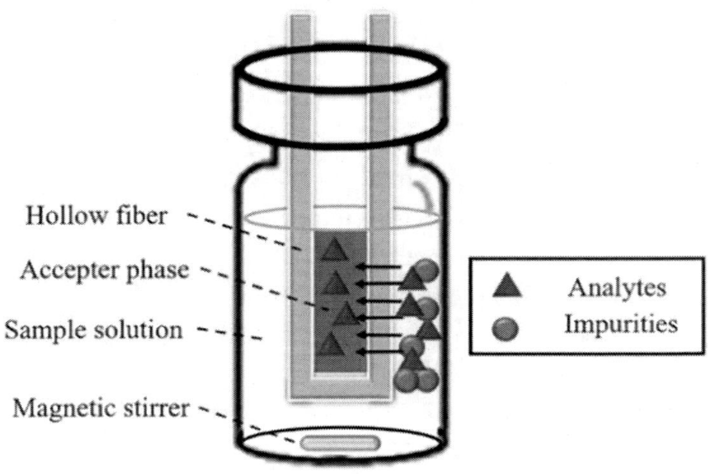

Figure 9. Hollow fiber liquid phase microextraction procedure.

Some of the recently developed works of LLE and its derivatives are accumulated in Table 3. The advantages and disadvantages of these techniques are summarized in Table 6.

Table 3. Applicability of recent LLE methods for the extraction of flavonoids

Extract. Tech.	**Matrix**	**Ext.**	**Extraction conditions**		**Recov./ Yield (%)**	**Refer.**
			Solvent (mL)	Time (min)		
LLE	Orange Juice	HES, NAR, NARIN, PO	300 (ethyl acet.	40	98-107	[33]
LLE	Chinese herbal products	FI, QU, IS, KA	0.37 (heptanol:IL:water)	4.3	51-95	[49]
LLE	*A.Heliant-hus* flower	TFC	1.30 g mL^{-1} ratio (75% EtOH)	30	27.8	[85]
LLME	Food samples	QU	200 μL (N,N-dimethyl-n-octylamine)	21	97-105	[87]
DLLME	Human milk	NARIN, HTIN, KA, QU, EP, EPG, EGCG, GE, DA	550 μL ethyl acet. (extractant) and 1100 μL ACN (disperser)	5.1	94.3-103	[99]
DLLME	Honey	BA, HES, FI, NARIN, CH, MY, QU, KA	1.5 mL (ACN, disperser) and 150 μL (chloroform, extractant)	1	80 - 111	[89]
HP-LPME	Faba beans	CA, RU	300 μL NaCl (2 M)	180	-	[100]
HP-LPME	*E.platyloba* DC. and *M. piperita*	QU, MO, NARIN, KA, AP, LU	30 μL (borate)	80	92-99	[94]
HP-LPME	Chinese medicines	RU, QU, GA	40 μL (NaOH)	80	90.0–106.3	[101]

Acet.: acetate; AP: apigenin; ACN: acetonitrile; BA: baicalein; CA: catechin; CH: chrysin; DA: daidzein; DLLME: dispersive liquid-liquid microextraction; EP: epicatechin; EPG: epicatechin gallate; EGCG: epigallocatechin gallate; Ext. extractant; Extract.: extraction; FI: fisetin; GA: galuteilin; GE: genistein; HES: hesperidin; HP-LPME: hollow fiber liquid-phase microextraction; HTIN: hesperitin; IL: ionic liquid; IS: isorhamnetin; KA: kaempferol; LLE: liquid-liquid extraction technique; LLME: liquid-liquid microextraction; LU: luteolin; MY: myricetin; MO: morin; NARIN: naringenin; NAR: naringin; PO: poncirin; QU: quercetin; RU: rutin; Recov.: recovery; TFC; total flavonoids content; Temp.: temperature, Tech. technique.

Solid-Phase Extraction

SPE is a good choice to concentrate, separate, purify, and clean-up flavonoids from a crud plant, food, pharmaceutical samples in one operational step. It was introduced as an alternative greener extraction technique to LLE. The simplest SPE provides some advanced features over LLE, such as being faster, requiring less samples and solvents volume, centrifugation is not required due to automation, there is consistency in the choice of extraction sorbents, suitability of used solvents to HPLC, and the provided recoveries are notably higher than LLE [102]. Frequently, flavonoids acidified samples are used in the SPE procedure to prevent the ionization of these substances that prevent reducing their retention [34]. However, C_{18} is the widest sorbent used for the extraction of flavonoids from all kinds of matrices due to its own properties such as strong hydrophobicity, non-polarity and its functionality on a wide range of polarity compounds [103]. Oasis HLB has also been used for flavonoids extraction due to its RP-capability with a special "polar hook" that enhances the capturing of polar flavonoids [102]. Straightforward, a molecular imprinted polymer (MIP) is a sorbent that offers a quick and selective extraction phase for certain active compounds in the studied matrix [104]. Nevertheless, MIP is not the ideal choice for flavonoids' extraction as the presence of several OH groups in their main structures and the source of hydrogen bonding and non-localized electrostatic interactions between the OH groups [105]. In this direction, some attempts have been developed. The most common one is quercetin's attempt that provides a high affinity to the same related flavonoids family (flavonol). This MIP attempt was widely used for the extraction of quercetin, rutin, kaempferol, isorhamnetin, and their related derivatives from ginkgo Biloba leaf samples. For instance, MIP sorbent was prepared from DES solvent using quercetin as a template and an ethylene glycoldimethacrylate as a crosslinker [104]. This MIP was successfully applied to extract the quercetin, kaempferol, and isorhamnetin from ginkgo Biloba leaves using 200 mg of MIP in the SPE column and 1 mL of methanol triplicate.

In spite of all these great features applying SPE, it still consumes more solvents and time. Its manual procedures also need many manual handling steps. The usual procedure steps of SPE are filling the extraction resin into a column or directly using commercial extraction cartridges. The conditioning, washing, sample loading, and eluting of the analytes are the necessary steps to success with the SPE approach (Figure 10). Therefore, sometimes the evaporation step is applied to concentrate the analytes in the case of having to repeat the eluting step more than one time. For example, the extraction of flavonoids from various food matrices used a SPE-C_{18} cartridge. The extraction phase was conditioned by 3 mL of methanol followed by 3 mL of water as a washing step, then 1 mL of the sample was aliquoted. The loaded cartridge was dried for 10 min and let stand for 15 min to improve the extraction efficiency and selectivity. Later, the sorbent was washed by 1 mL of 5% methanol and eluted three times by 2 mL of 80% methanol. The collected eluents (6 mL) were evaporated to dryness and reconstituted with 1 mL of 0.1 M formic acid in 80% methanol and injected into HPLC [106]. It is important to indicate that the total SPE procedure was achieved within 13 min.

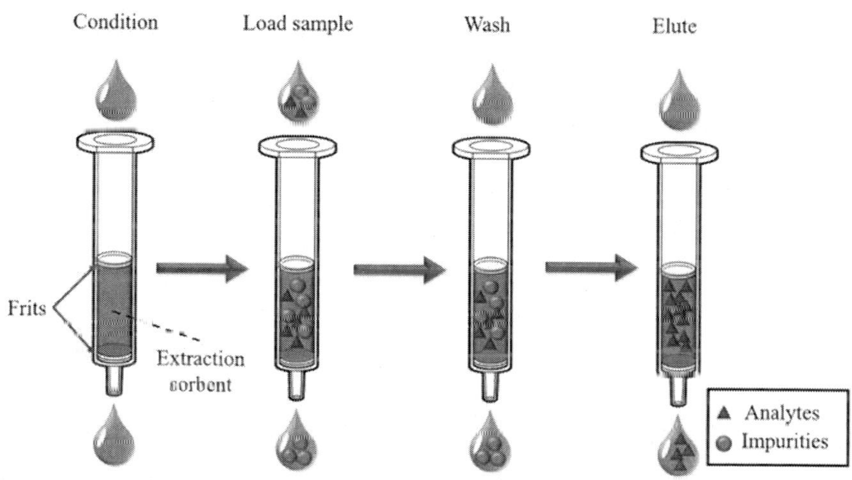

Figure 10. The most common solid phase extraction approach.

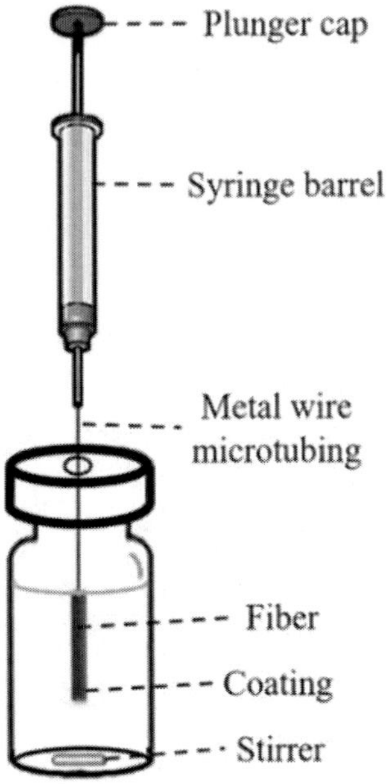

Figure 11. Solid-phase microextraction approach.

The evolution of SPE has led to the rising micro-SPE extraction techniques that successfully increase the extraction efficiency with less time, solvent, sample, and money consumptions. Solid-phase microextraction (SPME) is based on the adsorption of the target analytes on a suitable sorbent coated on fibre and fixed on a supporting wire that could be made of stainless steel [107], copper [108], aluminium [109], titanium [110], zinc [111], and gold [112]. Then, it was followed by thermal adsorption (solvent-free) or by a small quantity of a desorption solvent [113] (Figure 11). It is a useful technique for the effective extraction of flavonoids from food samples. This technique is simple for its ease of performance, solvent-free (eco-friendly), a short extraction time, and high sensitivity and preconcentration factors are observed [114]. On the other hand, the main drawbacks of this method come from the unavailability of the commercial stationary phase, its relatively high cost, and the stripping of coatings, the instability, swelling in organic solvents, bending of the needle, breakage of the fibre, and the short lifetime of the fibre [115, 116]. The extraction of 16 phenolic compounds, including five flavonoids, was carried out applying montmorillonite composites including ionic liquid to prepare a SPME fibre on the stainless steel wire as support [117]. The fibre was directly inserted into 20 mL of fresh fruit juice samples. Then the fibre was withdrawn into the needle after a certain extraction time, the adsorbed phenolic compounds were thematically eluted (solvent-free) by inserting the needle into the GC injection port after the derivatization process. In the case of HPLC, 500 μL of a mobile phase was used as an elution solvent before the HPLC determination.

Dispersive solid-phase micro-extraction (DSPME) has been recently developed to overcome the inherent limitations of SPME. It is considered as a non-fibre SPME method. In this technique, the sorbent is directly poured into the sample solution and not fixed onto the fibre core. Then, the sorbent desorption in the sample solution (extract the analytes) and in the acceptor solution (elute the adsorption analytes) are needed to increase mass transfer. The desorption could be done using a stirrer, ultrasonic, vortex, or even microwave-, ultrasonic-assisted system [113] (Figure 12). In this procedure, the extraction time, desorption time, and the quantity of sorbent are significantly reduced, providing a high enrichment factor, and low consumption of sample solution. While the difficulty of the sorbent separation and the need for many manual steps limit its automation [118]. In particular, the extraction of seven flavonoids (vitexin, luteolin, wogonoside, apigenin, chrysoeriol, acacetin, and pectolinarigenin) from *Veronicastrum latifolium* using DSPME assisted with microwave irradiation was reported. Zinc oxide and 500 μL of ethanol were attained as an adsorbent and eluent solvent, respectively. The mixing time between the analytes and the adsorbent in the sample solution was 120 s. The achieved recoveries were more than 98.1% with a higher achieved sensitivity ≤ 0.078 μg mL^{-1} [119]. Whereas, reducing the limitation of DSPME, micro- or even nano magnetic sorbents are the considerable novel sorbents in this method [120]. The ability to simplify the sorbent separation step helped with an external magnet providing a higher yield in a shorter extraction time due to the elimination of centrifugation or filtration steps. Additionally, it is easy to automate and maybe online hyphenated with the analytical technique [121]. In the literature, the use of magnetic sorbent in extraction procedures is classified as an individual technique, magnetic solid-phase extraction (MSPE) [122] or magnetic dispersive solid-phase extraction MDSPE [123].

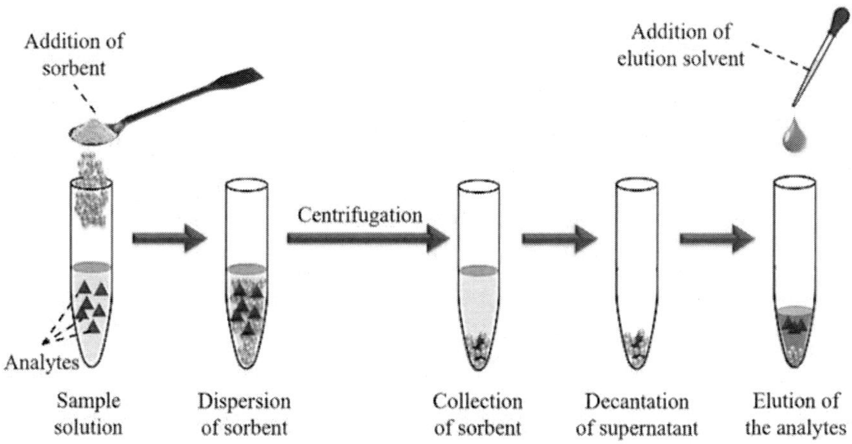

Figure 12. The main dispersive solid-phase microextraction approach.

Matrix solid-phase dispersive extraction (MSPDE) is the latest trend in flavonoids sample preparation of fruits, plants, herbs and other natural matrices. This procedure depends on the direct mixing of the sorbent with a solid or semi-solid sample; blending and homogenizing steps are implemented. Next, the mixture will be transferred to the extraction column and the analytes adsorbed onto the sorbent are eluted by applying a suitable organic solvent [124] (Figure 13). MSPDE provides the complete interaction between the sample and the sorbent, reducing the matrix interferences to the minimum, and increasing the extraction efficiency. Nevertheless, the handling, manual works are the main disadvantages of this technique such as blending, homogenizing, and filling the mixture into the extraction column or a syringe resulting in the difficulty of its automatization [124].

Table 4. Applicability of the recent SPE methods for the extraction of flavonoids

Extract. Tech.	Matrix	Ext.	Extraction conditions		LOD (μg mL⁻¹)	Recov./yield (%)	Refer.
			Solvent (mL)	Time (min)			
SPE	Food matrices	QU, HTIN, NARIN, NAR,	6 (MeOH)	13	≤ 29.2	72.5–104.2	[106]
SPE	Honey	RU, QU, KA	50 (MeOH)	-	≤ 0.1	40-85.4	[34]
MIP-SPE (QU templet)	Ginkgo biloba leaves	QU, KA, IS	3 (EtOH)	-	-	99.5-100.3	[126]
MIP-SPE (BI-A templet)	Laboratory prepared sample	DA, GE, BI-A	3 (MeOH:ACN 3:2)	-	-	89-101	[127]
SPME	Fresh fruits juice	QU, KA, RU, CA, EC	Thermally in GC, 0.5 (20:10:70 (v/v) MeOH: ACN:2% (v/v) acetic acid) in HPLC	5	$\leq 3.2 \times 10^{-3}$	86.6 – 98.4	[117]
MIP-SPME (QU templet)	Tea and coffee	QU	0.2 (MeOH: acetic acid) (9:1 V/V)	50	9.9×10^{3}	94.2 – 98.5	[128]
DSPME	Honey, and urine	QU, KA, IS	0.1 (MeOH)	2	$\leq 0.1 \times 10^{3}$	94.1 – 100.5	[129]
DSPME	*V. latifolium*	VI, LU, WO, AP, CH, AC, PE	0.5 (EtOH)	2	≤ 0.8	98.1 – 101.4	[119]
DSPME	Juice and smoothie	10 flavonoids	6.5 (MeOH: Water) (95:5 V/V)	30	-	57-102	[130]
MSPDE	Lime fruit	Neo-HES, NAR	0.4 (1-butyl-3-methylimidazolium tetrafluoroborate)	6	≤ 5.0	90.2–96.5	[131]
MDSPME	Dark tea, chocolate, vegetable, and fruit juice	MO, QU, KA	0.1 (DES, tetramethylammonium chloride and lactic acid)	7.1	$\leq 1.1 \times 10^{3}$	90.4-98.8	[132]

Table 4. (Continued)

Extract. Tech.	Matrix	Ext.	Extraction conditions		LOD ($\mu g\ mL^{-1}$)	Recov./yield (%)	Refer.
			Solvent (mL)	Time (min)			
MDSPME	Chinese medicine	AS, ISM, AST	4.5 (MeOH)	13	≤ 0.6	91.1 - 96.2	[123]
MSPDE	*M. cauliflora*	CA, QU, NARIN, MY	6 (MeOH)	6.8	-	-	[133]
MSPDE	Citrus fruits	NAR, HES, Neo-HES, NA	2.5 (MeOH)	11	$\leq 6.5\times10^{-3}$	96.8-104.7	[125]

AC: acacetin; ACN: acetonitrile; AP: apigenin; AS: astrapterocarpan; AST: astagaloside BI-A: biochanin A; CAT: catechin; CH: chrysoeriol; DA: DA: daidzein; DES: deep eutectic solvent; EC: epicatchin; EtOH: ethanol; Ext. extractant; Extract.: extraction; GE: genistein; HES: hesperidin; Neo-HES: neohesperedin; HTIN: hesperetin; IS: isorhamnetin; ISM: isomucronulatol; KA: kaempferol; LU: luteolin; MDSPME: magnetic dispersive solid-phase microextraction; MeOH: methanol; MO: morin; MSPDE: matrix solid-phase dispersive extraction; MDSPE: magnetic dispersive solid phase extraction; MY: myricetin; NAR: naringin; NARIN: naringenin; NA: narirutin; PE: pectolinarigenin; QU: quercetin; Recov.: recovery; SPME: solid-phase micro-extraction; VA-MSPD: vortex-assisted matrix soplid-phase micro-extraction; VI: vitexin; WO: wogonoside; Temp.: temperature; Tech. technique.

For instance, a simple and sensitive MSPDE method was established for the extraction of chiral flavonoids from citrus fruits using 30 mg of C_{18} as a dispersant mixed with 25 mg of sample in a mortar and grounded together along 1 min. The mixture was placed in a centrifuge tube and eluted in a vortex with 2.5 mL of methanol for 5 min [125]. The eluate was diluted and centrifuged again for another 5 min before the mass spectroscopic determination.

The comparison between all evaluated SPE methods concerning the extraction of flavonoids is illustrated in Table 4. The summarizing of SPE's advantages and disadvantages are illustrated in Table 6.

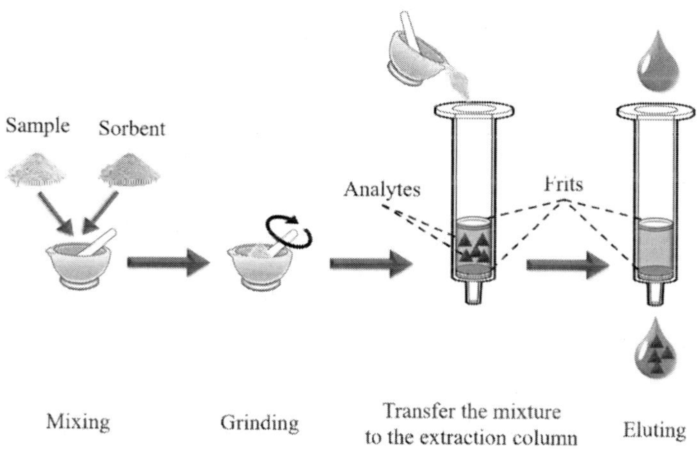

Figure 13. Matrix solid-phase dispersive extraction approach.

Other Modern Extraction Techniques

SFE is one of the modern and effective alternative techniques applied for a greener extraction of flavonoids from food and natural sources. This technique is similar to PLE, and the main difference is that SFE is based on the extraction of the analytes with solvents at temperature and pressure conditions above their corresponding critical points. Then these solvents will be at supercritical conditions, while PLE uses the temperature above the solvents boiling points and pressure enough to keep the extracting fluid in the liquid state. Regarding this point, the solvents acquire different unique physical properties such as the possessing of diffusivities of gases while maintaining densities of liquids. These features are responsible for owning a unique character of SFE compared with other extraction techniques. As can be inferred, carbon dioxide CO_2 is the most important extraction solvent implemented in SFE, even for the extraction of flavonoids. This gas possesses several very interesting characteristics for use as a supercritical fluid. For instance, but not limited to being cheaper than many other extraction solvents, it is environmentally friendly and since it is produced as a subproduct of different technological processes, no extra CO_2 emission is needed, and it's food-grade solvent. Additionally, the most important character regarding the analytical point of view is a solvent-free extract since it is a gas at atmospheric conditions. It totally evaporates due to the depressurization and leaves the extractant totally solvent-free and ready to be further analyzed, thus the ease of its being coupled with the analysis tools [134]. In the extraction of flavonoids, CO_2 is considered a non-powerful extraction solvent concerning its very low polarity. In this sense, the combination between CO_2 and other polar organic solvents such as methanol, ethanol [135], and ethyl acetate [136] help it to encourage the enhancing of the extraction yield of flavonoids. In practice, the comparison between the yields of total phenolic content, total flavonoid content, and antioxidant activity, applying different SFE extractions solvents for a Brazilian variety of *Humulus lupulus* using CO_2, CO_2 + ethanol, CO_2 + ethyl acetate, and compressed propane has been published. The results indicate that ethyl acetate increases the extraction yield with higher total phenolic content, total flavonoid content, and antioxidant activity values compared with other studied extraction solvents [136]. The extraction time ranged between 30 and 110 min depending on the kind of solvents used, resulting in the faster extraction in the combination of (CO_2 + ethyl acetate). On the contrary, pure CO_2 was the longest process. This clearly depends on the polarity and the affinity of the solvents to the extraction compounds. Worthwhile, the evaporation step of the adding co-solvents needed 30 h (a long processing time), while the pure CO_2 directly evaporated. Moreover, the comparison between the extraction of total flavonoids from *S.purpurea,*

S.viminalis, and *P. nigra* using pure CO_2, CO_2 + water, and pure water indicated that the highest yield was observed by using CO_2 at 9 h of the extraction time, which indicates the hydrophobic nature of this group of chemicals present in the analyzed raw materials [137].

The chosen suitable extraction solvent plays an important role in the improvement of the enrichment factor as well as the optimization of the attained temperature and pressure [138]. In the case of the elevating of these physical factors, the possibility of the flavonoids' degradation will be also increased. For that, these variables should be carefully optimized to enhance the sensitivity and the selectivity of the extraction. Otherwise, the main disadvantages of this technique are the higher maintenance cost, the requirement of special high-pressure vessels according to the high-pressure operation conditions, and the difficulty to reach the equilibrium status between solute and solvent.

HSCCC is a fast-growing separation and purification technique. It is similar to solid-phase chromatography (LC), but its support-free liquid-liquid partition chromatography, thus its functionality in principle is a LLE procedure. In other words, the separation can be realized according to the different partition coefficients (*K*) of the target compounds provided by a suitable solvent system. HSCCC has a number of advantages over these similar mentioned techniques, such as its being more efficient than LLE, belonging to a great number of exchanges that perform between liquid phases providing a higher recovery and higher capacity, no irreversible adsorption and eliminating the possibility of the extraction column to be clogged by the impurities located in the crude samples compared with LC, and low risk of sample denaturation [139].

Recently, HSCCC has been carried out in the analysis, separation and preparative procedures of flavonoids from natural samples [140]. For example, the preparative extraction of five flavonoids from *Caragana korshinskii* Kom was developed using ethyl acetate: n-butanol: water (0.5% glacial acetic acid) (4:6:10, v:v:v) at a flow rate of 35 mL min^{-1} as a suitable solvent system. The separation was done in 400 min and the obtained purities were higher than 98% [141]. As another example, HSCCC based DSE solvent was developed for the extraction and separation of five flavonoids from *Malus hupehensis*. Choline chloride: glucose-water-ethyl acetate (1:1:2, v/v) as separation liquid systems, 2 mL min^{-1} of flow rate, and 240 min of extraction time were the best-operating conditions [142].

From the literature, the most developed HSCCC have been carried out at a long extraction time that can reach up to 400 min applying a high flow rate [141], resulting in a high solvent and time consumptions. These points are the main drawback of the technique along with the non-commercially available instrument, its low separation efficiency, and the difficulty to maintain the liquid stationary phase in a steady-state while the mobile phase is pushed through. The change in the composition of one liquid may induce a change in the other depending on the mechanism of HSCCC that is based on the diffusion process between liquids and the efficiency of the mixing phase. Therefore, the selecting of the stationary phase and the mobile phase must be taken with a lot of care [143]. However, owing to the limited peak capacity of unidimensional HSCCC, two dimensional (2D-HSCCC) has been exploited advancing its power on the separation of the similar polarities of flavonoids in the complex samples. The inherent drawbacks of the 2D approach are the need for the double extraction time that is already considered very long (off-line mode), over the required two expensive HSCCC instruments in on-line mode [144].

The main parameters that should be optimized in HSCCC are the composition of the solvent system, the operating temperature, the flow rate of the mobile phase, and revolution speed [145].

Automation of Extraction Techniques

Due to the industrial revolution and increasing demands for quality control of the produced goods, the need for quick, costless, and effective sample preparation and analytical techniques have been the essential needs. These requirements aroused the beginning of another revolution to miniaturize and automate the analysis procedure. This term, reduces the required time and effort for the analysis of each production batch and results in minimizing the production cost. The sample preparation step is considered a long and tedious step in the analytical procedure that can consume 80% of the completely analytical operation. Moreover, some analytical techniques are mandatory to clean-up the matrices before the analysis, such as all chromatographic systems.

In this sense, the interest in the development of automated extraction procedures has recently been increased. However, the development of fully automated methods covering all processing steps, from the extraction to the detection stage, is not always possible and sometime needs intensive working-time for designing, creating, optimizing, programing, and also, in some cases needs additional expensive tools.

Regarding flavonoids analysis, the majority of sample pretreatment techniques have been manually implemented; hence, some works have been illustrated as semi-automated extraction techniques of flavonoids from fruit and plant matrices. Semi-automated techniques mean that some works must be manually handled, such as the off-line connection of the extraction and detection systems that the extractants should be collected and manually transferred to the analysis tool [146, 147], and filling and replacing the extraction column in case of SPE [103, 146, 148-150]. This can happen in either off-line or even on-line approaches.

In principle, Flow analysis techniques (FATs), in all their generations, attempted to automate and integrate the sample pretreatment step with detection or separation instruments. Many FATs systems have been developed since the 1980s [151]. To their credit, FATs are fast, economic, more robust, more flexible than their batch counterparts. They reduce analytical costs and waste production by using samples and reagents sparingly, thus minimizing human intervention and decreasing their exposure to the solvents [152].

Since the first development of the FATs, numerous works and reviews have been focusing on the automation of the sample pretreatment steps and to integrate them (off-, on-, in-, at-line) with the detection and/or analytical instruments. Many works have been found in environmental analysis, biological analysis, medical analysis, radiochemical analysis, among other analytical fields [153-160].

In flavonoids, determination-based FATs, the first trail in this field was the online connection between flow injection analytical system (FIA) and capillary electrophoresis (CE). This system has been successfully implemented for the extraction and determination of three flavonoids (catechin, epicatechin, and quercetin), among other phenolic acids in different wine samples. The method is based on the continuous flow connection assisted by peristaltic pumps to simultaneously pump up the sample through the SPE-C_{18} minicolumn. Then, after their elution by 2 mL of methanol, the analytes were transferred to the CE autosampler helped by a homemade programable arm [161]. 6.30 min was the required extraction time to reach the eluted analysts to CE. The achieved limits of detection and the recoveries were $\leq$ 0.36 mg L^{-1} and 92% - 110%, respectively.

The following work in 1999 by Perez et al. [162] was for the online determination of hesperidin in orange samples. It was based on a coupling of the FIA system to a fluorescence detector. The method depends on the online formation of a fluorescent complex between hesperidin and Al^{+3}. The detection limit was better than 79 μg L^{-1}.

It is worth mentioning that the most important and computerized FATs systems that can be benefit from the automation of sample pretreatments and on-line coupling to the analytical techniques are the multisyringe flow-injection analysis (MSFIA) and the lab-on-valve (LOV) tools. Much later, beginning in 2012, the interest in flavonoids-based FATs determination increased. Wang et al. successfully applied a LOV-SPE-voltammetry method for an effective extraction of quercetin from human urine and red wine samples coupled with voltammetric detection [50]. The consumed solvent and time were 0.45 mL and 57.3 min, respectively. The achieved sensitivity was 0.39×10^{-6} μg mL^{-1}. Another LOV-μSPE-HPLC-DAD has been developed by Sammani et al. in 2019 [53].

The on-line extraction of hesperidin and diosmin from citrus juice and their related products was the target of this work. LOV was on-line coupled to the HPLC system. In this method, the repeatability, sensitivity and total procedure time were notably enhanced. The extraction and the analysis were carried out simultaneously in 25 min consuming only 0.27 mL of extraction solvent. The achieved LODs were lower than 0.32 μg mL^{-1}.

Analyzing the previously mentioned data, several advantages belonging to LOV are observed. It is able to provide multi-communication of reagents and samples, allowing to hold one of its micro-channels (a grafted micro-channel) by a small amount of extraction resin achieving a micro-SPE system (Figure 14). Hence, it greatly facilitates the automation of SPE mandatory steps including the filling, conditioning, washing, loading, eluting, and even replacing of the extraction column. Further, directly transferring the analytes to the separation/detection tools is also carried out by the net command of the controlling software [160]. All of these automated steps have been done assisted by a MSFIA. It is a powerful aid system that includes a conventional automatic titration burette headed with a fast-switching solenoid valve, adapted with one motor that can simultaneously move a metal bar with up to 4 syringes (Figure 15). MSFIA provides a programable aspiration and dispensing of solvents and samples using simple and powerful controlling software such as the Autoanalysis software developed by Sciware Systems-SL [163]. All these observed features led to achieving higher analysis throughput, a much greater reduction of the consumption of solvents, the required amount of extracting resin and the total analysis time compared to the existing manual pretreatment methods that can reach up to a 99% reduction [53]. Furthermore, the extraction beads can be used several times before their replacement, thus reducing the analysis time, analysis cost, waste generation and increasing the precision of the analysis [53].

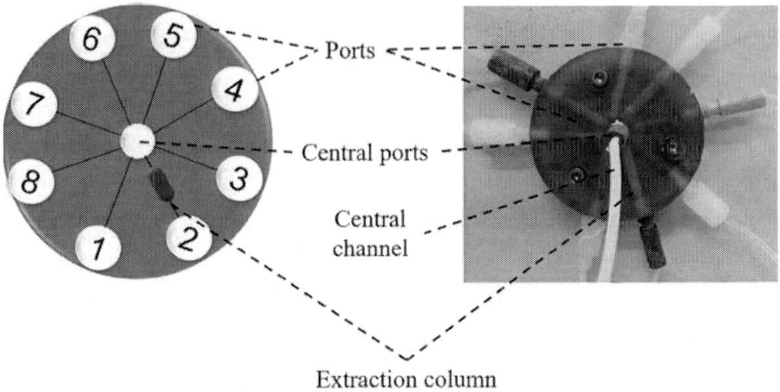

Figure 14. The diagram and real lab-on-valve (LOV) device.

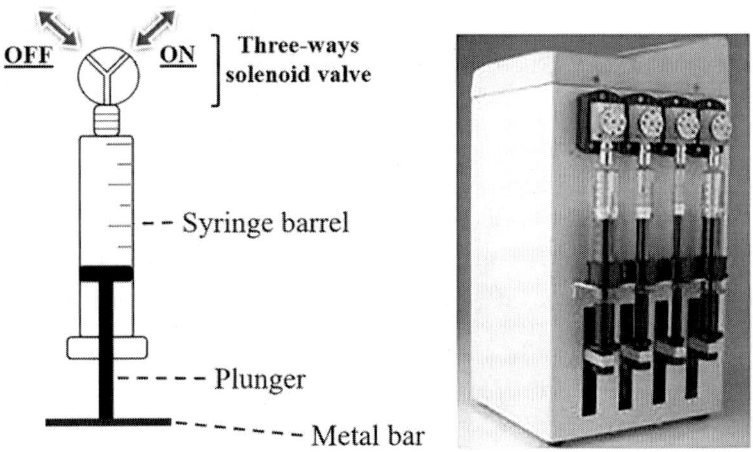

Figure 15. Multisyringe flow injection analysis system (MSFIA).

Microfluidics is another dynamic scientific-technology constructed on a solid substrate (glass or plastic) containing a network of micro-sized channels, which are connected to the reagents or sample reservoirs allowing the manipulation of fluid volumes in the order of nano- or picoliters. This technology holds superior advantages compared to conventional macroscale technical systems. Microfluidics is an attractive system in many fields such as biological analysis (the most microfluidic applications have been done in this field concerning the separation and determination of DNA and RNA), food analysis, and medicinal analysis.

In the field of food chemistry, it plays an important role in the miniaturization and automatization of not only the sample pretreatment and the extraction of the analytes, but also the integration of the separation and detection tool in one small device. In addition, it greatly minimizes the consumption of solvents and time, increases the sensitivity, selectivity, and analytical throughput. Its valveless flow control and small scales provide the advantage to be used in the field. To date, there are a countless number of developed microfluidic systems with the purpose of the automation of the extraction stage such as lab-off-chip, lab-on-chip and paper-based microfluidic system [164]. In sample preparation based microfluid, the use of off-chip sample pretreatment is very common in microfluidic analytical methods. This option is time consuming, leads to an undesirable loss of analytes, may introduce contaminants, and does not achieve a fully automated system [165]. To avoid these limitations and to automate the whole analytical process, on-chip sample pretreatment methods are implemented integrating all of the required analytical processes, by loading, filtrating, mixing, extracting, separating, and making the final detection. The LLE-based microfluidic system offers the possibility of performing the separation without the need of stirring or agitation, owing to the high surface-to-volume and short diffusion distances typical in microfluidic environments [166]. The SPE-based microfluidic system has been implemented in two ways, coating channel walls with a solid sorbent, but in this case, the loading capacity of this approach was relatively low or filling the microchannels with a solid sorbent that leads to increase the capacity of the extraction. Figure 16 shows an example of the solid-phase extraction based on a lab-on-chip device. Nevertheless, it is necessary to localize these particles in targeted regions of microchips and prevent their loss by the solvents flow using physical barriers, such as a sol-gel structure [167], two-weir design [168], and photopolymerized frit [169]. In this sense, the use of functionalized magnetic beads has shown its usefulness when applying SPE in microfluidic systems, making the packing, removal, and replenishment of the extraction beads straightforward. On the other hand, the main disadvantages of using these systems are the integration of all analysis-required stages in one device, which is not always possible and sometimes impossible to achieve a functional device that can provide the fully automated tool. For this purpose, tedious work for designing, formatting, creating, testing, optimizing the factors in microscale is usually required. The complexity of food, biological, and environmental samples, does not always allow one to directly apply in a microfluidic device and needs some purification and filtration steps to avoid the clogging of the microchannels [170]. Another aspect that needs to be improved is the limited extraction length and optical path which led to a limited extraction capacity and detection limits [171].

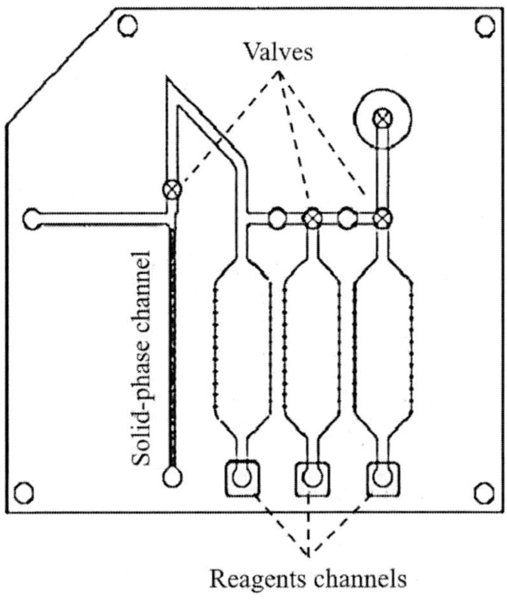

Figure 16. The example of solid-phase extraction-based lab-on-chip.

The analysis of flavonoids using microfluidic systems is quite limited. A lab-on-a-chip device coupled with a chemiluminescence detector has been developed for the simultaneous determination of total flavonoids and total phenolic acid in tea and honey samples using 20 mg of a magnetic zinc-imidazole framework (ZIF-4) as extraction sorbents [54].

The developed microfluidic device has notably reduced the consumption of solvents (0.7 mL) and time (21.1 min) providing good accuracy (95.6%-104%). Another developed work is based on the application of a porous polymer monolithic column for SPME of catechin in green tea using a fabricated lab-on-chip device with chemiluminescence detection [172]. Solvent-free elution, high recovery (90%-100%), high sensitivity (0.29×10^{-6} μg mL^{-1}), and high enrichment factor (20 times) of catechin were achieved due to a short extraction time (5 min).

A paper-based microfluidic system is a portable alternative compatible microfluidic analytical technique. In this technique, some zones (chromatographic or filtration paper) are placed onto a fixing tape. The flowing of the analytes between the zones is carried out with a power-free fluid transport via capillary forces [173]. Some of its advantages include low cost, easy to use, low reagent consumption, portability, biocompatibility, and biodegradability. This system has been recently developed for the determination of four flavonoids (rutin, quercetin, morin, and taxifolin) in hawthorn tincture and onion peel after their filtration, purification, and preconcentration onto the used paper in one step. The method consists of three applying paper zones (loading, preconcentration and detection). The sorbent (Hypercrosslinked polystyrene) was fixed on the preconcentration zone using a strong cylindrical magnet, which was put under the preconcentration paper. Flavonoid solutions were trickled down into the loading zone, a paper drain leading to a beaker was inserted instead of a detection zone for the elimination of the excess solution out of the paper, then this paper was replaced with a detection one and flavonoids were eluted with 50 μL of 0.1 mol L^{-1} KOH ethanol solution. After that, elution profiles of flavonoids on magnets were determined by HPLC [174]. The methods were run out in 12 min and achieved LODs lower than 14 μg mL^{-1} and recoveries between 60% and 90%.

Finally, the automated sample pretreatment strategies draw serious attention in the field of flavonoids extraction and determination. Unfortunately, these techniques are still in their beginning stages regarding flavonoids matter and have not yet been widely applied in this field. Reviewing the literature, FATs have a future outcome as powerful alternative techniques in front of the conventional ones. It is also true for the microfluidic systems due to their ability to simplify and automatize the whole sample pretreatment of flavonoids within one commercially available instrument assembly benefiting from advantages of the coupling analytical techniques, especially the separative ones that can effectively separate and determine the desired extracted flavonoids in the complex samples. On the contrary, the developed lab-on-chip devices are based on the extraction of total flavonoid content [54] or one compound [172], being unable to separate and determinate flavonoids individually. Even though, paper-based microfluidic system's developed work depends on the preconcentration of the analytes by drying the sample spot, nonetheless, the method achieved low recovery (60% – 90%) and poor sensitivity (14 $\mu g\ mL^{-1}$) compared with the lab-on-chip or developed flow analysis methods, but it is still attractive if it is improved. On the other hand, the difficulty of integrating the extraction part into the microfabricated chip limited the applicability of these devices. That has caused most development microfluidic systems to focus on the determination of total flavonoids, total phenolic content or the total antioxidant property of the matrices by spectroscopy or chemiluminescent detection directly without relying on the extraction process [175-177].

The developed automated sample pretreatment methods applied to flavonoids extraction from food and plant matrices are illustrated in Table 5. The main advantages and drawbacks of all mentioned extraction techniques belonging to flavonoids extraction are summarized in Table 6.

Table 5. Semi- and fully automated methods developed for the extraction of flavonoids

Extract. Tech.	Matrix	Ext.	Extraction conditions			LOD ($\mu g\ mL^{-1}$)	Recovery/ yield (%)	Refer.
			Solvent (mL)	Temp (°C)	Time (min)			
On-line-PLE-SPE	Apple pomace	TFC	2 mL min^{-1} (EtOH)	60	70	-	≤ 0.97	[146]
Off-line- MAE-SP	*S. medusa*	TFC	2L (EtOH)	80	60	-	4.97	[147]
On-line-PLE-SPE-UV	Yerba mate	4 flav.	Variety of flow rates (water and EtOH)	40	-	-	-	[148]
On-line-UAE-SPE-HSCCC-DAD	*R. rosea*	4 flav.	0.5 mL (IL)	-	485	-	0.02 – 0.2	[149]
On-line- SPE-HPLC	Chinese medicines	14 flav.	MeOH- acetic acid (8:2, v/v) in a dynamic mode	-	0.82 (elution)	≤ 0.05	93.2-94	[133]
On-line-SPE-HPLC	Environment water samples	LU, AP, CHR	1.0 mL min^{-1} (MeOH: ACN: 0.02 M H_3PO_4, 35:35:30) (v/v/v)	-	12	≤ 0.11× 10^3	79.8- 105.3	[150]
On-line- VA-MSPD-HPLC	Tomato, onion, grapes, red wine	MY, QU	Citrate buffer (0.001 M): ACN (70:30) in a dynamic mode	-	55	≤ 48.3 × 10^{-6}	96–100	[178]
On-line-FIA-CE	Wine	EP, QU, CA	2 mL (MeOH)	-	6.3	≤ 0.36× 10^3	92 - 110	[161]
On-line-SI-LOV-SPE-V	Human urine, red wine	QU	0.45 mL (30% (v/v) MeOH–phosphoric buffer)	-	57.3	≤ 0.39× 10^{-6}	97.7-103.8	[59]
On-line LOV-µSPE	Orange, lemon, mandarin Juices	HES, DIO	0.27 mL (ACN)	-	25	≤ 0.1	99.4-105.5	[53]

Table 5. (Continued)

Extract. Tech.	Matrix	Ext.	Extraction conditions			LOD ($\mu g\ mL^{-1}$)	Recovery/ yield (%)	Refer.
			Solvent (mL)	Temp (°C)	Time (min)			
Lab-on-a-chip-LC	Tea, honey	TFC	0.7 mL	-	21.1	≤ 0.1	95.6 - 104	[54]
Lab-on-a-chip-CL	Honey	TFC	20 mL min^{-1} (0.5 mmol L^{-1} potassium permanganate)	-	10	12.0×10^{-9} mol L^{-1}	$\leq 765.4 \times 10^4$	[177]
On-chip-CL	Green tea	CA	Without elution step	-	5	0.29×10^{-6}	90-110	[172]
μPADs-SP	Hawthorn tincture, onion peel	QU, MO, RU, TA	100 μL (0.1 mol L^{-1} KOH EtOH)	-	12	≤ 14	60-90	[174]

ACN: acetonitrile; AP: apigenin; CE: capillary electrophoresis; CA: catechin; CHR: chrysin; CL: chemiluminescence; DIO: diosmin; EP: epicatechin; EtOH: ethanol; Ext. extractant; Extract.: extraction; Flav.: flavonoids; FIA: Flow injection analysis; HSCCC: counter-current chromatography; HES: hesperidin; LU: luteolin; LOV: lab-on-valve, MAE: microwave-assisted extraction; MeOH: methanol; MO: morin; MY: myricetin; μPADs: paper-based analytical devices; PLE: pressurized liquid extraction; QU: quercetin; RU: rutin; Recov.: recovery; V: voltammetry; VA-MSPD: vortex-assisted matrix solid-phase microextraction; SP: spectrophotometer; SI-LOV: sequential injection lab-on-valve; TFC: total flavonoids content; TA: taxifolin, UAE: ultrasound-assisted extraction; Temp.: temperature; Tech. technique.

Table 6. The summarizing advantages and disadvantages of extraction techniques of flavonoids

Extract. Tech.	Advantages	Disadvantages
MAE	Ease of use, reducing the extraction time, and solvents than CETs, increasing the extraction yield.	High-processing temperature may decompose the flavonoids, microwave radiation negatively affects the biological properties, low selectivity, consuming solvent and time.
UAE	Less degradation effect than MAE, does not affect the biological properties of flavonoids, increase the extraction yield from the plant materials.	Elevated temperature produces flavonoids degradation, is difficult to maintain a constant temperature along the process, requires big solvent quantity and time, has low selectivity of flavonoids.
PLE	Increase the mass transfer rate, provides a high extraction yield, better flavonoids solubility, has the possibility of its automation.	High pressure and temperature processing conditions lead to the degradation of flavonoids and decrease the extraction yield, high equipment's cost, non-selective.
SWAE	Prevents the flavonoids degradation, very short extraction time, small extraction solvents' quantity, eco-friendly, observes high extraction efficiency.	Low sensitivity, unavailability.
LLE	Less solvent, sample, and time consumption than CETs, it has simple operation, does not requires complex apparatus, effective for some flavonoids' subclasses.	Tedious, time-consuming, large amount of toxic and expensive organic solvents are required, non-environmentally friendly, low enrichment factor, low selectivity of some flavonoids' subclasses, the difficulty of phase separation.
LLME	More eco-friendly than traditional LLE, higher recovery and higher enrichment factor than LLE, simple and rapid of operation, low operational cost.	As LLE, is difficult to be applied to many sub-groups of flavonoids extraction, the difficult phase separation, requires centrifugation, the difficult its automation.
DLLME	Simple and rapid operation, effective in the extraction of flavonoids from aqueous matrices, achieves high enrichment factor, eco-friendly, not required evaporation before eluent's analysis.	Difficulty to completely separate the extraction solvents, low selectivity, limitation of solvents choice for many sub-groups of flavonoids, requires centrifugation, not appropriate for complex matrices.

Table 6. (Continued)

Extract. Tech.	Advantages	Disadvantages
HF-LPME	Simple, inexpensive, reducing a lot the consumption of solvents, environmentally friendly, tolerates a wide pH range, excellent sample clean up, high enrichment factor, easy of automation.	Relatively long extraction time, possibility of blocked fibre pores, fibres commercially unavailable, fibre with a relatively high cost.
SPE	Less solvent and sample consumption than LLE, relatively short extraction time, diversity availability of sorbents, ease of automation, high reproducibility, high enrichment factor.	Time and solvents consumption regarding the required operational steps, clogging the pores of the solid phase after several extraction times thus the efficiency is reduced and the using of new extraction beads is essential.
SPME	Simple, the separation is fast, high sensitivity, elution solvent-free, eco-friendly, easy of automation.	Lack of robustness, fibres commercially unavailable, relatively high cost, polymer is fragile and has a limited lifetime, the small coating surface reduces the extraction capacity.
DSPME	SPE column is not required, reducing the required amount of sorbent, the consumption solvents, the operational time, increasing the selectivity and the enrichment factor.	Reducing the extraction capacity due to the limitation of the extractant amount, its automation is challenging since it requires centrifugation.
MSPDE	Centrifugation and filtration steps are not required, the solid or semi-solid sample can be used directly, dramatically reduces the consumption of solvents and time, the maximum contacting between the sample and the analytes is performed.	Many manual steps are required, lack of its automation, further sample clean-up is almost required.
SFE	Eco-friendly, reducing the waste generation, unique physical features of using mobile phase in a corresponding critical point, high mass transfer, and high extraction yield.	Required high temperature and pressure conditions,-long processing time led to flavonoids degradation, high cost of required high-pressure vessels, high periodic maintenance cost, expensive equipment, difficult solvents equilibration.
HSCCC	Completely recovering sample, can be used for industrial extraction amounts, much less noise and the stability is effective, enhancing the throughput performance.	Expensive equipment, unavailability, long processing time, low separation efficiency, difficulty to maintain the liquid stationary phase in a steady state.

Extract. Tech.	Advantages	Disadvantages
FTAs	Robust, closed solvents protecting loss of volatile organic solvents, non-transfer loss of analyses, higher recovery, highly reducing the required time and solvents, eco-friendly, increasing the analysis throughput, the extraction and separation procedures are performed simultaneously, effective automation of the whole analysis, provides a solution to the limitation of several procedures such as SPE and DLLME.	Less flexible since the final elution conditions need to be compatible with the detection system due to the difficulty to apply the evaporation and centrifugation steps, high back pressure when using SPE extraction columns which could be led the leakage of the solvents from the system.
MFS	Microfabricated device, ease to handle, suitable for fieldwork, small lab's space is required, reducing of the solvent and time consumptions, saving energy, eco-friendly, enhancing the method throughput.	Tedious work is required, difficulty to include all required processes in the same chip (the main objective of MFS), each analysis objective needs a new appropriate chip, low selectivity and sensitivity regarding the limited pathway.

DLLME: dispersive liquid-liquid microextraction; DSPME: dispersive solid-phase microextraction; FATs: flow analysis technique, HSCCC: high-speed counter-current chromatography; HF-LPME: hollow fiber liquid phase microextraction; LLE: liquid-liquid extraction; LLME: liquid-liquid microextraction; MSPDE: matrix solid-phase dispersive extraction; MAE: microwave assisted extraction; MFS: microfluidic systems; PLE: pressurized liquid extraction; SPE: solid-phase extraction; SPME: solid-phase microextraction; SFE: supercritical fluid extraction; UAE: ultrasound-assisted extraction.

Analytical Methodologies of Flavonoids

The determination of flavonoids in their natural sources and the knowledge of their exact concentration and the estimation of the quality of the origin has utmost importance regarding their remarkable therapeutic and pharmacological behaviours. On the other hand, concerning the vast differences' characterization of flavonoids compounds in nature, their countless derivatizations, and the numerous linkage groups to the main framework of flavonoids, differentiate their ability to dissolve in aqueous and organic solvents, polarity, the complexity of the matrices, and their low concentrations in the natural samples, several analytical methodologies have been developed to fulfil this purpose. Therefore, the objective of the analysis and the concentration of the target flavonoids in the sample play a significant role in the choosing of the appropriate analysis method. In general, if the determination of total flavonoids or the estimation of the matrix's biological or antioxidant behaviour is a matter of concern, spectroscopic techniques are the one of choice belonging on their fast, simple, and low-cost. When the individual estimation of flavonoids in the plants, food, dietary supplements, and pharmaceutical samples are the aim, separation analysis techniques are focused on regarding their high sensitivity and selectivity in the separation and determination of individual flavonoids.

In the following paragraphs, the classification of the analytical techniques used in the field of flavonoids determination in plants, fruits, foods, and commercial products will be detailed. The benefits and the drawbacks of each technique will be also explained. The recent applications for this purpose will be mentioned.

Spectrophotometric Assay

Several absorptions and vibrational spectroscopic methods have been developed for the determination of flavonoids in their different sources. They include direct ultraviolet radiation (UV) [179], visible radiation (Vis) [180], fluorescence technique (FL) [181], nuclear magnetic resonance spectroscopic technique (NMR) [182], X-ray crystallography (XR) [183], and circular dichroism spectroscopy (CD) [184] as an absorption spectroscopic technique. On the other hand, the vibrational ones are near-infrared (NIR) [185], Raman spectroscopy (RS) [186], terahertz (THz) [187], and hyperspectral imaging spectroscopy (HSI) [188]. Also, the interesting applications of mass spectroscopy in the area of flavonoids' determination cannot be forgotten [189].

In UV-VIS radiation techniques, flavonoids are well-defined compounds for their powerful absorption of UV radiation. In principle, the main skeleton of flavonoids provides two important absorption bands I (240-285 nm) and II (300-560 nm) from the absorption of ring A and B, respectively (Figure 17) [183]. Almost all flavonoids have at least one maximum absorption wavelength in the range of these bands. In this sense, UV spectroscopy is the technique of interest to evaluate the flavonoids in different matrices. In terms of quickness, cost-effectiveness and environmentally friendly method comparing with chromatographic techniques, UV spectroscopy is the major technique for the evaluation of total flavonoids, total antioxidant power, and total biological behaviours in the natural samples and it has been widely accepted for these purposes during decades [184].

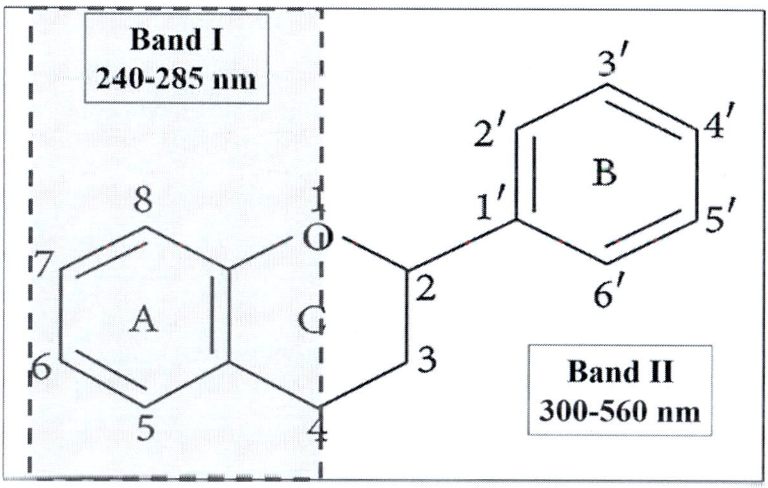

Figure 17. The main ultraviolet absorption bands of flavonoids' main framework.

However, this technique is considered as a non-selective detection technique of flavonoids because of the maximum wavelength for most flavonoids located in narrow two range bands. Additionally, the linkage molecules to the flavonoid's skeleton such as methyl, methoxy, non-dissociated hydroxyl and sugar molecules, in the case of glycosides, have also overlapped the absorptions and effect negatively on the accuracy and the sensitivity of the analysis. On the other hand, natural sample endogenous' unexpected absorptions have limited the application of this technique. Whereas, some works were found for direct determination of total flavonoids in their sources [179]. Whilst the colouration methods are better than the direct UV estimation, it will not be disturbed by other unwanted components, as well as increase the selectivity to flavonoid groups [190], but still is not able to differentiate between the different flavonoids' compounds located in the sample which affected the accuracy and reliability of the results [191]. For example, all anthocyanidins can form intensive blue compounds chelating with Al^{3+} [192]. Due to these properties, it is only used for total flavonoids determination in natural samples. The coloured flavonoids are usually done by complex formation of flavonoids with shift reagents such as 2.4-dinitrophenylhydrazine [193], 4-aminoantipyrine [194], oxidation with Cu (IV) [195], and aluminium chloride ($AlCl_3$) [192]. In practice, $AlCl_3$ is the most widely implemented reagent for total flavonoids evaluation. The principal procedure is based on the mixing of methanolic [196], ethanolic [197], or alkaline [191] sample medium with 1.5% to 10% of $AlCl_3$ [198-200], then the mixture stands for incubation for the time ranging between 2 and 60 min [198, 199] and then the mixture was directly measured at the appropriate visible-light wavelength depending on the solvent used.

Otherwise, some flavonoids' groups such as isoflavonoids, polymethoxylated flavones and flavonoids with a free hydroxyl in the C_3 position show a slight natural fluorescent behaviour [201]. In these groups, the FL technique provides a selective and useful choice for their total determination in natural complex matrices. In general, flavonoids are considered as good chelating compounds. Taking advantage of this feature, the enhancing of fluorescent behaviour of these groups as well as the other flavonoids' groups that do not exhibit a native fluorescence, FL can be applied after the formation of the chelating complex with highly fluorescent chelating ions such as Al^{+3} [181], Tb^{+3} [202] and Cu^{+2} [203]. Of note fluorescently, Naturstoff reagent A (diphenylboric acid 2-aminoethyl ester) has been reacted with some flavonoids to visualizing them in plant tissues and inhuman in vitro cell cultures [204]. This complexation is enhanced significantly by the fluorescent signal of studied flavonoids such as apigenin [205], kaempferol, naringenin and quercetin [204], but it has not been implemented yet outside the scope of tracking flavonoids at the cellular level. Hence, FL has notably improved the sensitivity and the selectivity of the detection flavonoids' groups but still required additional preparation time to form a chelating complex that can reach up to 60 min, plus it is poor selectivity for the determination of individual flavonoids. It should be noted here that these detection techniques have been deeply used for decades for the determination of total flavonoids in plants, foods, and human fluids, but nowadays, many other powerful tools are replacing these techniques in the determination of flavonoids.

In the field of flavonoids' structural elucidation, NMR [206], mass spectroscopy (MS) [189], XR [183], and CD [184], in addition to vibrational spectroscopic techniques are the common analytical techniques that have been implemented for this purpose.

NMR is usually used in quality control and research for determining the content and purity of a sample along with its molecular structure. It is increasingly used to provide insight into a mixture of natural products simply and without time-consuming chemical degradations and syntheses [207]. NMR can quantitatively analyze known compounds. For unknown ones, it can either be used to match against spectral libraries or to infer the basic structure directly. The main application of NMR in flavonoid research is the structural elucidation of novel compounds, but the potential for its quantitative ability (qNMR) has recently increased. The qNMR method could simultaneously detect multiple components in a very short time (1–5 mins) using a very cheap internal standard reference [208]. On the other hand, the application of NMR for naturally occurring flavonoids, most of which are glycosides is limited, because of their low solubility in most organic solvents, common attached structures, such as methoxy and acetyl, that are generally not suitable for the NMR analysis because the signal pattern of the natural product is often obscured, in part, by the signals of the additional groups [209]. Regarding the limited dissolving properties of flavonoids in many aqueous and organic solvents, DMSO-d_6 is the optimal solvent among others to perform NMR of flavonoids that almost all kinds of flavonoids could be well dissolved in it, and solvent peaks rarely overlap the resonance signals of flavonoids [210]. In the implementation, NMR was used for the structural analysis of flavonoids in herbs and medicinal plants matrices [206, 211]. Quantitative NMR (qNMR) was successfully applied for the quantitative analysis of three flavonoids (liquiritin, liquiritigenin, and isoliquiritinin glycyrrhizin) in *Glycyrrhiza uralensis* Fisch extract using DMSO as the solvent, and dichloromethane as the internal standard. The obtained LOD was less than 0.023 mg mL^{-1}, the recovery ranged between 100% and 108.8% [212]. qNMR is also utilized for the simultaneous determination of four flavonoids (baicalin, baicalein, wogonin, and wogonoside) in *Scutellaria baicalensis* Georgi Extract using 3,4,5-trichloropyridine as an internal standard [213].

Another important spectroscopic technique that has also been utilized for flavonoids identification is MS. This technique is a physicochemical method that has proved to be one of the most effective techniques for the identification and structural determination in biomedical research, especially when complex matrices must be analyzed. MS provides high sensitivity and high specificity, as it can separate molecules of the same molecular weight but different atom composition [214]. In the case of flavonoids, especially in glycoside forms, these compounds are polar, nonvolatile and thermally labile, which limit the application of MS without a prior derivatization process. Consequently, some MS techniques such as electron impact and chemical ionization require the analytes to be in gas form for ionization. In desorption ionization techniques, for instance, field desorption, direct chemical ionization, fast atom bombardment, and liquid secondary ions mass spectrometry, the analysis of flavonoid glycosides without derivatization became possible. However, the tedious preparation work and the low selectivity of MS toward flavonoids compounds seriously limited its applicability. Especially by using a single MS technique that will not be able to provide all necessary structural information allowing an unambiguous assignment of the structure to an unknown compound regarding flavonoids' structure (the number of OH bonds, and the diversity of attached molecules) that increase the generated fragmented ions together with the overlapping and noisy resulting data [214]. These limitations can be prevented by the coupling of MS with chromatographic techniques [215], which will be discussed later.

In works of literature, the application of MS as a single analysis stage for the determination of flavonoids is quite low. Therefore, most application works are assisted by logarithm data analysis programs for the filtration, processing, and elimination of the overlapping in raw data, thus allowing for the accurate determination of studied flavonoids [189, 216].

Other lower frequency absorption spectroscopic techniques that have been applied for flavonoids' structural elucidations are XR and CD. XR is the technique that is able to provide crystal structures in challenging powders by detecting the arrangement of atoms within a crystal by the atom-induced diffraction of X-rays. XR reported works of flavonoids are very low because these substances only form crystals in sporadic conditions and the crystallization state is essential in XR to be performed [209]. As an example, the identification of the intermolecular π- π interactions of stacking parallel aglycones that form supramolecular layers, thus identify aminoacyl residues involved in the formation of protein-flavonoid complexes [217]. More recently, X-ray powder diffraction has been utilized as an important tool for obtaining structural data of solid-phase flavonoids such as catechins [218] and the identification of co crystallization of flavonoids with caffeine [183].

CD allows the analysis of the differential absorption of left and right circularly polarized light. It can be used to determine the contribution of individual chromophores and to reach their possible substitution patterns that are confined to the narrow absorption range of each individual chromophore (the main strength over optical rotation measurements) [219]. Some CD configuration characterization of flavonoids has been reviewed [220]. In practice, CD has been used for the determination of the absolute configuration of flavonoid molecules, such as isoflavan-4-ols stereoisomers [184] and flavan-3.4-diols [221], in addition, to the interaction study of flavonoids with biomolecules in terms of biomolecules-drug studies, such as the interaction with DNA [222], proteins [223, 224], and haemoglobin [225] in biological samples. On the other side, the CD is only used for qualitative analysis and the spectrum data cannot be able to provide quantitative information [226].

Concluding the applicability of previous techniques, UV-VIS is the most common, worldwide analysis technique for the determination of total flavonoids along with studying their antioxidant capacity in biological and medicinal plant samples. While the other absorption spectroscopic techniques have proved several limitations, relative long sample preparation time, or only used for concrete analysis objectives.

Vibrational spectroscopic techniques have been effectively utilized for the qualitative and quantitative analysis of flavonoids in food samples. Generally, it is a non-invasive, non-destructive, very rapid, accurate, chemical-free, and environmentally friendly method. The additional main advantage of these techniques is that most do not require previous sample treatments [201], which are considered as the essential requirement before absorption spectroscopies or chromatographic techniques.

IR, RS, and THz vibration spectroscopies can only provide point-based spectral information. They are more often associated with the structural elucidation and furthermore with qualitative analysis of flavonoids [227-229] rather than it is with quantitative analysis [230]. The absolute band intensities are dependent on the analyte concentration and a number of other factors, such as emission source power intensity, position or band shape, temperature, impurities, and crystalline size [231]. By adjusting these factors and using a standard substance, the well-resolved spectra can be used for quantitative analysis. The quantitative assay through vibrational techniques is based on the predicting concentration (indirect estimation) from the extracted and processed raw data by special chemometric algorithms applications expressed by the correlation regression (r^2) of the numerous detecting samples. How much r^2 is closed to 1 expressed to the real predicted concentration? Anyway, this is still a big challenge and requires special and expensive chemometric applications. Also required are specialists in the field of vibrational spectroscopy who are able to read, extract, and compare the bands in the spectral data, being that this is the main disadvantage of vibrational spectroscopic quantitative analysis.

RS and IR spectroscopy are referred to as sisters since both are complementary techniques and usually are measuring the vibrational modes of a molecule. IR relies on the measurement of the interaction of infrared radiation with the matter by absorption, while RS is based on inelastic scattering from the interaction of incident radiations with vibrating molecules [232, 233]. Though the similarity between RS and IR, RS is the best for symmetric vibrations of non-polar groups, whereas IR is the best at the asymmetric vibrations of polar groups [233]. Both techniques are able to detect the flavonoids in all sample states, solid, liquid, or gas states, however, the absorption bands caused by OH stretching vibration are so intense in IR that these bands interfere with those of other vibrations. Therefore, IR spectroscopy cannot be applied to water-rich samples, as well as flavonoids are also reached by OH bonds, so the detecting of multi-components samples is difficult by IR. On the contrary, RS is non-sensitive to OH vibration, so it is more useful for the analysis of flavonoids in aqueous samples and multi-component samples over other analytical techniques, such as IR and UV-VIS absorption spectroscopies [234]. On the other hand, another advantage of RS over IR is that RS detects the sample directly without the need for previous preparation comparing to IR, thus a very fast detection method can be observed [235]. Moreover, the smallest particle that can be identified using IR is in the 35–50μm range, particles as small as 1–2 μm in diameter can be identified by the RS method.

HIS is an innovative technique that integrates conventional imaging and spectroscopy to attain both spatial and spectral information from an object. It is often carried out in VIS/NIR range, it combines the main features of vibrational spectroscopy and imaging technique into one system, and it is the widely used approach for the analysis of food quality and safety such as contaminant detection, defect identification, constituent analysis and quality evaluation [236]. The main advantages of HIS over the previous vibrational techniques are the ability to provide both spatial and spectral information simultaneously, hence, giving extensive information about both externals and internals of the sample, and it provides multi-constituent information and higher sensitivity to minor components [237]. On the contrary, its high cost, and relatively lengthy times necessary for hypercube image acquisition, processing, and classification compared with other vibrational techniques that can range from 2 to 4 min, while the others need seconds for the same purpose, are disadvantages. Likewise, processing and classification time requires better computer hardware and software capabilities to reduce the required processing time, which means additional cost [238].

THz spectroscopy exploits a part of the electromagnetic spectrum in far-infrared vibrational modes between the microwave and infrared region. To date, it has emerged for the analysis of biological molecules, but, generally, its application in food and flavonoids monitoring is very slight. THz is very sensitive to H bonds, thus when it is applied in rich-water samples as food samples that have high moisture content high noiselevel is obtained [239]. This might be the main restriction point in food analysis.

In practical applications of vibrational spectroscopy, RS has recently been gaining a lot of interest in the field of quantitative analysis of flavonoids in different food and beverages samples. The quality analytical study of the Bulgarian wine using RS has been established. Total flavonoids content, anthocyanins, and catechins in red and white wine samples have been used as the indicator of wine's quality. r^2 was 0.87 (526.6 mg dm^{-3} catechin equivalents) [240]. The comparison between IR, RS, and the integration between them IR-RS for the determination of total phenolic content and total antioxidant capacity of Chinese rice wine was also reported [241]. Good r^2 ranged between 0.83 and 0.96 was observed. Only 20s is the required time of each spectrum.

IR was also widely used for total flavonoids' prediction in herbs, foods, vegetables, fruits, and medicinal plants. The determination of total flavonoid, flavanols as the indicator quality of kiwifruit after their extraction by organic methanol was reported. r^2 were relatively high (from 0.82 to 0.99) [242]. Total flavonoids' content was estimated in coffee grounds by NIR directly in ethanolic extract. Good linearity with

r^2 of 0.95 was observed [243]. Numerous other applications using NIR for the determination of total flavonoids content in several natural samples with high correlation regressions have been established, such as in honey [244], green tea [245], black tea [246], Chinese tea [247], matcha [248] rice grain [249], Ginkgo biloba [250], herbs [251], Goji berries [252], lyophilized durian, mango and avocado samples [253], and blueberry, grape, and blackberry [254].

Elsewhere, HIS was also attained for predicting total flavonoids content in *Chrysanthemum morifolium.* The method used 3g of dried sample in this evaluation, and a high correlation regression was obtained being over 0.9 [188].

Analyzing the data from all of the above-mentioned works, the best application of vibration spectroscopic techniques was for the determination of total flavonoids content and/or total antioxidant activity. Therefore, these techniques have been also utilized for the determination of individual flavonoids in food samples. The detection of quercetin in onion peel after its three hours extraction with heat ethanol and methanol was reported employing RS. The predicted quantity was 73 mg 100 g^{-1}, LOD was 5×10^{-5} mol L^{-1} with a very high r^2 of 0.9999 [255]. The determination of three flavonoids (chrysin, apigenin and luteolin) in Weld textile were obtained using citrate reduced Ag colloids was obtained. The sample was just hydrolyzed and dissolved in 50% ethanol then directly detected by RS [256]. IR and RS have been carried out for the predicting analysis of 13 catechins in tea powder. The samples were directly measured and the comparison between theoretical and practical regression results showed that these two techniques are implemented successfully for the analysis of catechins [257].

NIR was successfully utilized for the direct detection of nine catechins in green tea. The calibration models for the studied catechins had high r^2 (more than 0.90) [258]. Moreover, NIR was applied to determine the content of rutin in Tartary buckwheat. The predicted concentration was compared with this determined in HPLC. 15 g of sample powder was directly used in this detection, a good correlation was observed $r^2 = 0.76$ [259].

THz was performed for the quantitative prediction of 10 common flavonoids, including baicalein, baicalin, apigenin, quercetin, naringenin, hesperetin, daidzein, genistein, puerarin, and gastrodin, in a ternary mixture. The laboratory prepared mixtures of ternary flavonoids were mixed in a mortar with high-density polyethylene powder in a ratio of 1:2. Finally, a slice with a thickness of 1 mm was prepared under a tablet press then detected by THz. r^2 was over 0.99 [187].

As mentioned before, vibrational spectroscopic techniques are successfully and widely implemented for the structural elucidation and quantitative prediction of flavonoids concentration in their natural sources. RS and NIR are the most common and effective techniques that have been carried out for flavonoids' determination with high correlation regression observed in almost all developed works in the kinds of literature, while THz and HIS have been in the earlier stage application in this area and have not yet been widely applied. Nevertheless, these techniques are a predictive matter and not direct detection methods, but all attained techniques showed robust, high accuracy and validity testing, quickness, and minimizing the using and the contacting with the harmful organic solvents and protect the environment.

Chromatographic Techniques

CTs are the most powerful separation techniques of flavonoids in almost all of their matrices. The coupling of CTs with detection techniques are required for the evaluation of separated substances individually. Generally, these techniques are strongly able to separate the different flavonoids, also the most similar molecule structures, isomers, and the closest derivatives of the same aglycone form from the complex matrices. Therefore, it provides the best solution of the facing barriers in spectroscopic determination techniques, constancy, increase the selectivity to identify flavonoids individually, and increasing the sensitivity of detection methods. In addition, it provides high reliability and accurate results, and the automation of most analysis steps. Almost all existing CTs have been employed to determine flavonoids in their different formats. Some examples are; paper chromatography (PC), thin-layer chromatography (TLC), gas chromatography (GC), capillary electrophoresis (CE), supercritical fluid chromatography (SFC), and high-performance liquid chromatography (HPLC).

Paper and Thin-Layer Chromatography

PC is the simplest chromatographic shape that has been carried out for flavonoids' separation. Therefore, this technique is simple and can separate the flavonoids, but it loses the power to separate the complex forms along with affecting many environmental factors [260]. Furthermore, TLC is a more powerful technique than PC for flavonoids separation. It is easy to use, has inexpensive equipment, and several stationary phases can be used for this purpose. In literature, there are hundreds of developed works in this area [261, 262]. Silica gel layers are the most stationary phase that has been used to separate the flavonoids coupling with UV-light detector at 250-260 or 350-365 nm [263]. Nevertheless, nowadays, these techniques are not often relied on for flavonoids' determination, mainly because of used detectors, the detection limit is much higher compared to other CTs. It provides unreliable results due to its being affected by the environmental surrounding conditions such as temperature and humidity. Its coupling is not possible with other systems such as MS, NMR or FT-IR, which have problems of selectivity with difficulty differentiating between overlapping bands, and have a negative environmental impact regarding the quantity and the evaporation matter of used organic solvents [152].

Recently, two-dimensional-TLC (2D-TLC) [264] and high performance-TLC (HPTLC) [262] provide great solutions for difficult separations by improving the sensitivity and reducing the analysis time and solvents consumptions. More recently, the innovative combination of TLC and vibration spectroscopic techniques provides rapid, accurate and sensitive determination technique of flavonoids. In workable applications, TLC-NIR has been applied for methoxylated flavones predicted amounts in phytomedicine samples taking advantages of compound libraries for cluster analysis and to perform quantitative analysis of the same recorded spectrum [265]. 2D-TLC-Raman has also been employed for 14 citrus flavonoids determination, achieving better sensitivity from traditional TLC, as well as better detection efficiency and less analysis time than HPLC-UV applied for the same purpose [266].

Gas Chromatography

Generally, GC is a very powerful separation technique of natural compounds regarding the higher given resolution power, especially when it combines with MS spectroscopic detection [267]. GC requires the analytes to be in a gas stage for their separation and detection. This is not the ideal condition in the case of flavonoid fields regarding their non-volatile properties and their high boiling points (more than 300°C), accordingly, a derivatization process for converting flavonoids to volatile derivative shapes and improve their thermal stability is almost essential prior to sample GC injection [268].

Therefore, there are few works that run directly without the derivatization process [269], but several derivatization reagents have been employed in the most developed ones. In this sense, silylating or methylating reagents are the most famous, for example, trimethylchlorosilane and hexamethyldisilazane [268], N,O-bis(trimethylsilyl) trifluoroacetamide) [270], *N*-methyl-*N*-(trimethylsilyl) trifluoroacetamide [271], *N*-methyl-*N*-(tert-butyldimethylsilyl) trifluoroacetamide and trimethylanilinium hydroxide.

This process is tedious, consuming time and solvents, leading to form multiple derivatives corresponding to flavonoids hydroxyl groups which add some difficulty in detection and increase the analysis errors [272]. For instance, the determination of 24 phenolic compounds in five different aromatic plants by GC-MS and HPLC-UV was established [273]. This work achieved better selectivity and sensitivity in a shorter analysis time (50 min) in GC compared to the HPLC method (120 min), but it still consumed more sample preparation time than HPLC, about 45 min for the derivatives formation requirement.

The most widely utilized separation columns in GC for flavonoids analysis are 100% dimethyl polysiloxane [269], 5% phenyl 95% dimethyl polysiloxane [270], proprietary phase [274], 1,4-bis(dimethylsiloxy) phenylene dimethyl polysiloxane [275] as non-polar fused capillary columns, 50% cyanopropyl 50% dimethyl polysiloxane as polar column, and polyethylene glycol as high polar column [276]. In the detection stage, the MS detector is the most powerful used detection method coupled with GC for the identification of separated flavonoids, followed by a flame ionization detector [269].

Capillary Electrophoresis

CE is a competent technique to separate and identify charged low and medium molecular weight substances in the presence of the electrical field effectively, rapidly, with a high-resolution power, and low sample and solvents volume consumptions [277]. According to its unique advantages and the charging of flavonoids molecules in usually basic pH range, CE is one of the most important separation methods that has been used smoothly and rapidly for flavonoids analysis in natural product research [278]. From the six CE modes, only three have been developed to analyze the flavonoids in their natural sources, capillary zone electrophoresis (CZE) [279], micellar electrokinetic capillary electrophoresis (MECE) [280] and capillary electrochromatography (CEC) [281].

UV [282], electrochemistry detector (ED) [283], fluorescence detector [284], amperometric detector (AD) [285], and MS detector [286] are the applicable ones with CE for flavonoids detection matter. Therefore, the benefits of CE, nonetheless, requires a long time for the equilibration of the separation column, low sensitivity trend to sample overload, and less quantitatively reproducible than HPLC. Additionally, the HPLC instruments are more widely used than CEmaking it the strongest separation technique of flavonoids [287]. As a comparing application between CE and HPLC for the separation of flavonoids, the determination of 9 flavonoids in grape wine was carried out using CZE-UV and HPLC-UV [288]. In this work, the CE method achieved a better detection limit and less analysis time (6 min) than HPLC (15 min), however, CE needs 80 min for column equilibration before analysis.

Supercritical Fluid Chromatography

SFC is the method that relies on CO_2 as an essential component in the mobile phase and NP-beads as a stationary phase. It performs a highly efficient chromatographic platform allowing higher linear velocities due to faster diffusion and lower viscosities [289]. The use of CO_2 as a unique component in the mobile phase due to flavonoids separation is very limited because of its highly non-polar behaviour against the polarity of target analytes [290]. To overcome this limitation, the adding of a polar organic solvent modifier is performed, such as methanol [291], acetonitrile [292], and ILs [293]. Some researchers have compared the power of SFC against HPLC for the separation of flavonoids. In practice, the comparison between ultra-SFC and ultra-HPLC for the separation of ten flavonoids in *Radix hedysari* was demonstrated. The required analysis time for successful separation of the target compound was 22 min in SFC, while HPLC needed 103 min to achieve the same objective [292]. The SFC's required time for the separation of six flavonoids in Citri Reticulate Pericarpium was 10 min, while the HPLC that it was being compared to for the sample purpose needs 16 min to completely separate these six target flavonoids [291]. However, the non-availability of the wide polarity of separation column is another disadvantage of SFC, in this term, some works have studied the separation power of many column types. As for example, the comparison between four different stationary phases, including bridged ethyl hybrid, BEH, the same hybrid phase modified with 2-ethyl pyridine, CSH fluorophenyl, and HSS C_{18} SB. The analytes included seven flavonoids. The CSH fluorophenyl column was observed as the most suitable with the fastest separation within 17 min using gradient elution with CO_2 and methanol as a modifier. This column achieved a good resolution between all provided asymmetric peaks of 1.41 [294]. In another work, six different separation columns have been compared: ZORBAX SB-C_{18} column, ZORBAX RX-SIL column, Venusil HILIC column, Venusil Imidazolyl column, Venusil ASB Phenyl column, and Venusil NP column. The work showed that the ZORBAX RX-SIL column effectively separated 12 flavonoids within 20 min using CO_2 and acidified methanol (0.1 phosphoric acid) in a gradient mode [295].

Analyzing the data showed that despite the limitation of SFC, it can effectively implement the separation of flavonoids from natural matrices realizing good resolution in a shorter analysis time than traditional HPLC, fulfilling high selectivity and reliable results.

High-Performance Liquid Chromatography

Among all chromatographic techniques, HPLC is the most convenient technique, which enables both separation and identification of flavonoids in the complex matrices, such as plants extracts, fruits, vegetables, honey, propolis, beverages, food supplements and pharmaceutical formulations accurately, sensitively, and selectively.

The workability of the HPLC tool in the separation of flavonoids lead the concerned organizations to approve it and adopt it for adjusting the quality of flavonoids' products, such as the world health organization (WHO) [296], U.S. Food and Drug Administration (FDA) [297], pharmacopoeias: United States [298], British [299] and European pharmacopoeia [300]. Further, the successfully developed techniques usually should be compared to HPLC to appear their advantages.

HPLC is continuously improving to be more effective and reliable in the determination of flavonoids, many HPLC significant factors are being studied practically. To date, there are thousands of developed works, reviews, and books in this field.

In the separation stage, the separation of flavonoids has been carried out on both normal (NP) and reversed (RP) phases. In spite of this there are some works that used silica column [301], polystyrene divinylbenzene [302], diol [303] and cyanopropyl-bonded phase [304] as a NP-column for separating flavonoids. The most separation approaches have been done on RP-columns, taking into account, the higher polarity of these flavonoids besides the highly-polarity of linkage groups as saccharide molecules and the probability of their retaining inside NP-column, besides the required long analysis time [301].

RP-C_{18} [31] and RP-C_8 [305] columns are the most demonstrated stationary phase that have been used for flavonoids separation regarding the ability to separate deferent polarities of flavonoids effectively, with high-resolution power, and due to an acceptable analysis time usually less than 40 minutes [306]. As usual, 5 μm of separation beads' diameter are used in column length between 100 mm and 250 mm [307]. More recently, both Fused-Core® (FC) and monolithic column technologies have been developed attaining effective flavonoids separation. FC depends on a fused-core silica substrate with a layer of coated porous silica, while MC consists of a porous silica rod. Both have been designed to offer the improvement in reducing diffusion path peak for allowing fast mass transfer, therefore, providing more selectivity, separation efficiency and high reproducibility in less analysis time and less backpressure than traditional columns. Practically, FC-C_{18} was carried out to separate 11 flavonols and flavones in Ginkgo biloba, Betula pendula, and a variety of Sorbus species at 30 min of analysis time achieving a high resolution between the pair peaks in plant materials [308]. Whereas FC-amid has been used for the separation of rutin, troxerutin, diosmin and hesperidin in food supplements at 5 min of analysis time [309]. Further, the MC-C_{18} column has been widely applied for flavonoids separation in different sources [310-313]. The employed MC-C_{18} column is usually done on 100 × 4.6 mm of column dimensions using high flow rates ranged between 1 and 2.1 mL min^{-1}, in either isocratic [310, 313] or gradient modes [311, 312]. The observed results note that these conditions are able to effectively separate up to 17 monomeric compounds in different complex foodstuffs due to short analysis time ranged between 7 and 30 min.

On the other hand, the mobile phase typically comprises a binary solvent, both aqueous and organic phases. Less polar organic solvents are usually conducted, methanol and/or acetonitrile are the most famous organic mobile phase part [314, 315]. Whilst, aqueous phase often includes acidic modifier such as glacial acetic acid, formic acid, phosphoric acid or trifluoroacetic acid for the purpose of maintaining the non-charged flavonoids molecules. Isocratic mode is rarely employed for flavonoids separation [316], while gradient mode is more suitable for the separation of closed flavonoids molecules in complex profiles [317].

In the detection stage, HPLC has been combined with several detection systems in terms of the quantification of the separated flavonoids. In fact, this combination can produce a powerful analytical technique supporting the advantages of used spectroscopic tools and overcoming many of their drawbacks, furthermore, increasing the selectivity of the detection step. In this sense, UV-VIS and its diode array (DAD) are the most usable detector of flavonoids compared with other detection systems [318], benefiting from the familiarity of flavonoids to the absorption of UV radiations as mentioned before in the spectroscopic essay paragraph. Following by MS detector regarding its advantages uses, in special when trace flavonoids in complex matrices should be analyzed and/or when the reference standards are not available [319].

Recently, the combination with NMR has significantly enhanced the prediction detection and elucidation perspective of separated flavonoids. NMR is indispensable for this objective. Overviewing, NMR is hindered by cost, complexity, throughput limitation, and the requirement of sufficient concentrations of relatively pure analysts that can be circumvented by the coupling with HPLC due to the separation of the closest compounds prior to their detection individually [320].

The less dependable implemented detection systems in this area are FL detector because of the low native fluorescence of flavonoids. It may require a prior treatment to form a fluorescent complex carried out post derivatization column located before the detection stage [303, 321]. Other detectors are rarely used for flavonoids' detection such as coulometric [322], amperometric [323], and electrochemical detectors [324].

Otherwise, as the developing product of HPLC, hydrophilic interaction liquid chromatography (HILIC) has been introduced as an alternative HPLC mode that depends on the separation of polar compounds using both polar stationery and mobile phases. Unlike the mechanism separation on NP-HPLC, the stationary phase should be more polar than the mobile phase, thus the solubility of polar analytes could be insufficient and lead to a strong retention behaviour of the analytes onto the more polar stationary phase, and vice versa in case of RP-HPLC that leads to less familiarity to retaining on the nonpolar stationary phase, hence less selectivity. HILIC employs traditional polar stationary phases, but the mobile phase is similar to those are used in RP-HPLC, which could solve the drawbacks of traditional RP- and NP- HPLC [325]. HILIC was increasingly applied for the effective separation of flavonoids from complex samples. Many investigations have been done in this area regarding the study of the separation efficiency and the retention of flavonoids in the HILIC polar stationary phase [326-328]. These pieces of literature observed that HILIC provides a powerful separation of many sub-groups of flavonoids due to an adequate analysis time of less than 30 min. The main disadvantage of HILIC is the requirement of a high organic solvent content in the mobile phase to enhance the sensitivity and column retention of polar analytes that could harm the environment.

On the other hand, the mandatory advanced extraction and purification procedure from the complex matrices prior to HPLC analysis encourages the emerging of two-dimensional HPLC (2DLC). 2DCL functionality depends on the elution of unresolved analytes in the sample by 1D then the separation of target analytes through 2D based on the differences in selectivity of the two dimensions [325]. This mode enhances the separation power of isomeric and enantiomeric compounds, improving their peak capacity, solving the overlapping peaks of similar structures, and decreasing the need for advanced sample pretreatment techniques before chromatographic separation. In the area of flavonoids, 2DCL is carried out off-line [329] or on-line [330] and can consist of the combination between RP-HPLC×RP-HPLC [331], NP-HPLC×RP-HPLC [332], HILIC×RP-HPLC [333, 334], and HILIC×NP-HPLC [335]. The main drawbacks of 2DLC, especially in off-line mode, are the relatively time-consuming, labour-intensive multi-step procedure, losing or evaporating of the collected fractions, which lead to less accuracy and fewer precision results [336]. For instance, 51 min of analysis time is required for the separation of five related flavonoids impurities of rutin in drug tablets using 2DLC-MS/MS [337].

The applicability, flexibility, and the diversity of separation columns and detection systems that can be coupled to HPLC technique in the objective of the separation of flavonoids put the HPLC on the top of the hierarchy of technical choices to separate and determinate the flavonoids in real-natural and artificial sources with highly accurate, reliable, and precious results.

The advantages and disadvantages of using the analytical methods for the analysis of flavonoids are summarized in Table 7.

The characterizations and the validation figures of merit of the developed analytical methods of flavonoids in food, beverages, plant, herbs, and commercial products are also collected in Table 8, in the term of the comparison between them.

Table 7. Advantages and disadvantages of some used techniques for flavonoids determination

Anal. tech.	Advantages	Disadvantages
UV	Quick and cost-effectiveness, can be applied for all flavonoid's groups	Non-selective, less accurate results, can be employed just for total flavonoids detection
VIS	Higher sensitivity, and selectivity than UV detection	Non-selective for individual flavonoids, limitation of the accuracy and reliability, coloured complexation formation is required, long preparation time
FL	Selective to some flavonoid's groups, high sensitivity	Limited number of flavonoids exhibit native fluorescence, limited for total flavonoids determination, fluorescent complexation formation is required, long preparation time
NMR	No required reference standard, detect unknown flavonoids, structure elucidation, quick set up, fast analysis, ease of use, applied for solid and liquid samples, required minimal pretreatment procedure	Lack of sensitivity, costly, sample should be stable, not volatile and does not include water, pure samples are needed to achieve an acceptable signal-to-nose level, not suitable for glycoside determination due to the linkage groups

Table 7. (Continued)

Anal. tech.	Advantages	Disadvantages
MS	High sensitivity, specificity, and selectivity, structure elucidation	Required volatile and thermally stable compounds, required derivatization process, tedious preparation work, single MS not able to provide all necessary structure information
XR	Relatively cheap and simple, structure elucidation into the atomic level, can yield a high atomic resolution	The sample must be crystallizable, flavonoids only form crystals in sporadic conditions, tedious sample preparation and crystallization
CD	Examining different aspects of optically active flavonoids, molecules of any size can be studied, determines the absolute configuration of flavonoid molecules, useful for biomedicine interaction study, fast data collection.	Only provide qualitative information, several aqueous buffers are not compatible due to their absorption in the studying range
IR	High and constant resolution, simple, rapid, non-destructive, quick sample preparation is needed, applied for structure elucidation and quantitative analysis, eco-friendly	Required standard substance and a complicated and expensive chemometric algorithms application for quantitative analysis, required a specialist to analyze the raw spectral data, sample should be stable, very sensitive to OH bands in water content, that interfere with analyte vibrational bands, not applicable for aqueous samples
RS	Suitable for any type, size or sample shape, direct analysis, no sample preparation is needed, solvent-free, non-destructive, not sensitive to OH bands, can be used for aqueous solutions, fast analysis	The sample can be damaged by laser light, background fluorescence notorious that can even mask the spectra, low sensitivity, equipment cost and requires a chemometric algorithms application for quantitative analysis, required specialist to analyze the raw spectral data
HSI	High sensitivity, gives both spatial and spectral information simultaneously, provides an extensive information of the external and internal of the sample, multi-constituent information, quick sample preparation, non-destructive, eco-friendly	High cost, long hypercube image acquisition time, longer processing and classification time than other vibrational techniques, high computer hardware and software capabilities, requires a chemometric algorithms application for quantitative analysis, required a specialist to analyze the raw spectral data

Anal. tech.	Advantages	Disadvantages
THz	Environmentally friendly, non-destructive, high sensitivity, not required a complex samples treatment, robust, accurate, quick analysis	Very sensitive to OH bands in water content, highly noising data for moisture food samples, not applicable for aqueous samples, requires a chemometric algorithms application for quantitative analysis, required specialist to analyze the raw spectral data
PC/ TLC	Inexpensive equipment, simple, ease of use, diversity of stationary phase	Low selectivity and sensitivity, limit of used detectors, the difficulty to hyphenate it with other systems, low-reliability results
GC	Sensitive, selective, high-resolution power	A laborious derivatization process is required; thus, the errors of the analysis will be increased, difficulty in detection due to multiple derivatives fractions formation by flavonoids' hydroxyl groups
CE	High-resolution power, low analysis time and low sample and volume consumptions	Long column equilibration time, low sensitivity, sample overload, less quantitatively reproducibility than HPLC, non-availability
SFC	Fast separation, low solvent volume, high efficiency, good resolution between the peaks, eco-friendly	CO_2 is highly non-polar, polar organic solvent as a modifier in the mobile phase is required, non-availability of the wide polarity of the separation column
HPLC	Flexibility, diversity of separation column and coupling detectors, separation efficiency, high sensitivity, selectivity, reproducibility, resolution, accuracy and extremely precise, not requires a preliminary derivatization	Complex to use, costly, required large quantities of expensive organic solvents, non-environmentally friendly, coelution (depending on the used column and the polarity of the studied flavonoids)

Anal. tech.: analytical technique; UV: ultraviolet radiation spectroscopy; VIS: visible radiation spectroscopy; FL: fluorescence technique; NMR: nuclear magnetic resonance spectroscopic technique; MS: mass spectroscopy; XR: X-ray crystallography; CD: circular dichroism spectroscopy; IR: infra-red spectroscopy; RS: Raman spectroscopy; THz: terahertz spectroscopy, HSI. hyperspectral imaging spectroscopy PC, paper chromatography; TLC: thin layer chromatography; GC: gas chromatography; CE: capillary electrophoresis; SFC: supercritical fluid chromatography, HPLC: high-performance liquid chromatography

Table 8. Recent applied analytical techniques for flavonoids determination in natural samples

Spectroscopic methods									
Anal. Tech.	**Pretreat. proc.**	**Matrix**	**Target flav.**	**Determination conditions**		**LOD (μg mL^{-1})**	**Recov. (%)**	**RSD%**	**Refer.**
				Solvent (mL)	Time (min)				
UV	UAE	*S. barbata*	TFC	60% EtOH (25 mL)	12.5	-	92.2-107.6	≤ 6.9	[179]
VIS	Complex (Al^{+3})	Plant/food	Antho.	EtOH (830 μL)	15	0.18	98.8-103.3	≤ 1.5	[180]
VIS	Complex (Al^{+3})	*P. oleracea*	TFC	$NaNO_2$ (0.6 mL)	27	-	101.1-103.8	1.13	[190]
VIS	Complex (DNH)	Propolis	TFC	MeOH	50	-	-	-	[193]
VIS	Complex (AAP)	Tablets	DIO	NaOH (5 mL)	0.25	0.1	101.6-101.9	≤ 2.2	[194]
FL	Complex (Al^{+3})	Human plasma, drugs	HES	70% MeOH	-	≥ 0.3	9.3–99.7	1.3	[181]
FL	Complex (Tb^{+3})	Tablets	HES, DIO	Tris buffer (2 mL)	15	1.8×10^{-5}	97.8-99.1	≤ 0.9	[338]
NMR	Vortex (8.5 min)	*Crataegus*	4 flav.	MeOH-d_4 (1 mL) + DSS (internal standard)	0.67	0.5×10^{-3}	-	-	[182]
NMR	Reflex (2h)	*G.uralensis*	3 flav.	DMSO + DCM (internal standard)	0.17	23	100.0-110.8	≤ 4.4	[212]
NIR	non	Propolis	TFC	EtOH (2 mL)	-	-	$r^2 \geq 0.74$	-	[185]
RS	non	Wine	TFC, Antho., Cat.	-	0.33	-	$r^2 - 0.87$	-	[240]

Spectroscopic methods									
Anal. Tech.	**Pretreat. proc.**	**Matrix**	**Target flav.**	**Determination conditions**		**LOD (μg mL^{-1})**	**Recov. (%)**	**RSD%**	**Refer.**
				Solvent (mL)	Time (min)				
RS	F-CC	Rice wine	TPC	Folin–Ciocalteu reagent (1.5 mL)	-	-	$r^2 \geq 0.83$	-	[241]
RS	MeOH, EtOH (3h)	Onion	QU	MeOH/ EtOH	0.02	1.5×10^{-2}	$r^2 = 0.99$	0.3	[255]
IR	MeOH extract (3h)	kiwifruit	TFC, flavonols	MeOH	-	-	$r^2 \geq 0.99$	-	[242]
NIR	EtOH extract (5h)	Coffee grounds	TFC	EtOH	-	-	$r^2 = 0.95$	2.6 mg	[243]
NIR	EtOH extract (2h)	Green tea	9 Cat.	EtOH	0.33	-	$r^2 \leq 0.90$	-	[258]
THz	non	Lab. mix	10 flav.	-	-	-	$r^2 \geq 0.99$	-	[187]
HSI	non	*C.morifolium*	TFC	75% EtOH (2 mL)	-	-	$r^2 \geq 0.90$ (recovery 97.5)	2.35	[188]
Chromatographic methods									
Anal. Tech.	Pretreat. proc.	Matrix	Target flav.	Separation conditions		LOD (μg mL^{-1})	Recov.(%)	RSD%	Refer.
				Stationary/mobile phase	Time (min)				
2DTLC-UV	EtOH extract (15 min)	Propolis	9 flav.	Silica gel/n-hexane: EAT: AAC, 31:14:5 (1D), CF: MeOH: FAC, 44:3.5:2.5 (2D)	-	-	-	-	[264]

Table 8. (Continued)

Chromatographic methods									
Anal. Tech.	**Pretreat. proc.**	**Matrix**	**Target flav.**	**Separation conditions**		**LOD (µg mL^{-1})**	**Recov.(%)**	**RSD%**	**Refer.**
				Stationary/mobile phase	Time (min)				
HPTLC-UV	EtOH (24h)	G. senegalensis	RU, QU	C_{18}/ ACN:water, 4:6	20	≤ 33.0 ng/band	98.6-101.1	≤ 1.0	[262]
HPTLC-UV	EtOH (24h)	G. senegalensis	NARIN	Silica gel/ toluene: EAT: FC, 6:4:0.8	20	≤ 35.6 ng/band	98.6-99.9	≤ 1.5	[262]
TLC-NIR	EtOH	Plants	5 MTF	Al_2O_3/n-hexane:EAT 7:3	-	-	$r^2 \geq 0.99$	-	[265]
2DTLC-RS	MeOH (35 min)	Citrus	14 flav.	Silica gel/ DCM: MeOH 20: 1(1D), PE: acetone 6: 4 (2D)	10	≤ 16.7 µM	91.5-121.7	≤ 20.8	[266]
GC-MS	TMS deriv. (14h, 35°C)	*G. glabra*	3 deriv. of GL	5% phenylmethylsilicone/ temp. (40-250°C)	17	-	-	-	[268]
GC-FID	NaOH (18h), SPE	*L. sericeus*	21 flav., 6 Antho.	100% dimethylpolysiloxane/ Temp 230°C, N_2	22	-	-	-	[269]
GC-MS	MSTFA deriv. (1h, 60°C), SPE	Human urine	17 flav.	(5%-phenyl)-methylpolysiloxane/ Temp (160-270°C)	27	$\leq 9.7 \times 10^{-3}$	70.2-99.6	≤ 5.0	[271]
CE-UV	Sonicate (48 min)	*C. siliqua*	5 flav.	Fused silica capillary/ borate buffer	50	≤ 0.66	99.0	≤ 1.3	[339]
CE-UV	70% MeOH (1h)	*F. Chrysanthemi*	9 flav.	Fused silica capillary/ borate buffer	35	≤ 0.5	95.1-110.7	≤ 6.4	[279]
CE-UV	48% MeOH (182 min)	Tomato	5 flav.	Fused silica capillary/ borax	42	3.8	77-106	≤ 3.7	[280]

Chromatographic methods									
Anal. Tech.	**Pretreat. proc.**	**Matrix**	**Target flav.**	**Separation conditions**		**LOD (µg mL^{-1})**	**Recov.(%)**	**RSD%**	**Refer.**
				Stationary/mobile phase	Time (min)				
SFC	MeOH (70 min)	C.Reticulatae	6 flav.	Zorbax RX-SIL/ CO_2: MeOH (gradient mode)	10	≤ 4.5	98.7-104.3	≤ 2.6	[291]
UHPSFC-UV	75% EtOH (25 min), SPE	*R. hedysari*	4 flav.	C_{18}/ CO_2: 0.1% FC in MeOH (gradient), Temp 70°C	22	≤ 0.4	96.6-104.5	≤ 4.1	[292]
SFC-UV	MeOH	Lab. mix.	6 flav.	C_{18}/ CO_2: MeOH +ILs (gradient), Temp 45°C	6	≤ 17.1	-	-	[293]
UHPLC-UV	75% EtOH (25 min), SPE	*R. hedysari*	4 flav.	C_{18}/ 0.1% FC in water: 0.1% FC in MeOH, Temp 30°C	103	≤ 0.4	96.6-104.5	≤ 4.1	[292]
HPLC-UV	MeOH (15 min)	Drugs	4 flav.	C_{18}/ MeOH: ACN: acidified water (gradient), Temp 22°C	37	≤ 0.02	95.2- 107.9	≤ 3.4	[31]
HPLC-UV	MeOH (75 min)	Drugs, F.S.	4 flav.	C_{18}/ MeOH: ACN: acidified water (gradient), Temp 22°C	46	≤ 0.03	96- 107	≤ 5.1	[308]
HPLC-UV	-	Drugs	2 flav.	C_8/ MeOH: 0.05% FC in water (gradient), Temp 30°C	6	-	982-102	≤ 1.5	[305]
HPLC-UV	Reflux (95% MeOH), (1h)	Plants	10 flav.	FC-C_{18}/ ACN: THF (gradient), Temp 30°C	30.3	≤ 0.5 × 10^{-3}	95.3-100.2	≤ 7.5	[303]
HPLC-UV	DMSO (10 min)	F.S.	4 flav.	FC-Amid/ ACN: acidified water, Temp 50°C	5	≤ 0.5	96.2-104.4	≤ 3.5	[309]

Table 8. (Continued)

Chromatographic methods									
Anal. Tech.	**Pretreat. proc.**	**Matrix**	**Target flav.**	**Separation conditions**		**LOD (μg mL^{-1})**	**Recov.(%)**	**RSD%**	**Refer.**
				Stationary/mobile phase	Time (min)				
HPLC-UV	MAE (35 min)	Tea	8 Cat.	MC- C_{18}/ MeOH: ACN: Water	7	≤ 0.29	94.2-105.2	≤ 2.5	[313]
HPLC-UV	MeOH (35 min)	Citrus	14 flav.	Amid/water: ACN: THF (75:20:5)	45	≤ 1.5 μM	-	-	[266]
HILIC-UV	Filter	Wine	QU	ZIC-HILIC/ACN: acetate, Temp. 35°C	6	0.01	-	1.3	[340]
2DHPLC-MS	MeOH (2h), SPE	Plants	14 flav.	1D, CD/ water: ACN: AF, 2D, C_{18}/ 0.2% FC–water; 0.2% FC–ACN	45	-	-	-	[331]

2DTLC: two-dimensional thin layer chromatography;AAC: acidic acid; ACN: acetonitrile; AF: ammonium formate; Anal. Tech.: analytical technique; AAP: 4-aminoantipyrine; Antho.: Anthocyanidins; Cat.: catechins; CF: chloroform; derv.: derivatives; DCM: dichloromethane; DIO: diosmin; DMSO: dimethyl sulfoxide; DNH: 2,4-dinitrophenylhydrazine; DSS: 4,4-dimethyl-4-silapentane-1-sulfonic acid; EAT: ethyl acetate; EtOH: ethanol; Flav.: flavonoids; FC: fused-core column; FAC: formic acid; F.S.: food supplements; F-CC: Folin–Ciocalteu colorimetric reagent; FID: flame ionization detector; FL: fluorescence technique; GL: glabridin; HES: hesperidin; HIS: hyperspectral imaging spectroscopy; HILIC: hydrophilic interaction liquid chromatography; HPTLC: high performance thin layer chromatography; ILs: ionic liquids; IR: infrared spectroscopy; MTF: methoxylated flavones; MSTFA: N-methyl-N-(trimethylsilyl)trifluoroacetamide; MAE: microwave-assisted extraction; MC: monolithic column:; MeOH: methanol; NIR: near-infrared spectroscopy; NMR: nuclear magnetic resonance spectrometry; PE: petroleum ether; Pretreat. Proc.: pretreatment procedure; QU: quercetin; Recov.: recovery; RT: room temperature; RS: Raman spectroscopy; TFC: total flavonoids content; TPC: total phenolic content; TCP: 3,4,5-trichloropyridine; THF: tetrahydrofuran; THz: terahertz spectroscopy; UV: ultraviolet radiation spectroscopy; UHPSFC: ultra-high performance supercritical fluid chromatography; UHPLC: ultra-high performance liquid chromatography; VIS: visible radiation spectroscopy.

Conclusion

Flavonoids as secondary metabolism natural compounds are diverse. The attaching chemical groups to the main flavonoids' framework make them more complex to be extracted from their natural matrices, as well adding more difficulties to differentiate between them and detect the target ones. Plus, the diversity of their solubility in organic and aqueous solutions and the different chemical behaviours that they show, lead to there being no ideal extraction and an analytical method for the determination of all kinds of flavonoids in the matrices. In this sense, the developing, modifying and implementing the ideal sample pretreatment and analytical method depend on the kind of target flavonoids, their concentrations, their chemical behaviour, the complexity of the matrix, and the goal of the analysis.

Achieving simple, rapid, more environmentally friendly, and lower analysis cost methods to provide factories with more sensible and workable determination methods for their routine quality control are the main objective of developing a new sample pretreatment and analytical methods.

Regarding these objectives, the simplest sample pretreatment procedure of flavonoids is direct dissolving in an appropriate solvent, filtering, and directly analyze the target flavonoids. This procedure is commonly utilized in pharmaceutical industries concerning the high concentration of flavonoids as active pharmaceutical ingredients among the simpler composition of the pharmaceutical forms compared to the natural samples or if the analysis objective is the determination of total flavonoids and/or total antioxidant capacity of these compounds. In more complex matrices, such as plants, food, food supplements, beverages, and biological samples, advanced sample preparation methods are required. LLE and SPE are the most common alternative sample pretreatment techniques used for flavonoids extraction. These techniques notably minimize the consumption of solvents and time compared to other extraction techniques, gained the advantages of working at an ambient temperature that protects the flavonoids from degradation. The microextraction techniques of LLE and SPE have been taking the attention in the extraction of flavonoids from their complex matrices, greatly improving the extraction efficiency, increasing the sensitivity, the throughput of detection methods, decreasing the extraction time much more, and generating less waste.

The automation of these techniques is the target of scientists these days. The automated procedures add more robustness, reliability, and accuracy to the manual ones. In this context, flow analysis and microfluidic techniques are the ones of concern. They may be online coupled with the analysis tools, thus providing more dynamic integrated instruments to fully automate the sample pretreatment step and online transferring of analytes to detection or analytical systems, achieving much fewer solvents and time consumption, high injection throughputs, plus reducing exposure under hazardous solvents, and protecting workers and the environment.

Unfortunately, online coupled approaches have not yet been widely used for flavonoids determination in their biological, natural, and commercial sources lacking the benefit of clearly unique features of these coupling systems. Hence, these tools are continuously developing and applying in this field to miniaturize and automate the analysis methods of flavonoids. Covering more flavonoids and matrices that are still treated manually. Thus, providing workable and powerful tools for required routine quality control analysis of food and pharmaceutical factories and reducing their analysis cost and increasing environmental sustainability.

On the other hand, the analytical step is the next important stage after the extraction of flavonoids. Spectroscopic systems are usually employed for a simple and effective determination of total flavonoids and their antioxidant capacity in biological and natural samples. Toward this objective, spectroscopic tools such as UV VIS or FL (in the case of naturally fluorescence characters of some flavonoids' groups) are quick, reliable and can be relied on for this purpose. Additionally, the power in the structure elucidation of new flavonoids molecules such as MS, RS, NIR, THz, and HIS cannot be missed. Furthermore, chromatographic techniques are the most effective analytical tools that have been used for individual flavonoids determination of sensitively and selectively. HPLC is worldwide, flexible, and shows the diversity of its stationary phase and coupling detectors such as MS, NMR, and RS detections, making it the first-choice tool as an effective analytical technique for flavonoids' determination by the ability to fulfil the requirements of separation and determination of most flavonoids in all their sources. The modern HPLC developed approaches such as SFC, HILIC and 2DCL have gained interest because of their powerful role in overcoming the difficulties of conventional chromatographic methods, decreasing the necessity to apply advanced extraction techniques before the analysis and consuming less solvents and time than traditional systems.

In the analytical stage, the development of suitable methods that give a fast and effective approach for the determination of interested flavonoids sensitively and selectively continues to realize the best determination conditions with the minimum time and solvents consumptions regarding the goal of each analysis.

REFERENCES

[1] Benthsath, A., S. T. Rusznyak, and A. Szent-Gyorgyi. 1937. "Vitamin P." *Nature* 139 (3512): 326-327. https://doi.org/10.1038/139326b0.

[2] Renaud, S., and M. de Lorgeril. 1992. "Wine, alcohol, platelets, and the French paradox for coronary heart disease." *Lancet* 339 (8808): 1523-6. https://doi.org/10.1016/0140-6736(92)91277-f.

[3] Wang, Yechun, Steven Chen, and Oliver Yu. 2011. "Metabolic engineering of flavonoids in plants and microorganisms." *Appl Microbiol Biotechnol* 91 (4): 949-56. https://doi.org/10.1007/s00253-011-3449-2.

[4] Grotewold, Erich. 2006. *The Science of Flavonoids*. USA: Springer Science Business Media, Inc.

[5] Sandhar, Harleen Kaur, Bimlesh Kumar, Sunil Prasher, Prashant Tiwari, Manoj Salhan, and Pardeep Sharma. 2011. "A review of

phytochemistry and pharmacology of flavonoids." *International Pharmaceutica Sciencia* 1 (1): 25-41. https://www.academia.edu/8181572/A_Review_of_Phytochemistry_and_Pharmacology_of_Flavonoids_INTRODUCTION.

[6] Harborn, J. B., and H.Baxter. 1999. *The Handbook of Natural Flavonoids*. Vol. 2. USA: John Wiley and son.

[7] Terahara, Norihiko. 2015. "Flavonoids in foods: A review." *Natural Product Communications* 10 (3): 521-528. https://journals.sagepub.com/doi/pdf/10.1177/1934578X1501000334.

[8] Samanta, Amalesh, Gouranga Das, and Sanjoy Kumar Das. 2011. "Roles of flavonoids in plants." *International Journal of Pharmaceutical Science and Technology* 6 (1): 12-35. https://www.researchgate.net/publication/279499208_Roles_of_flavonoids_in_Plants.

[9] Agrawal, A. D. 2011. "Pharmacological activities of flavonoids: A review." *International Journal of Pharmaceutical Sciences and Nanotechnology* 4 (2): 1394-1398. https://doi.org/10.37285/ijpsn.2011.4.2.3.

[10] Tapas, AR, DM Sakarkar, and RB Kakde. 2008. "Flavonoids as nutraceuticals: A review." *Tropical Journal of Pharmaceutical Research* 7 (3): 1089-1099. https://doi.org/10.4314/tjpr.v7i3.14693.

[11] Zhang, XinFeng, Tran Manh Hung, Phuong Thien Phuong, Tran Minh Ngoc, Byung-Sun Min, Kyung-Sik Song, Yeon Hee Seong, and KiHwan Bae. 2006. "Anti-inflammatory activity of flavonoids from Populus davidiana." *Archives of Pharmacal Research* 29 (12): 1102-8. https://doi.org/10.1007/bf02969299.

[12] Alcaraz, L. E., S. E. Blanco, O. N. Puig, F. Tomas, and F. H. Ferretti. 2000. "Antibacterial activity of flavonoids against methicillin-resistant *Staphylococcus aureus* strains." *Journal of Theoretical Biology* 205 (2): 231-40. https://doi.org/10.1006/jtbi.2000.2062.

[13] Ait Lahcen, S., L. El Hattabi, R. Benkaddour, N. Chahboun, M. Ghanmi, B. Satrani, M. Tabyaoui, and A. Zarrouk. 2020.

"Chemical composition, antioxidant, antimicrobial and antifungal activity of Moroccan *Cistus Creticus* leaves." *Chemical Data Collections* 26: 100346. https://doi.org/10.1016/j.cdc.2020.100346.

[14] Lyu, Su Yun, Jee Young Rhim, and Won Bong Park. 2005. "Antiherpetic activities of flavonoids against herpes simplex virus type 1 (HSV-1) and type 2 (HSV-2) in vitro." *Archives of Pharmacal Research* 28 (11): 1293-1301. https://doi.org/10.1007/BF02978215. https://doi.org/10.1007/BF02978215.

[15] Xu, H. X., M. Wan, H. Dong, P. P. But, and L. Y. Foo. 2000. "Inhibitory activity of flavonoids and tannins against HIV-1 protease." *Biol Pharm Bull* 23 (9): 1072-6. https://doi.org/10.1248/bpb.23.1072.

[16] Jo, Seri, Suwon Kim, Dong Hae Shin, and Mi-Sun Kim. 2020. "Inhibition of SARS-CoV 3CL protease by flavonoids." *Journal of Enzyme Inhibition and Medicinal Chemistry* 35 (1): 145-151. https://doi.org/10.1080/14756366.2019.1690480.

[17] Kumar, G. Suresh, R. Manivannan, and B. Niveth. 2021. "In silico identification of flavonoids from *Corriandrum Sativum* seeds against coronavirus covid-19 main protease." *Journal of Drug Delivery and Therapeutics* 11 (2): 145-152. https://doi.org/10.1248/bpb.23.1072.

[18] Jain, Anisha S., P. Sushma, Chandan Dharmashekar, Mallikarjun S. Beelagi, Shashanka K. Prasad, Chandan Shivamallu, Ashwini Prasad, Asad Syed, Najat Marraiki, and Kollur Shiva Prasad. 2021. "In silico evaluation of flavonoids as effective antiviral agents on the spike glycoprotein of SARS-CoV-2." *Saudi Journal of Biological Sciences* 28 (1): 1040-1051. https://doi.org/10.1016/j.sjbs.2020.11.049.

[19] Liu, Renhao, Xiaoze Bao, Xuanrong Sun, Yue Cai, Tianwei Zhang, Xinyi Ye, and Xing-Nuo Li. 2020. "Cytisine-flavonoid conjugates: Synthesis and antitumor structure-activity relationship research." *Tetrahedron Letters* 61 (17): 151803. https://doi.org/10.1016/j.tetlet.2020.151803.

[20] Nie, Jiang-Ping, Zhen-Ni Qu, Ying Chen, Jia-Hao Chen, Yue Jiang, Mei-Na Jin, Yang Yu, Wen-Yan Niu, Hong-Quan Duan, and Nan Qin. 2020. "Discovery and anti-diabetic effects of novel isoxazole based flavonoid derivatives." *Fitoterapia* 142: 104499. https://doi.org/10.1016/j.fitote.2020.104499.

[21] Park, Ki Hun, Yong-Dae Park, Jong-Min Han, Kyung-Ran Im, Byong Won Lee, Il Yun Jeong, Tae-Sook Jeong, and Woo Song Lee. 2006. "Anti-atherosclerotic and anti-inflammatory activities of catecholic xanthones and flavonoids isolated from *Cudrania tricuspidata*." *Bioorganic & Medicinal Chemistry Letters* 16 (21): 5580-3. https://doi.org/10.1016/j.bmcl.2006.08.032.

[22] Kamboh, Asghar Ali, Muhammad Awais Arain, Muhammad Jameel Mughal, Ali Zaman, Zeeshan Mahmood Arain, and Abdul Hameed Soomro. 2015. "Flavonoids: health promoting phytochemicals for animal production-a Review." *Journal of Animal Health and Production* 3 (1): 6-13. https://doi.org/10.14737/journal.jahp/2015/3.1.6.13.

[23] Chun, Ock Kyoung, Sang Jin Chung, and Won O. Song. 2007. "Estimated dietary flavonoid intake and major food sources of U.S. adults." *J Nutr* 137 (5): 1244-52. https://doi.org/10.1093/jn/137.5.1244.

[24] Skibola, Christine F., and Martyn T. Smith. 2000. "Potential health impacts of excessive flavonoid intake." *Free Radical Biology and Medicine* 29 (3-4): 375-383. https://doi.org/10.1016/s0891-5849(00)00304-x.

[25] Gattuso, Giuseppe, Davide Barreca, Claudia Gargiulli, Ugo Leuzzi, and Corrado Caristi. 2007. "Flavonoid composition of Citrus juices." *Molecules* 12 (8): 1641-73. https://doi.org/10.3390/12081641.

[26] DrugBank. 2020. *Online database*. Canada: DrugBank.ca.

[27] Kozlowska, Aleksandra, and Dorota Szostak-Wegierek. 2017. "Plant flavonoids in health, prevention, and treatment of chronic diseases." In *Nutritional Antioxidant Therapies: Treatments and*

Perspectives, edited by Kais Hussain Al-Gubory and Ismail Laher, 347-376. Cham: Springer International Publishing.

[28] Koch, Wojciech. 2019. "Dietary polyphenols-important non-nutrients in the prevention of chronic noncommunicable diseases. A systematic review." *Nutrients* 11 (5): 1039. https://doi.org/10.3390/nu11051039.

[29] Cerda, Víctor, Laura Ferrer, Jessica Avivar, and Amalia Cerda. 2014. "Flow analysis. A practical guide." *Analytical and Bioanalytical Chemistry* 406 (30): 7761-7762. https://doi.org/10.1007/s00216-014-8245-2.

[30] Martha J. M. Wells, Somenath Mitra, and Roman Brukh. 2003. *Sample Preparation Techniques in Analytical Chemistry*, ed SOMENATH MITRA. USA: John Wiley & Sons, Inc.

[31] Sammani, Mohamad Subhi, Sabrina Clavijo, Lindomar Portugal, Ruth Suarez, Hassan Seddik, and Víctor Cerda. 2017. "Use of multiresponse statistical techniques to optimize the separation of diosmin, hesperidin, diosmetin and hesperitin in different pharmaceutical preparations by high performance liquid chromatography with UV-DAD." *Talanta* 167: 695-702. https://doi.org/10.1016/j.talanta.2017.02.069.

[32] Sharma, Veena, and Pracheta Janmeda. 2017. "Extraction, isolation and identification of flavonoid from *Euphorbia neriifolia* leaves." *Arabian Journal of Chemistry* 10 (4): 509-514. https://doi.org/10.1016/j.arabjc.2014.08.019.

[33] Silva, Lidercia C. R. Cerqueira e, Jorge M. David, Rafael dos S. Q. Borges, Sergio L. C. Ferreira, Juceni P. David, Pedro S. dos Reis, and Roy E. Bruns. 2014. "Determination of flavanones in orange juices obtained from different sources by HPLC/DAD." *Journal of Analytical Methods in Chemistry* 2014: 5. https://doi.org/10.1155/2014/296838.

[34] Michalkiewicz, Anna, Magdalena Biesaga, and Krystyna Pyrzynska. 2008. "Solid-phase extraction procedure for determination of phenolic acids and some flavonols in honey."

Journal of Chromatography A 1187 (1): 18-24. https://doi.org/10.1016/j.chroma.2008.02.001.

[35] Tsao, Rong, and Jason McCallum. 2009. "Chemistry of Flavonoids." In *Fruit and Vegetable Phytochemicals*, 131-153.

[36] Oyvind M. Andersen, and Kenneth R. Markham. 2006. *Flavonoids: Chemistry, Biochemistry and Applications*. USA: Taylor & Francis Group.

[37] Ren, Feiyue, Yingqun Nian, and Camila A. Perussello. 2020. "Effect of storage, food processing and novel extraction technologies on onions flavonoid content: A review." *Food Research International* 132: 108953. https://doi.org/10.1016/j.foodres.2019.108953.

[38] Somers, Bernard, Fekadu F. Dinssa, Qingli Wu, and James E. Simon. 2020. "Elemental micronutrients, antioxidant activity, total Polyphenol, and total flavonoid content of selected spider plant accessions (*Cleome gynandra*) grown in eastern Africa and the eastern United States." *Journal of Medicinally Active Plants* 9 (3): 157-165. https://doi.org/10.7275/jnrp-3y92.

[39] Bondonno, Nicola P., Frederik Dalgaard, Cecilie Kyro, Kevin Murray, Catherine P. Bondonno, Joshua R. Lewis, Kevin D. Croft, Gunnar Gislason, Augustin Scalbert, Aedin Cassidy, Anne Tjonneland, Kim Overvad, and Jonathan M. Hodgson. 2019. "Flavonoid intake is associated with lower mortality in the Danish Diet Cancer and Health Cohort." *Nature Communications* 10 (1): 3651. https://doi.org/10.1038/s41467-019-11622-x. https://doi.org/10.1038/s41467-019-11622-x.

[40] Knekt, Paul, Jorma Kumpulainen, Ritva Jarvinen, Harri Rissanen, Markku Heliövaara, Antti Reunanen, Timo Hakulinen, and Arpo Aromaa. 2002. "Flavonoid intake and risk of chronic diseases." *The American Journal of Clinical Nutrition* 76 (3): 560-568. https://doi.org/10.1093/ajcn/76.3.560. https://doi.org/10.1093/ajcn/76.3.560.

[41] Dianursanti, Pandu Nugroho, and Muhamad Bagus Prakasa. 2020. "Comparison of maceration and soxhletation method for flavonoid

production from Spirulina platensis as a sunscreen's raw material." *AIP Conference Proceedings* 2230 (1): 020006. https://doi.org/10.1063/5.0002806.

[42] Fiamegos, Yiannis C., Christos G. Nanos, Jacques Vervoort, and Constantine D. Stalikas. 2004. "Analytical procedure for the in-vial derivatization-extraction of phenolic acids and flavonoids in methanolic and aqueous plant extracts followed by gas chromatography with mass-selective detection." *Journal of Chromatography A* 1041 (1): 11-18. https://doi.org/10.1016/j.chroma.2004.04.041.

[43] Hui, Chong Kah, Nurul Izwanie Majid, Mohamad Khairi Mohd Zainol, Habsah Mohamad, and Zamzahaila Mohd Zin. 2018. "Preliminary phytochemical screening and effect of hot water extraction conditions on phenolic contents ad antioxidant capacities of *Morinda citrifolia* leaf." *Malaysian Applied Biology* 47 (4): 13-24. https://www.academia.edu/37954400/PRELIMINARY_PHYTOCHEMICAL_SCREENING_AND_EFFECT_OF_HOT_WATER_EXTRACTION_CONDITIONS_ON_PHENOLIC_CONTENTS_AND_ANTIOXIDANT_CAPACITIES_OF_Morinda_citrifolia_LEAF.

[44] Zhang, J.G., and C.P. Lin. 2006. "Extracting of total flavone from *Astragalus membranaceus* by alcohol." *Mod. Food Sci. Technol.* 22: 158-162.

[45] Ahmad, Islamudin, Berna Elya, Yayuk Isneyni Ismail, Arikadia Noviani, Hadi Kuncoro, Maryono, and Neneng Siti Silfi Ambarwati. 2020. "Design of ionic liquid-based microwave-assisted extraction of flavonoids from *Cyclea barbata* Miers. And its lipoxygenase inhibitory activity." *Journal of Applied Pharmaceutical Science* 10 (4): 1-7. https://doi.org/10.7324/JAPS.2020.104001.

[46] Pham, Dinh-Chuong, Hoang-Chinh Nguyen, Thanh-Hang Le Nguyen, Hoang-Linh Ho, Thien-Kim Trinh, Jirawat Riyaphan, and Ching-Feng Weng. 2020. "Optimization of ultrasound-assisted extraction of flavonoids from *celastrus hindsii* leaves using

response surface methodology and evaluation of their antioxidant and antitumor activities." *BioMed Research International* 2020: 3497107. https://doi.org/10.1155/2020/3497107.

[47] Huaman-Castilla, Nils Leander, María Salomé Mariotti-Celis, Maximiliano Martinez-Cifuentes, and Jose Ricardo Perez-Correa. 2020. "Glycerol as alternative co-solvent for water extraction of polyphenols from carmenere pomace: hot pressurized liquid extraction and computational chemistry calculations." *Biomolecules* 10 (3): 474. https://doi.org/10.3390/biom10030474.

[48] Molina, Gustavo A., Fanny Gonzalez-Fuentes, Achim M. Loske, Francisco Fernandez, and Miriam Estevez. 2020. "Shock wave-assisted extraction of phenolic acids and flavonoids from *Eysenhardtia polystachya* heartwood: A novel method and its comparison with conventional methodologies." *Ultrasonics Sonochemistry* 61: 104809. https://doi.org/10.1016/j.ultsonch.2019.104809.

[49] Zhang, Li-Sha, Shuang Hu, Xuan Chen, Xiao-Hong Bai, and Qing-Shan Li. 2013. "A new ionic liquid–water–organic solvent three phase microextraction for simultaneous preconcentration flavonoids and anthraquinones from traditional Chinese prescription." *Journal of Pharmaceutical and Biomedical Analysis* 86: 36-39. https://doi.org/10.1016/j.jpba.2013.07.007.

[50] Wang, Yang, Lu Wang, Tian Tian, Xiaoya Hu, Chun Yanga, and Qin Xu. 2012. "Automated solid-phase extraction hyphenated to voltammetry for the determination of quercetin using magnetic nanoparticles and sequential injection lab-on-valve approach." *Analyst* 137 (10): 2400-2405. https://doi.org/10.1039/c2an35300h

[51] Yang, Chun, Yan-Rong Xu, and Wei-Xi Yao. 2002. "Extraction of pharmaceutical components from *Ginkgo biloba* leaves using supercritical carbon dioxide." *Journal of Agriculture and Food Chemistry* 50 (4): 846-849. https://doi.org/10.1021/jf010945f.

[52] Zhu, Yindi, Yue Liu, Ying Zhan, Lin Liu, Yajuan Xu, Tunhai Xu, and Tonghua Liu. 2013. "Preparative isolation and purification of five flavonoid glycosides and one benzophenone galloyl glycoside

from *Psidium guajava* by high-speed counter-current chromatography (HSCCC)." *Molecules* 18 (12): 15648-15661. https://doi.org/10.3390/molecules181215648.

[53] Sammani, Mohamad Subhi, Sabrina Clavijo, Alba Gonzalez, and Victor Cerda. 2019. "Development of an on-line lab-on-valve micro-solid phase extraction system coupled to liquid chromatography for the determination of flavonoids in citrus juices." *Analytica Chimica Acta* 1082: 56-65. https://doi.org/10.1016/j.aca.2019.06.032.

[54] Bagheri, Nafiseh, Haider A. J. Al Lawati, and Javad Hassanzadeh. 2021. "Simultaneous determination of total phenolic acids and total flavonoids in tea and honey samples using an integrated lab on a chip device." *Food Chemistry* 342: 128338. https://doi.org/10.1016/j.foodchem.2020.128338.

[55] Arora, Sumit, and Prakash Itankar. 2018. "Extraction, isolation and identification of flavonoid from *Chenopodium album* aerial parts." *Journal of Traditional and Complementary Medicine* 8 (4): 476-482. https://doi.org/10.1016/j.jtcme.2017.10.002.

[56] Damian-Reyna, A.A., J.C. Gonzalez-Hernandez, and M.C. Chavez-Parga. 2016. "Current procedures for extraction and purification of citrus flavonoides." *Rev. colomb. biotecnol.* 18 (1): 135-147. http://www.scielo.org.co/scielo.php?script=sci_arttext&pid=S0123-34752016000100015&nrm=iso.

[57] Hartonen, Kari, Jevgeni Parshintsev, Kati Sandberg, Eija Bergelin, Linda Nisula, and Marja-Liisa Riekkola. 2007. "Isolation of flavonoids from aspen knotwood by pressurized hot water extraction and comparison with other extraction techniques." *Talanta* 74 (1): 32-38. https://doi.org/10.1016/j.talanta.2007.05.040.

[58] Nandan, M. P., and V. Meena. 2015. "Extraction, modelling and purification of flavonoids from *citrus medica* peel." *International Journal of Applied Sciences and Biotechnology* 3 (4): 588-591. https://doi.org/10.3126/ijasbt.v3i4.13360.

[59] Pan, Yingming, Chunhuan He, Hengshan Wang, Xiaowen Ji, Kai Wang, and Peizhen Liu. 2010. "Antioxidant activity of microwave-assisted extract of *Buddleia officinalis* and its major active component." *Food Chemistry* 121 (2): 497-502. https://doi.org/10.1016/j.foodchem.2009.12.072.

[60] Aspe, Estrella, and Katherina Fernandez. 2011. "The effect of different extraction techniques on extraction yield, total phenolic, and anti-radical capacity of extracts from *Pinus radiata* Bark." *Industrial Crops and Products* 34 (1): 838-844. https://doi.org/10.1016/j.indcrop.2011.02.002.

[61] Gao, Min, and Chun-Zhao Liu. 2005. "Comparison of techniques for the extraction of flavonoids from cultured cells of *Saussurea medusa* Maxim." *World Journal of Microbiology and Biotechnology* 21 (8): 1461-1463. https://doi.org/10.1007/s11274-005-6809-1

[62] Yedhu Krishnan, R., M. Neelesh Chandran, Vellingiri Vadivel, and K. S. Rajan. 2016. "Insights on the influence of microwave irradiation on the extraction of flavonoids from *Terminalia chebula*." *Separation and Purification Technology* 170: 224-233. https://doi.org/10.1016/j.seppur.2016.06.039.

[63] Lin, Xue, Lingfeng Wu, Xiong Wang, Linling Yao, and Lu Wang. 2021. "Ultrasonic-assisted extraction for flavonoid compounds content and antioxidant activities of India *Moringa oleifera* L. leaves: Simultaneous optimization, HPLC characterization and comparison with other methods." *Journal of Applied Research on Medicinal and Aromatic Plants* 20: 100284. https://doi.org/10.1016/j.jarmap.2020.100284.

[64] Yi Ling, Yeong, Pang Sook Fun, Afiqah Yeop, Mashitah M. Yusoff, and Jolius Gimbun. 2019. "Assessment of maceration, ultrasonic and microwave assisted extraction for total phenolic content, total flavonoid content and kaempferol yield from *Cassia alata* via microstructures analysis." *Materials Today: Proceedings* 19: 1273-1279. https://doi.org/10.1016/j.matpr.2019.11.133.

[65] Chan, Chung-Hung, Rozita Yusoff, Gek-Cheng Ngoh, and Fabian Wai-Lee Kung. 2011. "Microwave-assisted extractions of active ingredients from plants." *Journal of Chromatography A* 1218 (37): 6213-6225. https://doi.org/10.1016/j.chroma.2011.07.040.

[66] Alara, O. R., N. H. Abdurahman, S. K. Abdul Mudalip, and O. A. Olalere. 2018. "Microwave-assisted extraction of Vernonia amygdalina leaf for optimal recovery of total phenolic content." *Journal of Applied Research on Medicinal and Aromatic Plants* 10: 16-24. https://doi.org/10.1016/j.jarmap.2018.04.004.

[67] Chaves, Jaisa Oliveira, Mariana Correa de Souza, Laise Capelasso da Silva, Daniel Lachos-Perez, Paulo Cesar Torres-Mayanga, Ana Paula da Fonseca Machado, Tania Forster-Carneiro, Mercedes Vazquez-Espinosa, Ana Velasco Gonzalez-de-Peredo, Gerardo Fernandez Barbero, and Mauricio Ariel Rostagno. 2020. "Extraction of flavonoids from natural sources using modern techniques." *Frontiers in Chemistry* 8 (864): 507887. https://doi.org/10.3389/fchem.2020.507887.

[68] Akbari, Sweeta, Nour H. Abdurahman, Rosli M. Yunus, and Fahim Fayaz. 2019. "Microwave-assisted extraction of saponin, phenolic and flavonoid compounds from Trigonella foenum-graecum seed based on two level factorial design." *Journal of Applied Research on Medicinal and Aromatic Plants* 14: 100212. https://doi.org/10.1016/j.jarmap.2019.100212.

[69] Chavez-Gonzalez, Monica L., Leonardo Sepulveda, Deepak Kumar Verma, Hugo A. Luna-Garcia, Luis V. Rodriguez-Duran, Anna Ilina, and Cristobal N. Aguilar. 2020. "Conventional and emerging extraction processes of flavonoids." *Processes* 8 (4): 434. https://doi.org/10.3390/pr8040434.

[70] Tzanova, Milena, Vasil Atanasov, Zvezdelina Yaneva, Donika Ivanova, and Toncho Dinev. 2020. "Selectivity of current extraction techniques for flavonoids from plant materials." *Processes* 8 (10): 1222. https://doi.org/10.3390/pr8101222.

[71] Piana, Fabrizio, Marco Ciulu, Rosa Quirantes-Pine, Gavino Sanna, Antonio Segura-Carretero, Nadia Spano, and Alberto Mariani.

2018. "Simple and rapid procedures for the extraction of bioactive compounds from Guayule leaves." *Industrial Crops and Products* 116: 162-169. https://doi.org/10.1016/j.indcrop.2018.02.057.

[72] Belwal, Tarun, Farid Chemat, Petras Rimantas Venskutonis, Giancarlo Cravotto, Durgesh Kumar Jaiswal, Indra Dutt Bhatt, Hari Prasad Devkota, and Zisheng Luo. 2020. "Recent advances in scaling-up of non-conventional extraction techniques: Learning from successes and failures." *TrAC Trends in Analytical Chemistry* 127: 115895. https://doi.org/10.1016/j.trac.2020.115895.

[73] Pereira, Debora Tamires Vitor, Adriana Gadioli Tarone, Cinthia Bau Betim Cazarin, Gerardo Fernandez Barbero, and Julian Martinez. 2019. "Pressurized liquid extraction of bioactive compounds from grape marc." *Journal of Food Engineering* 240: 105-113. https://doi.org/10.1016/j.jfoodeng.2018.07.019.

[74] Loske, A.M. 2017. "Novel uses and potential applications." In *Medical and biomedical applications of shock waves, Shock wave and high pressure phenomena*, 251-301. Switzerland: Springer, Cham.

[75] Kuraya, Eisuke, Akiko Touyama, Shina Nakada, Osamu Higa, and Shigeru Itoh. 2017. "Underwater shockwave pretreatment process to improve the scent of extracted citrus junos Tanaka (Yuzu) juice." *International Journal of Food Science* 2017: 2375181. https://doi.org/10.1155/2017/2375181.

[76] Cassol, Liliana, Eliseu Rodrigues, and Caciano Pelayo Zapata Norena. 2019. "Extracting phenolic compounds from Hibiscus sabdariffa L. calyx using microwave assisted extraction." *Industrial Crops and Products* 133: 168-177. https://doi.org/10.1016/j.indcrop.2019.03.023.

[77] Tomaz, Ivana, Luna Maslov, Domagoj Stupic, Darko Preiner, Danijela Asperger, and Jasminka Karoglan Kontic. 2016. "Multi-response optimisation of ultrasound-assisted extraction for recovery of flavonoids from red grape skins using response surface methodology." *Phytochemical Analysis* 27 (1): 13-22. https://doi.org/10.1002/pca.2582.

[78] Wang, Xin-Hong, and Jing-Ping Wang. 2019. "Effective extraction with deep eutectic solvents and enrichment by macroporous adsorption resin of flavonoids from *Carthamus tinctorius* L." *Journal of Pharmaceutical and Biomedical Analysis* 176: 112804. https://doi.org/10.1016/j.jpba.2019.112804.

[79] Corazza, Gustavo Osmar, Denise Bilibio, Odivan Zanella, Anderson Luis Nunes, Joao Paulo Bender, Naira Carniel, Priscilla Pereira dos Santos, and Wagner L. Priamo. 2018. "Pressurized liquid extraction of polyphenols from Goldenberry: Influence on antioxidant activity and chemical composition." *Food and Bioproducts Processing* 112: 63-68. https://doi.org/10.1016/j.fbp.2018.09.001.

[80] Barrales, Francisco Manuel, Paula Silveira, Paula de Paula Menezes Barbosa, Amanda Roggia Ruviaro, Bruno Nicolau Paulino, Glaucia Maria Pastore, Gabriela Alves Macedo, and Julian Martinez. 2018. "Recovery of phenolic compounds from citrus by-products using pressurized liquids — An application to orange peel." *Food and Bioproducts Processing* 112: 9-21. https://doi.org/10.1016/j.fbp.2018.08.006.

[81] Leyva-Jimenez, Francisco Javier, Jesus Lozano-Sanchez, Isabel Borras-Linares, David Arráez-Roman, and Antonio Segura-Carretero. 2018. "Comparative study of conventional and pressurized liquid extraction for recovering bioactive compounds from *Lippia citriodora* leaves." *Food Research International* 109: 213-222. https://doi.org/10.1016/j.foodres.2018.04.035.

[82] Su, Yi, Shitou Xia, Ruozhong Wang, and Langtao Xiao. 2017. "13 - Phytohormonal quantification based on biological principles." In *Hormone Metabolism and Signaling in Plants*, edited by Jiayang Li, Chuanyou Li and Steven M. Smith, 431-470. Academic Press.

[83] Stalikas, Constantine D. 2007. "Extraction, separation, and detection methods for phenolic acids and flavonoids." *Journal of Separation Science* 30 (18): 3268-3295. https://doi.org/10.1002/jssc.200700261.

[84] Herrero, M., M. Plaza, A. Cifuentes, and E. Ibanez. 2012. "4.08 - Extraction techniques for the determination of phenolic compounds in food." In *Comprehensive Sampling and Sample Preparation*, edited by Janusz Pawliszyn, 159-180. Oxford: Academic Press.

[85] Ri, Hyon-il, Chol-song Kim, Un-hak Pak, Myong-su Kang, and Tae-mun Kim. 2019. "Purification of total flavonoids from *Aurea Helianthus* flowers and in vitro hypolipidemic effect." *Biomolecules*: 10. https://doi.org/arXiv:1906.12007v1.

[86] Bian, Yu, Yuan Zhang, Yu Zhou, Guo-Hui Li, and Xue-Song Feng. 2020. "Progress in the pretreatment and analysis of flavonoids: An update since 2013." *Separation & Purification Reviews*: 1-27. https://doi.org/10.1080/15422119.2020.1801469.

[87] Soylak, Mustafa, Bircan Ozdemir, and Erkan Yilmaz. 2020. "An environmentally friendly and novel aminc-based liquid phase microextraction of quercetin in food samples prior to its determination by UV–vis spectrophotometry." *Spectrochimica Acta Part A: Molecular and Biomolecular Spectroscopy* 243: 118806. https://doi.org/10.1016/j.saa.2020.118806.

[88] Quigley, Andrew, Wayne Cummins, and Damian Connolly. 2016. "Dispersive liquid-liquid microextraction in the analysis of milk and dairy products: A review." *Journal of Chemistry* 2016: 4040165. https://doi.org/10.1155/2016/4040165.

[89] Campillo, Natalia, Pilar Vinas, Gema Ferez-Melgarejo, and Manuel Hernandez-Cordoba. 2015. "Dispersive liquid–liquid microextraction for the determination of flavonoid aglycone compounds in honey using liquid chromatography with diode array detection and time-of-flight mass spectromctry." *Talanta* 131: 185-191. https://doi.org/10.1016/j.talanta.2014.07.083.

[90] Maya, Fernando, Burkhard Horstkotte, Jose M. Estela, and Victor Cerda. 2014. "Automated in-syringe dispersive liquid-liquid microextraction." *TrAC Trends in Analytical Chemistry* 59: 1-8. https://doi.org/doi.org/10.1016/j.trac.2014.03.009.

[91] Venson, Rafael, Ann-Sophie Korb, and Gail Cooper. 2019. "A review of the application of hollow-fiber liquid-phase microextraction in bioanalytical methods – A systematic approach with focus on forensic toxicology." *Journal of Chromatography B* 1108: 32-53. https://doi.org/10.1016/j.jchromb.2019.01.006.

[92] Soares da Silva Burato, Juliana, Deyber Arley Vargas Medina, Ana Lucia de Toffoli, Edvaldo Vasconcelos Soares Maciel, and Fernando Mauro Lancas. 2020. "Recent advances and trends in miniaturized sample preparation techniques." *Journal of Separation Science* 43 (1): 202-225. https://doi.org/10.1002/jssc.201900776.

[93] Alsharif, Ali Mohamed Ali, Guan-Huat Tan, Yeun-Mun Choo, and Abubakar Lawal. 2016. "Efficiency of hollow fiber liquid-phase microextraction chromatography methods in the separation of organic compounds: A review." *Journal of Chromatographic Science* 55 (3): 378-391. https://doi.org/10.1093/chromsci/bmw188.

[94] Hadjmohammadi, Mohammadreza, Hanieh Karimiyan, and Vahid Sharifi. 2013. "Hollow fibre-based liquid phase microextraction combined with high-performance liquid chromatography for the analysis of flavonoids in *Echinophora platyloba* DC. and *Mentha piperita*." *Food Chemistry* 141 (2): 731-735. https://doi.org/10.1016/j.foodchem.2013.02.083.

[95] Ali Redha, Ali. 2021. "Review on extraction of phenolic compounds from natural sources using green deep eutectic solvents." *Journal of Agricultural and Food Chemistry* 69 (3): 878-912. https://doi.org/10.1021/acs.jafc.0c06641.

[96] Ivanovic, Milena, Masa Islamcevic Razborsek, and Mitja Kolar. 2020. "Innovative extraction techniques using deep eutectic solvents and analytical methods for the isolation and characterization of natural bioactive compounds from plant material." *Plant Analytical Chemistry* 9 (11): 1428. https://doi.org/10.3390/plants9111428.

[97] Ali, Mohammad Chand, Jia Chen, Haijuan Zhang, Zhan Li, Liang Zhao, and Hongdeng Qiu. 2019. "Effective extraction of flavonoids from *Lycium barbarum* L. fruits by deep eutectic solvents-based ultrasound-assisted extraction." *Talanta* 203: 16-22. https://doi.org/10.1016/j.talanta.2019.05.012.

[98] Rodriguez De Luna, Sara Luisa, R. E. Ramirez-Garza, and Sergio O. Serna Saldivar. 2020. "Environmentally friendly methods for flavonoid extraction from plant material: Impact of their operating conditions on yield and antioxidant properties." *The Scientific World Journal* 2020: 6792069. https://doi.org/10.1155/2020/6792069.

[99] Nalewajko-Sieliwoniuk, Edyta, Marta Hryniewicka, Diana Jankowska, Anatol Kojlo, Monika Kamianowska, and Marek Szczepanski. 2020. "Dispersive liquid–liquid microextraction coupled to liquid chromatography tandem mass spectrometry for the determination of phenolic compounds in human milk." *Food Chemistry* 327: 126996. https://doi.org/10.1016/j.foodchem.2020.126996.

[100] Chaieb, Nadia, Montserrat Lopez-Mesas, Johannes Luis Gonzalez, Messaoud Marsb, and Manuel Valiente. 2015. "Hollow fibre liquid phase micro-extraction by facilitated anionic exchange for the determination of flavonoids in faba beans (*Vicia faba* L.)." *Phytochemical Analysis* 26: 346-352. https://doi.org/10.1002/pca.2569.

[101] Li, Shang, Shuang Hu, Xuan Chen, Xiaohong Bai, and Shuqiu Zhang. 2015. "Two-phase/three-phase hollow fibre liquid-phase simultaneous microextraction combined with HPLC for analysis of phenolic acids and flavonoids in traditional Chinese medicine." *Chromatographia* 78 (17): 1159-1167. https://doi.org/10.1007/s10337-015-2930-y.

[102] Shen, Shi, Jingbo Wang, Xi Chen, Tingting Liu, Yan Ma, Liyuan Wang, Wen Qin, Zhaolong Gong, and Qin Zhuo. 2020. "Determination of flavonoids in honey by solid phase extraction coupled with ultra-performance liquid chromatography-tandem

mass spectrometry method." *Wei Sheng Yan Jiu* 49 (2): 259-266. https://doi.org/10.19813/j.cnki.weishengyanjiu.2020.02.016.

[103] Wang, Dandan, Jinxiang Zeng, Wei Xiang, Manni Yin, Guoyue Zhong, and Zhining Xia. 2020. "Online coupling of the Ussing chamber, solid-phase extraction and high-performance liquid chromatography for screening and analysis of active constituents of traditional Chinese medicines." *Journal of Chromatography A* 1609: 460480. https://doi.org/10.1016/j.chroma.2019.460480.

[104] Li, Xiaoxia, and Kyung Ho Row. 2019. "Preparation of deep eutectic solvent-based hexagonal boron nitride-molecularly imprinted polymer nanoparticles for solid phase extraction of flavonoids." *Microchimica Acta* 186 (12): 753. https://doi.org/10.1007/s00604-019-3885-8.

[105] Turiel, Esther, and Antonio Martín Esteban. 2020. "8 - Molecularly imprinted polymers." In *Solid-Phase Extraction*, edited by Colin F. Poole, 215-233. Elsevier.

[106] Arvapally, Manasa, Ankita Asati, Narendra Kumar Nagendla, and Mohana Krishna Reddy Mudiam. 2021. "Development of an analytical method for the quantitative determination of multi-class nutrients in different food matrices by solid-phase extraction and liquid chromatography-tandem mass spectrometry using design of experiments." *Food Chemistry* 341: 128173. https://doi.org/10.1016/j.foodchem.2020.128173.

[107] Zhao, Tong, Xiujuan Guan, Wanjin Tang, Ying Ma, and Haixia Zhang. 2015. "Preparation of temperature sensitive molecularly imprinted polymer for solid-phase microextraction coatings on stainless steel fiber to measure ofloxacin." *Analytica Chimica Acta* 853: 668-675. https://doi.org/10.1016/j.aca.2014.10.019.

[108] Sun, Min, Juanjuan Feng, Yanan Bu, and Chuannan Luo. 2015. "Highly sensitive copper fiber-in-tube solid-phase microextraction for online selective analysis of polycyclic aromatic hydrocarbons coupled with high performance liquid chromatography." *Journal of Chromatography A* 1408: 41-48. https://doi.org/10.1016/j.chroma.2015.07.024.

[109] Djozan, Djavanshir, Bahram Ebrahimi, Mehrdad Mahkam, and Mir Ali Farajzadeh. 2010. "Evaluation of a new method for chemical coating of aluminum wire with molecularly imprinted polymer layer. Application for the fabrication of triazines selective solid-phase microextraction fiber." *Analytica Chimica Acta* 674 (1): 40-48. https://doi.org/10.1016/j.aca.2010.06.006.

[110] Liu, Hongmei, Daoai Wang, Li Ji, Jubai Li, Shujuan Liu, Xia Liu, and Shengxiang Jiang. 2010. "A novel TiO2 nanotube array/Ti wire incorporated solid-phase microextraction fiber with high strength, efficiency and selectivity." *Journal of Chromatography A* 1217 (12): 1898-1903. https://doi.org/10.1016/j.chroma.2010.01.080.

[111] Djozan, Dj, and L. Abdollahi. 2003. "Anodized zinc wire as a solid-phase microextraction fiber." *Chromatographia* 57 (11): 799-804. https://doi.org/10.1007/BF02491768.

[112] Hafez, Aly M., and Bernd W. Wenclawiak. 2013. "Development of a new porous gold SPME fiber for selective and efficient extraction of dodecanethiol followed by GC–MS analysis." *Analytical and Bioanalytical Chemistry* 405 (5): 1753-1758. https://doi.org/10.1007/s00216-012-6580-8.

[113] Ghorbani, Mahdi, Mohsen Aghamohammadhassan, Mahmoud Chamsaz, Hossien Akhlaghi, and Toktam Pedramrad. 2019. "Dispersive solid phase microextraction." *TrAC Trends in Analytical Chemistry* 118: 793-809. https://doi.org/10.1016/j.trac.2019.07.012.

[114] Sajid, Muhammad, Mazen Khaled Nazal, Malgorzata Rutkowska, Natalia Szczepanska, Jacek Namiesnik, and Justyna Plotka-Wasylka. 2019. "Solid phase microextraction: Apparatus, sorbent materials, and application." *Critical Reviews in Analytical Chemistry* 49 (3): 271-288. https://doi.org/10.1080/10408347.2018.1517035.

[115] Aulakh, J. S., A. K. Malik, Varinder Kaur, and Philippe Schmitt-Kopplin. 2005. "A review on solid phase micro extraction-high performance liquid chromatography (SPME-HPLC) analysis of

pesticides." *Critical Reviews in Analytical Chemistry* 35 (1): 71-85. https://doi.org/10.1080/10408340590947952.

[116] Merkle, Sybille, Kim Karen Kleeberg, and Jan Fritsche. 2015. "Recent developments and applications of solid phase microextraction (SPME) in food and environmental analysis—A review." *Chromatography* 2 (3): 293-381. https://doi.org/10.3390/chromatography2030293.

[117] Tashakkori, Paniz, Aylin Altınısık Tagac, and Melek Merdivan. 2021. "Fabrication of montmorillonite/ionic liquid composite coated solid-phase microextraction fibers for determination of phenolic compounds in fruit juices by gas chromatography and liquid chromatography." *Journal of Chromatography A* 1635: 461741. https://doi.org/10.1016/j.chroma.2020.461741.

[118] Ghorbani, Mahdi, Mohsen Aghamohammadhassan, Haniyeh Ghorbani, and Ali Zabihi. 2020. "Trends in sorbent development for dispersive micro-solid phase extraction." *Microchemical Journal* 158: 105250. https://doi.org/10.1016/j.microc.2020.105250.

[119] Yin, Lujun, Xue Zheng, Gen Wang, and Wei Wang. 2019. "Microwave irradiation followed by zinc oxide based dispersive solid-phase extraction coupled with HPLC for simultaneous extraction and determination of flavonoids in *Veronicastrum latifolium* (Hemsl.) Yamazaki." *Analytical and Bioanalytical Chemistry* 411 (5): 1029-1040. https://doi.org/10.1007/s00216-018-1529-1.

[120] Khezeli, Tahere, and Ali Daneshfar. 2017. "Development of dispersive micro-solid phase extraction based on micro and nano sorbents." *TrAC Trends in Analytical Chemistry* 89: 99-118. https://doi.org/10.1016/j.trac.2017.01.004.

[121] Yin, Shi-Jun, Jing Zhao, and Feng-Qing Yang. 2021. "Recent applications of magnetic solid phase extraction in sample preparation for phytochemical analysis." *Journal of Pharmaceutical and Biomedical Analysis* 192: 113675. https://doi.org/10.1016/j.jpba.2020.113675.

[122] Yilmaz, Erkan, and Mustafa Soylak. 2020. "Chapter 3 - Type of new generation separation and preconcentration methods." In *New Generation Green Solvents for Separation and Preconcentration of Organic and Inorganic Species*, edited by Mustafa Soylak and Erkan Yilmaz, 75-148. Elsevier.

[123] Li, Peng, Junfeng Ye, Yue Zhang, Ziqi Wang, Sujuan Ren, Yongri Jin, and Xuwen Li. 2019. "Magnetic dispersive solid-phase extraction for the determination of three different glycosides in the Kang'ai injection." *Microchemical Journal* 146: 575-581. https://doi.org/10.1016/j.microc.2019.01.037.

[124] Perez, Rosa Ana, Beatriz Albero, and Jose L. Tadeo. 2020. "19 - Matrix solid phase dispersion." In *Solid-Phase Extraction*, edited by Colin F. Poole, 531-549. Elsevier.

[125] Zheng, Hui, Xiao-Ting Zhen, Yan Chen, Si-Chen Zhu, Li-Hong Ye, Si-Wei Yang, Qiu-Yan Wang, and Jun Cao. 2021. "In situ antioxidation-assisted matrix solid-phase dispersion microextraction and discrimination of chiral flavonoids from citrus fruit via ion mobility quadrupole time-of-flight high-resolution mass spectrometry." *Food Chemistry* 343: 128422. https://doi.org/10.1016/j.foodchem.2020.128422.

[126] Li, Xiaoxia, and Kyung Ho Row. 2019. "Preparation of deep eutectic solvent-based hexagonal boron nitride-molecularly imprinted polymer nanoparticles for solid phase extraction of flavonoids." *Microchim. Acta* 186 (12): 753. https://doi.org/10.1007/s00604-019-3885-8. https://doi.org/10.1007/s00604-019-3885-8.

[127] Chrzanowska, Anna M., Anna Poliwoda, and Piotr P.Wieczorek. 2015. "Characterization of particle morphology of biochanin A molecularly imprinted polymers and their properties as a potential sorbent for solid-phase extraction." *Materials Science and Engineering C* 49: 793-798. https://doi.org/10.1016/j.msec.2015.01.069.

[128] Rahimi, Marzieh, Soleiman Bahar, Rouhollah Heydari, and S. Mojtaba Amininasab. 2019. "Determination of quercetin using a

molecularly imprinted polymer as solid-phase microextraction sorbent and high-performance liquid chromatography." *Microchemical Journal* 148: 433-441. https://doi.org/10.1016/j.microc.2019.05.032.

[129] Chu, Chu, Shan Wang, Huan Zhang, Caijing Liu, Luyi Jiang, Yiping Tang, and Jizhong Yan. 2019. "A sensitive dispersive micro solid-phase extraction coupled with high performance liquid chromatography for determination of three flavonoids in complex matrics by using crab shell as a sorbent." *Electrophoresis* 40 (9): 1298-1305. https://doi.org/10.1002/elps.201900003.

[130] Casado, Natalia, Sonia Morante-Zarcero, Damian Perez-Quintanilla, Jose S. Camara, and Isabel Sierra. 2019. "Dispersive solid-phase extraction of polyphenols from juice and smoothie samples using hybrid mesostructured silica followed by ultra-high-performance liquid chromatography-ion-trap tandem mass spectrometry." *Journal of Agricultural and Food Chemistry* 67 (3): 955-967. https://doi.org/10.1021/acs.jafc.8b05578.

[131] Xu, Jing-Jing, Rui Yang, Li-Hong Ye, Jun Cao, Wan Cao, Shuai-Shuai Hu, and Li-Qing Peng. 2016. "Application of ionic liquids for elution of bioactive flavonoid glycosides from lime fruit by miniaturized matrix solid-phase dispersion." *Food Chemistry* 204: 167-175. https://doi.org/10.1016/j.foodchem.2016.02.012.

[132] Majidi, Seyedeh Maedeh, and Mohammad Reza Hadjmohammadi. 2021. "Development of magnetic dispersive micro-solid phase extraction based on magnetic agarose nanoparticles and deep eutectic solvents for the isolation and pre-concentration of three flavonoids in edible natural samples." *Talanta* 222: 121649. https://doi.org/10.1016/j.talanta.2020.121649.

[133] Senes, Carlos Eduardo Rubio, Carina Alexandra Rodrigues, Antonio Eduardo Nicácio, Joana Schuelter Boeing, Liane Maldaner, and Jesuí Vergílio Visentainer. 2021. "Determination of phenolic acids and flavonoids from *Myrciaria cauliflora* edible part employing vortex-assisted matrix solid-phase dispersion (VA-

MSPD) and UHPLC-MS/MS." *Journal of Food Composition and Analysis* 95: 103667. https://doi.org/10.1016/j.jfca.2020.103667.

[134] Tyskiewicz, Katarzyna, Marcin Konkol, and Edward Roj. 2018. "The application of supercritical fluid extraction in phenolic compounds isolation from natural plant materials." *Molecules* 23 (10): 2625. https://doi.org/10.3390/molecules23102625.

[135] Arias, Julian, Jesica Mejia, Yuri Cordoba, Jairo Rene Martinez, Elena Stashenko, and Jose Manuel del Valle. 2020. "Optimization of flavonoids extraction from *Lippia graveolens* and *Lippia origanoides* chemotypes with ethanol-modified supercritical CO2 after steam distillation." *Industrial Crops and Products* 146: 112170. https://doi.org/10.1016/j.indcrop.2020.112170.

[136] Veiga, Bruno Aguilar, Fabiane Hamerski, Mathias Porsmose Clausen, Massimiliano Errico, Agnes de Paula Scheer, and Marcos Lúcio Corazza. 2021. "Compressed fluids extraction methods, yields, antioxidant activities, total phenolics and flavonoids content for Brazilian Mantiqueira hops." *The Journal of Supercritical Fluids* 170: 105155. https://doi.org/10.1016/j.supflu.2020.105155.

[137] Ostolski, Mateusz, Marek Adamczak, Bartosz Brzozowski, and Wiesław Wiczkowski. 2021. "Antioxidant activity and chemical characteristics of supercritical CO2 and water extracts from willow and poplar." *Molecules* 26 (3): 545. https://doi.org/10.3390/molecules26030545.

[138] Leyva-Jimenez, Francisco Javier, Jesus Lozano-Sanchez, Alvaro Fernandez-Ochoa, Maria de la Luz Cadiz-Gurrea, David Arraez-Roman, and Antonio Segura-Carretero. 2020. "Optimized extraction of phenylpropanoids and flavonoids from lemon verbena leaves by supercritical fluid system using response surface methodology." *Foods* 9 (7): 931. https://doi.org/10.3390/foods9070931.

[139] Gong, Yuan, Xin-Yi Huang, Dong Pei, Wen-Da Duan, Xia Zhang, Xiao Sun, and Duo-Long Di. 2020. "The applicability of high-speed counter current chromatography to the separation of natural

antioxidants." *Journal of Chromatography A* 1623: 461150. https://doi.org/10.1016/j.chroma.2020.461150.

[140] Berthod, Alain, Tatyana Maryutina, Boris Spivakov, Oleg Shpigun, and Ian A. Sutherland. 2009. "Countercurrent chromatography in analytical chemistry (IUPAC Technical Report) %J Pure and Applied Chemistry." *Pure and Applied Chemistry* 81 (2): 355-387. https://doi.org/10.1351/PAC-REP-08-06-05.

[141] Ma, Yumei, Xue Yang, Jilin Chen, Jingyang Zhao, Li Yang, Shuping Yan, Hongmei Li, Cheng Shen, Yangfei Wei, Shuo Wang, Tao Chen, Zhi Chen, and Yulin Li. 2020. "Separation of five flavonoids with similar polarity from *Caragana Korshinskii* Kom. by preparative high speed counter-current chromatography with recycling and heart cut mode." *Journal of Separation Science* 43 (19): 3748-3755. https://doi.org/10.1002/jssc.202000535.

[142] Cai, Xu, Mi Xiao, Xianwei Zou, Jintian Tang, Bisheng Huang, and Hongkun Xue. 2021. "Extraction and separation of flavonoids from Malus hupehensis using high-speed countercurrent chromatography based on deep eutectic solvent." *Journal of Chromatography A* 1641: 461998. https://doi.org/10.1016/j.chroma.2021.461998.

[143] Dash, Nihar, and Hemant Jain. 2016. "Surgical management of ulcerative colitis: A descriptive mini review." *Journal of Indira Gandhi Institute of Medical Sciences* 1 (1): 8-15. http://www.jigims.co.in/text.asp?2016/2/2/8/303394.

[144] Jiang, Shujing, Qi Liu, Yixi Xie, Hualiang Zeng, Li Zhang, Xinyu Jiang, and Xiaoqing Chen. 2015. "Separation of five flavonoids from tartary buckwheat (*Fagopyrum tataricum* (L.) Gaertn) grains via off-line two dimensional high-speed counter-current chromatography." *Food Chemistry* 186: 153-159. https://doi.org/10.1016/j.foodchem.2014.08.120.

[145] Cui, Hai-Yan, Xiao-Yan Jia, Xia Zhang, Jing Zhang, and Zhi-Qi Zhang. 2011. "Optimization of high-speed counter-current chromatography for separation of polyphenols from the extract of

hawthorn (*Crataegus laevigata*) with response surface methodology." *Separation and Purification Technology* 77 (2): 269-274. https://doi.org/10.1016/j.seppur.2010.12.016.

[146] da Silva, Laise C., Mariana C. Souza, Beatriz R. Sumere, Luiz G. S. Silva, Diogo T. da Cunha, Gerardo F. Barbero, Rosangela M. N. Bezerra, and Mauricio A. Rostagno. 2020. "Simultaneous extraction and separation of bioactive compounds from apple pomace using pressurized liquids coupled on-line with solid-phase extraction." *Food Chemistry* 318: 126450. https://doi.org/10.1016/j.foodchem.2020.126450.

[147] Gao, Min, Bao-Zhen Song, and Chun-Zhao Liu. 2006. "Dynamic microwave-assisted extraction of flavonoids from *Saussurea medusa* Maxim cultured cells." *Biochemical Engineering Journal* 32 (2): 79-83. https://doi.org/10.1016/j.bej.2006.09.004.

[148] Souza, Mariana C., Laise C. Silva, Jaisa O. Chaves, Mayara P. Salvador, Vitor L. Sanches, Diogo T. da Cunha, Tânia Foster Carneiro, and Mauricio A. Rostagno. 2021. "Simultaneous extraction and separation of compounds from mate (*Ilex paraguariensis*) leaves by pressurized liquid extraction coupled with solid-phase extraction and in-line UV detection." *Food Chemistry: Molecular Sciences* 2: 100008. https://doi.org/10.1016/j.fochms.2020.100008.

[149] Ma, Chaoyang, Liming Hu, Qianyun Fu, Xiaohong Gu, Guanjun Tao, and Hongxin Wang. 2013. "Separation of four flavonoids from *Rhodiola rosea* by on-line combination of sample preparation and counter-current chromatography." *Journal of Chromatography A* 1306: 12-19. https://doi.org/10.1016/j.chroma.2013.07.052.

[150] Liu, Yue, Jia Hu, Yan Li, Xiao-Shuang Li, and Zhong-Liang Wang. 2016. "Metal-organic framework MIL-101 as sorbent based on double-pumps controlled on-line solid-phase extraction coupled with high-performance liquid chromatography for the determination of flavonoids in environmental water samples." *Electrophoresis* 37 (19): 2478-2486. https://doi.org/10.1002/elps.201600118.

[151] Clavijo, Sabrina, Jessica Avivar, Ruth Suarez, and Victor Cerda. 2015. "Analytical strategies for coupling separation and flow-injection techniques." *TrAC Trends in Analytical Chemistry* 67: 26-33. https://doi.org/10.1016/j.trac.2014.11.019.

[152] Sammani, Mohamad Subhi, Sabrina Clavijo, and Victor Cerda. 2021. "Recent, advanced sample pretreatments and analytical methods for flavonoids determination in different samples." *TrAC Trends in Analytical Chemistry* 138: 116220. https://doi.org/10.1016/j.trac.2021.116220.

[153] Rodriguez, Rogelio, Jessica Avivar, Luz O. Leal, Victor Cerda, and Laura Ferrer. 2016. "Strategies for automating solid-phase extraction and liquid-liquid extraction in radiochemical analysis." *TrAC Trends in Analytical Chemistry* 76: 145-152. https://doi.org/10.1016/j.trac.2015.09.009.

[154] de Castro, M. D. Luque. 1986. "Flow injection analysis: a new tool to automate extraction processes." *Journal of Automatic Chemistry* 8: 382048. https://doi.org/10.1155/S1463924686000135.

[155] Silvestre, Cristina I. C., Joao L. M. Santos, Jose L. F. C. Lima, and Elias A. G. Zagatto. 2009. "Liquid–liquid extraction in flow analysis: A critical review." *Analytica Chimica Acta* 652 (1): 54-65. https://doi.org/10.1016/j.aca.2009.05.042.

[156] Rocha, Fabio R.P., Alex D. Batista, Wanessa R. Melchert, and Elias A.G. Zagatto. 2018. "Solid-phase extractions in flow analysis." *Anais da Academia Brasileira de Ciências* 90: 803-824. http://www.scielo.br/scielo.php?script=sci_arttext&pid=S0001-37652018000200803&nrm=iso.

[157] Facchin, Ileana, Jarbas J. R. Rohwedder, and Celio Pasquini. 1997. "An automated system for liquid-liquid extraction in monosegmented flow analysis." *Journal of Automatic Chemistry* 19: 891547. https://doi.org/10.1155/S1463924697000059.

[158] Alexovic, Michal, Burkhard Horstkotte, Ivana Sramkova, Petr Solich, and Jan Sabo. 2017. "Automation of dispersive liquid–liquid microextraction and related techniques. Approaches based on flow, batch, flow-batch and in-syringe modes." *TrAC Trends in*

Analytical Chemistry 86: 39-55. https://doi.org/10.1016/j.trac.2016.10.003.

[159] Mejia-Carmona, Karen, Marcela Jordan-Sinisterra, and Fernando M. Lancas. 2019. "Current trends in fully automated on-line analytical techniques for beverage analysis." *Beverages* 5 (1): 13. https://doi.org/10.3390/beverages5010013.

[160] Boonjob, Warunya. 2014. "An overview about recent advances of micro-solid phase extraction in flow based techniques." *Journal of Analytical and Pharmaceutical Chemistry* 2 (1): 1006. https://austinpublishinggroup.com/analytical-pharmaceutical-chemistry/fulltext/ajapc-v1-id1006.php.

[161] Arce, Lourdes, Mariia Teresa Tena, Angel Rios, and Miguel Valcarcel. 1998. "Determination of trans-resveratrol and other polyphenols in wines by a continuous flow sample clean-up system followed by capillary electrophoresis separation." *Analytica Chimica Acta* 359 (1): 27-38. https://doi.org/10.1016/S0003-2670(97)00668-5.

[162] Perez-Ruiz, T., C. Martinez-Lozano, V. Tomas, and J. Fenoll. 1999. "Spectrofluorometric determination of hesperidin by manual and flow-injection methods." *Fresenius' Journal of Analytical Chemistry* 364 (3): 279-283. https://doi.org/10.1007/s002160051338.

[163] Becerra, E., A. Cladera, and Victor Cerda. 1999. "Design of a very versatile software program for automating analytical methods." *Laboratory Robotics and Automation* 11 (3): 131-140. https://doi.org/10.1002/(SICI)1098-2728.

[164] Gomez-Hens, Agustina, and Juan M. Fernandez-Romero. 2017. "Microfluidic systems in analytical chemistry." In *Encyclopedia of Analytical Chemistry*, 1-20. USA: John Wiley & Sons, Ltd.

[165] Giordano, Braden C., Dean S. Burgi, Sean J. Hart, and Alex Terray. 2012. "On-line sample pre-concentration in microfluidic devices: A review." *Analytica Chimica Acta* 718: 11-24. https://doi.org/10.1016/j.aca.2011.12.050.

[166] de Mello, Andrew J., and Nigel Beard. 2003. "Focus. Dealing with 'real' samples: sample pre-treatment in microfluidic systems." *Lab on a Chip* 3 (1): 11N-20N. https://doi.org/10.1039/B301019H.

[167] Wolfe, Kelley A., Michael C. Breadmore, Jerome P. Ferrance, Mary E. Power, John F. Conroy, Pamela M. Norris, and James P. Landers. 2002. "Toward a microchip-based solid-phase extraction method for isolation of nucleic acids." *Electrophoresis* 23 (5): 727-733. https://doi.org/10.1002/1522-2683(200203)23:5<727::AID-ELPS727>3.0.CO;2-O.

[168] Oleschuk, Richard D., Loranelle L. Shultz-Lockyear, Yuebin Ning, and D. Jed Harrison. 2000. "Trapping of bead-based reagents within microfluidic systems: On-chip solid-phase extraction and electrochromatography." *Analytical Chemistry* 72 (3): 585-590. https://doi.org/10.1021/ac990751n.

[169] Ramsey, Jeremy D., and Greg E. Collins. 2005. "Integrated microfluidic device for solid-phase extraction coupled to micellar electrokinetic chromatography separation." *Analytical Chemistry* 77 (20): 6664-6670. https://doi.org/10.1021/ac0507789.

[170] Escarpa, Alberto. 2014. "Lights and shadows on Food Microfluidics." *Lab on a Chip* 14 (17): 3213-3224. https://doi.org/ 10.1039/C4LC00172A.

[171] Yang, Weichun, and Adam T. Woolley. 2010. "Integrated multiprocess microfluidic systems for automating analysis." *Journal of the Association for Laboratory Automation* 15 (3): 198-209. https://doi.org/10.1016/j.jala.2010.01.008.

[172] Lin, Ling, Hui Chen, Huibin Wei, Feng Wang, and Jin-Ming Lin. 2011. "On-chip sample pretreatment using a porous polymer monolithic column for solid-phase microextraction and chemiluminescence determination of catechins in green tea." *Analyst* 136 (20): 4260-4267. https://doi.org/10.1039/C1AN 15530J.

[173] Cate, David M., Jaclyn A. Adkins, Jaruwan Mettakoonpitak, and Charles S. Henry. 2015. "Recent developments in paper-based

microfluidic devices." *Analytical Chemistry* 87 (1): 19-41. https://doi.org/10.1016/j.jala.2010.01.008.

[174] Gutorova, Svetlana V., Vladimir V. Apyari, Vyacheslav I. Kalinin, Aleksei A. Furletov, Veronika V. Tolmacheva, Maria V. Gorbunova, and Stanislava G. Dmitrienko. 2021. "Composable paper-based analytical devices for determination of flavonoids." *Sensors and Actuators B: Chemical* 331: 129398. https://doi.org/10.1016/j.snb.2020.129398.

[175] Kovachev, Nikolay, Antonio Canals, and Alberto Escarpa. 2010. "Fast and selective microfluidic chips for electrochemical antioxidant sensing in complex samples." *Analytical Chemistry* 82 (7): 2925-2931. https://doi.org/10.1021/ac9029218.

[176] Al Haddabi, Buthaina, Haider A. J. Al Lawati, and FakhrEldin O. Suliman. 2017. "A comprehensive evaluation of three microfluidic chemiluminescence methods for the determination of the total phenolic contents in fruit juices." *Food Chemistry* 214: 670-677. https://doi.org/10.1016/j.foodchem.2016.07.119.

[177] Al Lawati, Haider A. J., Buthaina Al Haddabi, and FakhrEldin O. Suliman. 2014. "A novel microfluidic device for estimating the total phenolic/antioxidant level in honey samples using a formaldehyde/potassium permanganate chemiluminescence system." *Analytical Methods* 6 (18): 7243-7249. https://doi.org/10.1039/C4AY00550C.

[178] Kumar, Ashwini, Ashok Kumar Malik, and Dhananjay Kumar Tewary. 2009. "A new method for determination of myricetin and quercetin using solid phase microextraction–high performance liquid chromatography–ultra violet/visible system in grapes, vegetables and red wine samples." *Analytica Chimica Acta* 631 (2): 177-181. https://doi.org/https://doi.org/10.1016/j.aca.2008.10.038. http://www.sciencedirect.com/science/article/pii/S0003267008018126.

[179] Hu, Xiuli, Jingyan You, Changli Bao, Huarong Zhang, Xiangzhe Meng, Tingting Xiao, Kun Zhang, Yutang Wang, Huimin Wang, Hanqi Zhang, and Aimin Yu. 2008. "Determination of total

flavonoids in Scutellaria barbata D. Don by dynamic ultrasonic extraction coupled with on-line spectrophotometry." *Analytica Chimica Acta* 610 (2): 217-23. https://doi.org/10.1016/j.aca.2008.01.036.

[180] Bernal, Freddy A., Luisa L. Orduz-Diaz, and Ericsson Coy-Barrera. 2015. "Exploitation of the complexation reaction of ortho-dihydroxylated anthocyanins with aluminum(III) for their quantitative spectrophotometric determination in edible sources." *Food Chemistry* 185: 84-9. https://doi.org/10.1016/j.foodchem.2015.03.116.

[181] Pavun, Leposava A., Jasmina M. Dimitric Markovic, Predrag T. Durdevic, Milena D. Jelikic-Stankov, Daniela B. Dikanovic, Andrija R. Ciric, and Dusan L. Malesev. 2012. "Development and validation of a fluorometric method for the determination of hesperidin in human plasma and pharmaceutical forms." *Journal of the Serbian Chemical Society* 77 (11): 1625-1640. https://doi.org/10.2298/JSC111005060P.

[182] Lund, Jensen A., Paula N. Brown, and Paul R. Shipley. 2020. "Quantification of North American and European Crataegus flavonoids by nuclear magnetic resonance spectrometry." *Fitoterapia* 143: 104537. https://doi.org/10.1016/j.fitote.2020.104537.

[183] Xia, Yanming, Yuanfeng Wei, Hui Chen, Shuai Qian, Jianjun Zhang, and Yuan Gao. 2021. "Competitive cocrystallization and its application in the separation of flavonoids." *International Union of Crystallography* 8 (2): 195-207. https://doi.org/doi:10.1107/S2052252520015997.

[184] Kim, Mihyang, Dongho Won, and Jaehong Han. 2010. "Absolute configuration determination of isoflavan-4-ol stereoisomers." *Bioorganic & Medicinal Chemistry Letters* 20 (15): 4337-4341. https://doi.org/10.1016/j.bmcl.2010.06.074.

[185] Turco, Joao F., Camila L. do Nascimento, Vanderlei A. de Lima, and Yohandra R. Torres. 2020. "Could antioxidant capacity and flavonoid content of ethanolic extracts of geopropolis from

Brazilian native bees be estimated from digital photos and NIR Spectra?" *Microchemical Journal* 157: 105031. https://doi.org/10.1016/j.microc.2020.105031.

[186] Chih-Hsien, Wang, Huang Chia-Chi, Chen Wenlung, and Lai Yen-Shi. 2020. "The chemical aspects of Raman spectroscopy: Statistical structure-spectrum relationship in the analyses of bioflavonoids." *Journal of Food & Drug Analysis* 28 (2): 45-53. https://doi.org/10.38212/2224-6614.1058.

[187] Yin, Ming, Jianlin Wang, Haoliang Huang, Qiuping Huang, Zhengping Fu, and Yalin Lu. 2020. "Analysis of flavonoid compounds by terahertz spectroscopy combined with chemometrics." *ACS Omega* 5 (29): 18134-18141. https://doi.org/10.1021/acsomega.0c01706.

[188] He, Juan, Lidan Chen, Bingquan Chu, and Chu Zhang. 2018. "Determination of total polysaccharides and total flavonoids in *Chrysanthemum morifolium* using near-infrared hyperspectral imaging and multivariate analysis." *Molecules* 23 (9): 2395. https://doi.org/10.3390/molecules23092395.

[189] Nagy, Tibor, Gergo Roth, Akos Kuki, Miklos Zsuga, and Sandor Keki. 2021. "Mass spectral filtering by mass-remainder analysis (MARA) at high resolution and its application to metabolite profiling of flavonoids." *International journal of Molecular Sciences* 22 (2). 864. https://doi.org/10.3390/ijms22020864.

[190] Zhu, Hongbin, Yuzhi Wang, Yuxuan Liu, Yalin Xia, and Tian Tang. 2010. "Analysis of flavonoids in portulaca oleracea L. by UV–Vis spectrophotometry with comparative study on different extraction technologies." *Food Analytical Methods* 3 (2): 90-97. https://doi.org/10.1007/s12161-009-9091-2.

[191] Pekal, Anna, and Krystyna Pyrzynska. 2014. "Evaluation of aluminium complexation reaction for flavonoid content assay." *Food Analytical Methods* 7 (9): 1776-1782. https://doi.org/10.1007/s12161-014-9814-x.

[192] Fedenko, Volodymyr S., Sergiy A. Shemet, and Marco Landi. 2017. "UV–vis spectroscopy and colorimetric models for detecting

anthocyanin-metal complexes in plants: An overview of in vitro and in vivo techniques." *Journal of Plant Physiology* 212: 13-28. https://doi.org/10.1016/j.jplph.2017.02.001.

[193] Chang, Chia-Chi, Ming-Hua Yang, Hwei-Mei Wen, and Jiing-Chuan Chern. 2002. "Estimation of total flavonoid content in propolis by two complementary colorimetric methods." *Journal of Food and Drug Analysis* 10 (3): 178-182. https://doi.org/10.38212/2224-6614.2748.

[194] Moldovan, Zenovia, Andrei A. Bunaciu, Mohammed A. Al-Omar, and Hassan Y. Aboul-Enein. 2010. "A spectrophotometric method for diosmin determination." *The Open Chemical and Biomedical Methods Journal* 3: 123-127. https://doi.org/10.2174/187503890 1003010123.

[195] Pekal, Anna, Magdalena Biesaga, and Krystyna Pyrzynska. 2011. "Interaction of quercetin with copper ions: complexation, oxidation and reactivity towards radicals." *BioMetals* 24 (1): 41-49. https://doi.org/10.1007/s10534-010-9372-7.

[196] Christ, B., and K. H. Muller. 1960. "For the serial determination of the content of flavonol derivatives in drugs." *Archiv der Pharmazie* 293 (12): 1033-1042. https://doi.org/10.1002/ardp.19602931202.

[197] Ramos, Rhayanne T. M., Isabelle C. F. Bezerra, Magda R. A. Ferreira, and Luiz Alberto Lira Soares. 2017. "Spectrophotometric quantification of flavonoids in herbal material, crude extract, and fractions from leaves of *Eugenia uniflora* Linn." *Pharmacognosy research* 9 (3): 253-260. https://doi.org/10.4103/pr.pr_143_16.

[198] Ordonez, A. A. L., J. D. Gomez, M. A. Vattuone, and M. I. Isla. 2006. "Antioxidant activities of Sechium edule (Jacq.) Swartz extracts." *Food Chemistry* 97 (3): 452-458. https://doi.org/10.1016/j.foodchem.2005.05.024.

[199] Chen, Ligang, Lan Ding, Aimin Yu, Ruilan Yang, Xiupin Wang, Jiantao Li, Haiyan Jin, and Hanqi Zhang. 2007. "Continuous determination of total flavonoids in *Platycladus orientalis* (L.) Franco by dynamic microwave-assisted extraction coupled with on-line derivatization and ultraviolet–visible detection." *Analytica*

Chimica Acta 596 (1): 164-170. https://doi.org/10.1016/j.aca.2007.05.063.

[200] Baba, Shoib A., and Shahid A. Malik. 2015. "Determination of total phenolic and flavonoid content, antimicrobial and antioxidant activity of a root extract of Arisaema jacquemontii Blume." *Journal of Taibah University for Science* 9 (4): 449-454. https://doi.org/10.1016/j.jtusci.2014.11.001.

[201] Fossen, Torgils, and Oyvind Andersen. 2005. "Spectroscopic techniques applied to flavonoids." In *Flavonoids*, 37-142.

[202] Shaghaghi, M., J. Manzoori, D. Afshar, and A. %J DARU Jouyban. 2009. "Determination of flavonoids in pharmaceutical preparations using terbium sensitized fluorescence method." *DARU-Journal of Faculty of Pharmacy* 17 (4): 264-268. http://dspace.tbzmed.ac.ir:8080/xmlui/handle/123456789/50933.

[203] Fatma, Nisha, Mohan Singh Mehata, Nupur Pandey, and Sanjay Pant. 2020. "Flavones fluorescence-based dual response chemosensor for metal ions in aqueous media and fluorescence recovery." *Journal of Fluorescence* 30 (4): 759-772. https://doi.org/10.1007/s10895-020-02540-z.

[204] Ferrara, Billy T, and Elinor P Thompson. 2019. "A method for visualizing fluorescence of flavonoid therapeutics in vivo in the model eukaryote *Dictyostelium discoideum*." *Biotechniques* 66 (2): 65-71. https://doi.org/10.2144/btn-2018-0084.

[205] Vargo, Melissa A., Oliver H. Voss, Frantisek Poustka, Arturo J. Cardounel, Erich Grotewold, and Andrea I. Doseff. 2006. "Apigenin-induced-apoptosis is mediated by the activation of PKCδ and caspases in leukemia cells." *Biochemical Pharmacology* 72 (6): 681-692. https://doi.org/10.1016/j.bcp.2006.06.010.

[206] de Lima, Maria de Fatima Rocha, Luziene A. Cavalcante, Emily Cintia Tossi de Araujo Costa, Bruno Oliveira de Veras, Marcia Vanusa da Silva, Livia Nunes Cavalcanti, and Renata Mendonca Araujo. 2021. "Bioactivity flavonoids from roots of *Euphorbia tirucalli* L." *Phytochemistry Letters* 41: 186-192. https://doi.org/10.1016/j.phytol.2020.10.017.

[207] Mahrous, Engy A., and Mohamed A. Farag. 2015. "Two dimensional NMR spectroscopic approaches for exploring plant metabolome: A review." *Journal of Advanced Research* 6 (1): 3-15. https://doi.org/10.1016/j.jare.2014.10.003.

[208] Chauthe, Siddheshwar K., Ram Jee Sharma, Farrukh Aqil, Ramesh C. Gupta, and Inder Pal Singh. 2012. "Quantitative NMR: An applicable method for quantitative analysis of medicinal plant extracts and herbal products." *Phytochemical Analysis* 23 (6): 689-696. https://doi.org/10.1002/pca.2375.

[209] Pinheiro, Pedro F., and Goncalo C. Justino. 2012. "Structural analysis of flavonoids and related compounds - A review of spectroscopic applications." In *Phytochemicals - A Global Perspective of Their Role in Nutrition and Health*, edited by Venketeshwer Rao. InTech.

[210] De Souza, Leonardo A., Wagner M. G. Tavares, Ana Paula M. Lopes, Malucia M. Soeiro, and Wagner B. De Almeida. 2017. "Structural analysis of flavonoids in solution through DFT 1H NMR chemical shift calculations: Epigallocatechin, Kaempferol and Quercetin." *Chemical Physics Letters* 676: 46-52. https://doi.org/10.1016/j.cplett.2017.03.038.

[211] De Souza, Leonardo A., Haroldo C. Da Silva, and Wagner B. De Almeida. 2018. "Structural determination of antioxidant and anticancer flavonoid rutin in solution through DFT calculations of (1)H NMR chemical shifts." *ChemistryOpen* 7 (11): 902-913. https://doi.org/10.1002/open.201800209.

[212] Yu, Ping, Qian Li, Yanmei Feng, Yuying Chen, Sinan Ma, and Xiaoqin Ding. 2021. "Quantitative analysis of flavonoids in Glycyrrhiza uralensis Fisch by 1H-qNMR." *Journal of Analytical Methods in Chemistry* 2021: 6655572. https://doi.org/10.1155/2021/6655572.

[213] Su, Feng, Ji-Peng Chen, Pei-Xi Zhu, Yi-Fan Wang, Xian-Rui Liang, and Wei-Ke Su. 2021. "Development of a quantitative NMR method for direct and simultaneous determination of four flavonoids in *Scutellaria baicalensis* georgi extracts." *Current*

Pharmaceutical Analysis 17 (2): 194-200. https://doi.org/10.2174/1573412916666191218111610.

[214] Cuyckens, Filip, and Magda Claeys. 2004. "Mass spectrometry in the structural analysis of flavonoids." *Journal of Mass Spectrometry* 39 (1): 1-15. https://doi.org/10.1002/jms.585.

[215] Oliveira, Fernanda Granja da Silva, Bruno Oliveira de Veras, Ana Paula Sant Anna da Silva, Amanda Dias de Araujo, Deyzi Caroline da Silva Barbosa, Tayane de Cassia Mendes Silva, Elisabete Regina Fernandes Ramos Ribeiro, Marta Melissa Leite Maia, Umberto Pereira Souza Junior, Vera Lucia de Menezes Lima, Marcia Vanusa da Silva, Norberto Peporine Lopes, Larissa Araujo Rolim, and Jackson Roberto Guedes da Silva Almeida. 2021. "Photoprotective activity and HPLC-MS-ESI-IT profile of flavonoids from the barks of Hymenaea martiana Hayne (Fabaceae): development of topical formulations containing the hydroalcoholic extract." *Biotechnology & Biotechnological Equipment* 35 (1): 504-516. https://doi.org/10.1080/13102818.2021.1901607.

[216] Strzemski, Maciej, Slawomir Dresler, Ireneusz Sowa, Lukasz Kurach, Jozef Kovacik, Kamila Wojas-Krawczyk, and Magdalena Wojciak. 2020. "Direct spectroscopic and GC profiling combined with chemometric analysis as an alternative approach to investigate Hypericum species: Is it possible to replace HPLC?" *Industrial Crops and Products* 157: 112930. https://doi.org/10.1016/j.indcrop.2020.112930.

[217] Rolo-Naranjo, Alberto, Edelsys Codorniu-Hernandez, and Noel Ferro. 2010. "Quantum chemical associations ligand−residue: Their role to predict flavonoid binding sites in proteins." *Journal of Chemical Information and Modeling* 50 (5): 924-933. https://doi.org/10.1021/ci900358z.

[218] Harper, James K., Jennifer A. Doebbler, Elisabeth Jacques, David M. Grant, and Robert B. Von Dreele. 2010. "A combined solid-state NMR and synchrotron X-ray diffraction powder study on the structure of the antioxidant (+)-catechin 4.5-hydrate." *Journal of*

the American Chemical Society 132 (9): 2928-2937. https://doi.org/10.1021/ja907671p.

[219] Kurouski, Dmitry. 2017. "Advances of vibrational circular dichroism (VCD) in bioanalytical chemistry. A review." *Analytica Chimica Acta* 990: 54-66. https://doi.org/10.1016/j.aca.2017.08.014.

[220] Slade, Desmond, Daneel Ferreira, and Jannie P. J. Marais. 2005. "Circular dichroism, a powerful tool for the assessment of absolute configuration of flavonoids." *Phytochemistry* 66 (18): 2177-2215. https://doi.org/10.1016/j.phytochem.2005.02.002.

[221] Ferreira, D., J. P. Marais, D. Slade, and L. A. Walker. 2004. "Circular dichroic properties of flavan-3,4-diols." *Journal of Natural Products* 67 (2): 174-8. https://doi.org/10.1021/np030318f.

[222] Ahmadi, S. M., G. Dehghan, M. A. Hosseinpourfeizi, J. E. Dolatabadi, and S. Kashanian. 2011. "Preparation, characterization, and DNA binding studies of water-soluble quercetin--molybdenum(VI) complex." *DNA Cell Biol* 30 (7): 517-23. https://doi.org/10.1089/dna.2010.1205.

[223] Di Bari, L., S. Ripoli, S. Pradhan, and P. Salvadori. 2010. "Interactions between quercetin and warfarin for albumin binding: A new eye on food/drug interference." *Chirality* 22 (6): 593-6. https://doi.org/10.1002/chir.20794.

[224] Matei, I., and M. Hillebrand. 2010. "Interaction of kaempferol with human serum albumin: a fluorescence and circular dichroism study." *Journal of Pharmaceutical and Biomedical Analysis* 51 (3): 768-73. https://doi.org/10.1016/j.jpba.2009.09.037.

[225] Chaudhuri, S., S. Chakraborty, and P. K. Sengupta. 2011. "Probing the interactions of hemoglobin with antioxidant flavonoids via fluorescence spectroscopy and molecular modeling studies." *Biophysical Chemistry* 154 (1): 26-34. https://doi.org/10.1016/j.bpc.2010.12.003.

[226] Barron, Pavan M. V. Raja, and R. Andrew. 2021. *Circular dichroism spectroscopy and its application for determination of secondary structure of optically active species.*

[227] Jurasekova, Z., J. V. Garcia-Ramos, C. Domingo, and S. Sanchez-Cortes. 2006. "Surface-enhanced Raman scattering of flavonoids." *Journal of Raman Spectroscopy* 37 (11): 1239-1241. https://doi.org/10.1002/jrs.1634.

[228] Unsalan, Ozan, Yusuf Erdogdu, and M. Tahir Gulluoglu. 2009. "FT-Raman and FT-IR spectral and quantum chemical studies on some flavonoid derivatives: Baicalein and Naringenin." *Journal of Raman Spectroscopy* 40 (5): 562-570. https://doi.org/10.1002/jrs.2166.

[229] Zaffino, Chiara, Bianca Russo, and Silvia Bruni. 2015. "Surface-enhanced Raman scattering (SERS) study of anthocyanidins." *Spectrochimica Acta Part A: Molecular and Biomolecular Spectroscopy* 149: 41-47. https://doi.org/10.1016/j.saa.2015.04.039.

[230] Pelletier, M. J. 2003. "Quantitative analysis using Raman spectrometry." *Applied spectroscopy* 57 (1): 20a-42a. https://doi.org/10.1366/000370203321165133.

[231] Hou, P. Y., J. Ager, J. Mougin, and A. Galerie. 2011. "Limitations and advantages of Raman spectroscopy for the determination of oxidation stresses." *Oxidation of Metals* 75 (5): 229-245. https://doi.org/10.1007/s11085-011-9235-9.

[232] Bunaciu, A. A., and H. Y. Aboul-Enein. 2017. "Vibrational spectroscopy applications in drugs analysis." In *Encyclopedia of Spectroscopy and Spectrometry (Third Edition)*, edited by John C. Lindon, George E. Tranter and David W. Koppenaal, 575-581. Oxford: Academic Press.

[233] Lin, Xiaohui, and Da-Wen Sun. 2020. "Recent developments in vibrational spectroscopic techniques for tea quality and safety analyses." *Trends in Food Science & Technology* 104: 163-176. https://doi.org/10.1016/j.tifs.2020.06.009.

[234] Hedoux, Alain. 2016. "Recent developments in the Raman and infrared investigations of amorphous pharmaceuticals and protein formulations: A review." *Advanced Drug Delivery Reviews* 100: 133-146. https://doi.org/10.1016/j.addr.2015.11.021.

[235] Singh, Keisham S., Mahesh S. Majik, and Supriya Tilvi. 2014. "Chapter 6 - Vibrational Spectroscopy for Structural Characterization of Bioactive Compounds." In *Comprehensive Analytical Chemistry*, edited by Teresa Rocha-Santos and Armando C. Duarte, 115-148. Elsevier.

[236] Gowen, A. A., C. P. O'Donnell, P. J. Cullen, G. Downey, and J. M. Frias. 2007. "Hyperspectral imaging – an emerging process analytical tool for food quality and safety control." *Trends in Food Science & Technology* 18 (12): 590-598. https://doi.org/10.1016/j.tifs.2007.06.001.

[237] Liu, Yuwei, Hongbin Pu, and Da-Wen Sun. 2017. "Hyperspectral imaging technique for evaluating food quality and safety during various processes: A review of recent applications." *Trends in Food Science & Technology* 69: 25-35. https://doi.org/10.1016/j.tifs.2017.08.013.

[238] Kamruzzaman, M., and D. W. Sun. 2016. "Chapter 5 - Introduction to Hyperspectral Imaging Technology." In *Computer Vision Technology for Food Quality Evaluation (Second Edition)*, edited by Da-Wen Sun, 111-139. San Diego: Academic Press.

[239] Afsah-Hejri, Leili, Parvaneh Hajeb, Parsa Ara, and Reza J. Ehsani. 2019. "A comprehensive review on food applications of terahertz spectroscopy and imaging." *Comprehensive Reviews in Food Science and Food Safety* 18 (5): 1563-1621. https://doi.org/10.1111/1541-4337.12490.

[240] Deneva, Vera, Ivan Bakardzhiyski, Krasimir Bambalov, Daniela Antonova, Diana Tsobanova, Valentin Bambalov, Daniel Cozzolino, and Liudmil Antonov. 2019. "Using Raman spectroscopy as a fast tool to classify and analyze Bulgarian wines-A feasibility study." *Molecules (Basel, Switzerland)* 25 (1): 170. https://doi.org/10.3390/molecules25010170.

[241] Wu, Z., E. Xu, J. Long, X. Pan, X. Xu, Z. Jin, and A. Jiao. 2016. "Comparison between ATR-IR, Raman, concatenated ATR-IR and Raman spectroscopy for the determination of total antioxidant capacity and total phenolic content of Chinese rice wine." *Food Chemistry* 194: 671-9. https://doi.org/10.1016/j.foodchem.2015.08.071.

[242] Park, Yong Seo, Buk-Gu Heo, Kyung-Sik Ham, Seong-Gook Kang, Yang-Kyun Park, Alina Nemirovski, Zeev Tashma, Shela Gorinstein, Hanna Leontowicz, and Maria Leontowicz. 2019. "Analytical determination of bioactive compounds as an indication of fruit quality." *Journal of AOAC INTERNATIONAL* 95 (6): 1725-1732. https://doi.org/10.5740/jaoacint.12-130.

[243] Pascoa, Ricardo N. M. J., Luis M. Magalhaes, and Joao A. Lopes. 2013. "FT-NIR spectroscopy as a tool for valorization of spent coffee grounds: Application to assessment of antioxidant properties." *Food Research International* 51 (2): 579-586. https://doi.org/10.1016/j.foodres.2013.01.035.

[244] Escuredo, O., M. C. Seijo, J. Salvador, and M. I. Gonzalez-Martin. 2013. "Near infrared spectroscopy for prediction of antioxidant compounds in the honey." *Food Chemistry* 141 (4): 3409-14. https://doi.org/10.1016/j.foodchem.2013.06.066.

[245] Luypaert, J., M. H. Zhang, and D. L. Massart. 2003. "Feasibility study for the use of near infrared spectroscopy in the qualitative and quantitative analysis of green tea, *Camellia sinensis* (L.)." *Analytica Chimica Acta* 478 (2): 303-312. https://doi.org/10.1016/S0003-2670(02)01509-X.

[246] Zareef, Muhammad, Quansheng Chen, Qin Ouyang, Muhammad Arslan, Md Mehedi Hassan, Waqas Ahmad, Annavaram Viswadevarayalu, Pingyue Wang, and Wang Ancheng. 2019. "Rapid screening of phenolic compounds in congou black tea (*Camellia sinensis*) during in vitro fermentation process using portable spectral analytical system coupled chemometrics." *Journal of Food Processing and Preservation* 43 (7): 13996. https://doi.org/10.1111/jfpp.13996.

[247] Wang, Jiahua, Yifang Wang, Jingjing Cheng, Jun Wang, Xudong Sun, Shuang Sun, and Zhenya Zhang. 2018. "Enhanced cross-category models for predicting the total polyphenols, caffeine and free amino acids contents in Chinese tea using NIR spectroscopy." *LWT* 96: 90-97. https://doi.org/10.1016/j.lwt.2018.05.012.

[248] Wang, Jingjing, Muhammad Zareef, Peihuan He, Hao Sun, Quansheng Chen, Huanhuan Li, Qin Ouyang, Zhiming Guo, Zhengzhu Zhang, and Delian Xu. 2019. "Evaluation of matcha tea quality index using portable NIR spectroscopy coupled with chemometric algorithms." *Journal of the Science of Food and Agriculture* 99 (11): 5019-5027. https://doi.org/10.1002/jsfa.9743.

[249] Zhang, Caiya, Yun Shen, Jian Chen, Peng Xiao, and Jinsong Bao. 2008. "Nondestructive prediction of total phenolics, flavonoid contents, and antioxidant capacity of rice grain using near-infrared spectroscopy." *Journal of Agricultural and Food Chemistry* 56 (18): 8268-8272. https://doi.org/10.1021/jf801830z.

[250] Shi, Ji-Yong, Xiao-Bo Zou, Jie-Wen Zhao, and Mel Holmes. 2012. "Near infrared quantitative analysis of total flavonoid content in fresh Ginkgo Biloba leaves based on different wavelength region selection methods and partial least squares regression." *Journal of Near Infrared Spectroscopy* 20 (2): 295-305. https://doi.org/10.1255/jnirs.989.

[251] Arslan, Muhammad, Zou Xiaobo, Haroon Elrasheid Tahir, Jiyong Shi, Muhammad Zareef, Allah Rakha, and Muhammad Bilal. 2021. "Rapid screening of phenolic compounds from wild *Lycium ruthenicum* Murr. Using portable near-infrared (NIR) spectroscopy coupled multivariate analysis." *Analytical Letters* 54 (3): 512-526. https://doi.org/10.1080/00032719.2020.1772807.

[252] Tingting, Shen, Zou Xiaobo, Shi Jiyong, Li Zhihua, Huang Xiaowei, Xu Yiwei, and Chen Wu. 2016. "Determination geographical origin and flavonoids content of Goji Berry using near-infrared spectroscopy and chemometrics." *Food Analytical Methods* 9 (1): 68-79. https://doi.org/10.1007/s12161-015-0175-x.

[253] Gorinstein, S., R. Haruenkit, S. Poovarodom, S. Vearasilp, P. Ruamsuke, J. Namiesnik, M. Leontowicz, H. Leontowicz, M. Suhaj, and G. P. Sheng. 2010. "Some analytical assays for the determination of bioactivity of exotic fruits." *Phytochemical Analysis* 21 (4): 355-62. https://doi.org/10.1002/pca.1207.

[254] Lam, Henry S., Andrew Proctor, Luke Howard, and Mi Jin Cho. 2005. "Rapid fruit extracts antioxidant capacity determination by fourier transform infrared spectroscopy." *Industrial Crops and Products* 70 (9): C545-C549. https://doi.org/10.1111/j.1365-2621.2005.tb08303.x.

[255] Numata, Yasushi, and Hiroyuki Tanaka. 2011. "Quantitative analysis of quercetin using Raman spectroscopy." *Food Chemistry* 126 (2): 751-755. https://doi.org/10.1016/j.foodchem.2010.11.059.

[256] Corredor, Charlie, Tatyana Teslova, Maria Vega Cañamares, Zhanguo Chen, Jie Zhang, John R. Lombardi, and Marco Leona. 2009. "Raman and surface-enhanced Raman spectra of chrysin, apigenin and luteolin." *Vibrational Spectroscopy* 49 (2): 190-195. https://doi.org/10.1016/j.vibspec.2008.07.012.

[257] Xia, Jing, Dan Wang, Pei Liang, De Zhang, Xiaoqing Du, Dejiang Ni, and Zhi Yu. 2020. "Vibrational (FT-IR, Raman) analysis of tea catechins based on both theoretical calculations and experiments." *Biophysical Chemistry* 256: 106282. https://doi.org/10.1016/j.bpc.2019.106282.

[258] Lee, Min-Seuk, Young-Sun Hwang, Jinwook Lee, and Myoung-Gun Choung. 2014. "The characterization of caffeine and nine individual catechins in the leaves of green tea (*Camellia sinensis* L.) by near-infrared reflectance spectroscopy." *Food Chemistry* 158: 351-357. https://doi.org/10.1016/j.foodchem.2014.02.127.

[259] Yang, Nan, and Guixing Ren. 2008. "Application of near infrared reflectance spectroscopy to the evaluation of rutin and D-chiro-Inositol contents in Tartary Buckwheat." *Journal of Agricultural and Food Chemistry* 56 (3): 761-764. https://doi.org/10.1021/jf072453u.

[260] Nambiar, Vanisha S., Mammen Daniel, and Parul Guin. 2010. "Characterization of polyphenols from Coriander leaves (*Coriandrum sativum*), red amaranthus (*A. paniculatus*) and green amaranthus (*A. frumentaceus*) using paper chromatography: and thier health implications." *Journal of Herbal Medicine and Toxicology* 4 (1): 173-177. https://www.academia.edu/25822564/Characterization_of_polyphenols_from_coriander_leaves_Coriandrum_sativum_red_amaranthus_A_paniculatus_and_green_amaranthus_A_frumentaceus_using_paper_chromatography_and_their_health_implications.

[261] Shawky, Eman. 2016. "Multivariate analyses of NP-TLC chromatographic retention data for grouping of structurally-related plant secondary metabolites." *Journal of Chromatography B* 1029–1030: 10-15. https://doi.org/10.1016/j.jchromb.2016.06.037.

[262] Alam, Perwez, Mohammad K. Parvez, Ahmed H. Arbab, and Mohammed S. Al-Dosari. 2017. "Quantitative analysis of rutin, quercetin, naringenin, and gallic acid by validated RP- and NP-HPTLC methods for quality control of anti-HBV active extract of Guiera senegalensis." *Pharmaceutical Biology* 55 (1): 1317-1323. https://doi.org/10.1080/13880209.2017.1300175.

[263] Sherma, Joseph. 2000. "Thin-layer chromatography in food and agricultural analysis." *Journal of Chromatography A* 880: 129-147. https://doi.org/10.1016/s0021-9673(99)01132-2.

[264] Medic-Saric, Marica, Ivona Jasprica, Ana Mornar, Asja Smolcic-Bubalo, and Petra Golja. 2004. "Quantitative analysis of flavonoids and phenolic acids in propolis by two-dimensional thin layer chromatography." *Journal of Planar Chromatography - Modern TLC* 17: 459-463. https://doi.org/10.1556/JPC.17.2004.6.12.

[265] Mattle, Christian, Nico Heigl, Gudrun Abel, Gunther K. Bonn, and Christian W. Huck. 2010. "Near-Infrared diffuse reflection spectroscopy and multivariate calibration hyphenated with thin-Layer chromatography for quality control of a phytomedicine and simultaneous quantification of methoxylated flavones." *JPC –*

Journal of Planar Chromatography – Modern TLC 23 (5): 348-352. https://doi.org/10.1556/JPC.23.2010.5.9.

[266] Li, Yuzhi, Chengying Zhao, Chang Lu, Shuaishuai Zhou, Guifang Tian, Lili He, Yuming Bao, Marie-Laure Fauconnier, Hang Xiao, and Jinkai Zheng. 2021. "Simultaneous determination of 14 bioactive citrus flavonoids using thin-layer chromatography combined with surface enhanced Raman spectroscopy." *Food Chemistry* 338: 128115. https://doi.org/10.1016/j.foodchem.2020.128115.

[267] Medeiros, Patricia M. 2018. "Gas chromatography–mass spectrometry (GC–MS)." In *Encyclopedia of Geochemistry: A Comprehensive Reference Source on the Chemistry of the Earth*, edited by William M. White, 530-535. Cham: Springer International Publishing.

[268] Denisova, S. B., E. G. Galkin, and Yu. I. Murinov. 2006. "Isolation and GC-MS determination of flavonoids from Glycyrrhiza glabra root." *Chemistry of Natural Compounds* 42 (3): 285-289. https://doi.org/10.1007/s10600-006-0100-z.

[269] Oyedeji O. A, Azeez L, Adewuyi S.O, Osinfade B.G, and Bamimore M.O. 2015. "Flavonoid profile, anthocyanin, carotenoid, sugar and vitamin compositions of Lonchocarpus sericeus seeds" *African Journal of Food Science and Technology* 6 (5). 131-135. https://doi.org/10.14303/AJFST.2015.051.

[270] Zhang, Kai, and Yuegang Zuo. 2004. "GC-MS determination of flavonoids and phenolic and benzoic acids in human plasma after consumption of Cranberry juice." *Journal of Agricultural and Food Chemistry* 52 (2): 222-227. https://doi.org/10.1021/jf035073i.

[271] Magiera, S., C. Uhlschmied, M. Rainer, W. Huck Ch, I. Baranowska, and G. K. Bonn. 2011. "GC-MS method for the simultaneous determination of beta-blockers, flavonoids, isoflavones and their metabolites in human urine." *Journal of Pharmaceutical and Biomedical Analysis* 56 (1): 93-102. https://doi.org/10.1016/j.jpba.2011.04.024.

[272] Liu, E-Hu, Lian-Wen Qi, Jun Cao, Ping Li, Chang-Yin Li, and Yong-Bo Peng. 2008. "Advances of modern chromatographic and electrophoretic methods in separation and analysis of flavonoids." *Molecules* 13: 2521-2544. https://doi.org/10.3390/molecules1310 2521.

[273] Proestos, C., D. Sereli, and M. Komaitis. 2006. "Determination of phenolic compounds in aromatic plants by RP-HPLC and GC-MS." *Food Chemistry* 95 (1): 44-52. https://doi.org/10.1016/ j.foodchem.2004.12.016.

[274] M.P. Fernandez, P.A. Watson, and C. Breuil. 2001. "Gas chromatography–mass spectrometry method for the simultaneous determination of wood extractive compounds in quaking aspen." *Journal of Chromatography A* 922: 225-233. https://doi.org/10. 1016/s0021-9673(01)00948-7.

[275] Grun, Christian H., Ferdi A. van Dorsten, Doris M. Jacobs, Marie Le Belleguic, Ewoud J. J. van Velzen, Max O. Bingham, Hans-Gerd Janssen, and John P. M. van Duynhoven. 2008. "GC–MS methods for metabolic profiling of microbial fermentation products of dietary polyphenols in human and in vitro intervention studies." *Journal of Chromatography B* 871 (2): 212-219. https://doi.org/10.1016/j.jchromb.2008.04.039.

[276] Cheng, Ka-Wing, Chi Chun Wong, Chi Kong Cho, Ivan K. Chu, Kong Hung Sze, Clive Lo, Feng Chen, and Mingfu Wang. 2008. "Trapping of phenylacetaldehyde as a key mechanism responsible for naringenin's inhibitory activity in mutagenic 2-amino-1-methyl-6-phenylimidazo [4,5-b]pyridine formation." *Chemical Research in Toxicology* 21 (10): 2026-2034. https://doi.org/10. 1021/tx800220h.

[277] Morzunova, T. G. 2006. "Capillary electrophoresis in pharmaceutical analysis (A review)." *Pharmaceutical Chemistry Journal* 40 (3): 158-170. https://doi.org/10.1007/s11094-006-0083-x.

[278] Liu, E-Hu, Lian-Wen Qi, Jun Cao, Ping Li, Chang-Yin Li, and Yong-Bo Peng. 2008. "Advances of modern chromatographic and

electrophoretic methods in separation and analysis of flavonoids." *Molecules* 13 (10): 2521-2544. https://doi.org/10.3390/molecules 13102521.

[279] Wu, Tingni, Changzhu Yu, and Rong Li. 2017. "Determination of flavonoids in Flos Chrysanthemi and Flos Chrysanthemi Indici by capillary electrophoresis." *Instrumentation Science & Technology* 45 (4): 412-422. https://doi.org/10.1080/10739149.2016.1258572.

[280] Marti, Raul, Mercedes Valcarcel, Jose Manuel Herrero-Martinez, Jaime Cebolla-Cornejo, and Salvador Rosello. 2017. "Simultaneous determination of main phenolic acids and flavonoids in tomato by micellar electrokinetic capillary electrophoresis." *Food Chemistry* 221: 439-446. https://doi.org/10.1016/j.foodchem.2016.10.105.

[281] Chen, X. J., J. Zhao, Q. Meng, S. P. Li, and Y. T. Wang. 2009. "Simultaneous determination of five flavonoids in licorice using pressurized liquid extraction and capillary electrochromatography coupled with peak suppression diode array detection." *Journal of Chromatography A* 1216 (43): 7329-7335. https://doi.org/10.1016/j.chroma.2009.08.034.

[282] Senem Sanlı, and Craig Lunte. 2014. "Determination of eleven flavonoids in chamomile and linden extracts by capillary electrophoresis." *Analytical Methods* 6: 3858-3864. https://doi.org/10.1039/C3AY41878B.

[283] Xu, Xueqin, Lishuang Yu, and Guonan Chen. 2006. "Determination of flavonoids in Portulaca oleracea L. by capillary electrophoresis with electrochemical detection." *Journal of Pharmaceutical and Biomedical Analysis* 41 (2): 493-499. https://doi.org/10.1016/j.jpba.2006.01.013.

[284] Xu, Jing-Jing, Mingrui An, Rui Yang, Jun Cao, Li Hong Ye, and Li-Qing Peng. 2016. "Trace amounts of poly-beta-cyclodextrin wrapped carbon nanotubes for the microextraction of flavonoids in honey samples by capillary electrophoresis with light-emitting diode induced fluorescence detection." *Electrophoresis* 37 (13): 1891-1901. https://doi.org/10.1002/elps.201600016.

[285] Wang, Wei, Ping Lin, Lihong Ma, Kaixuan Xu, and Xiuli Lin. 2016. "Separation and determination of flavonoids in three traditional chinese medicines by capillary electrophoresis with amperometric detection." *Journal of Separation Science* 39 (7): 1357-1362. https://doi.org/10.1002/jssc.201501287.

[286] Johnson, Ryan T., and Craig E. Lunte. 2016. "A capillary electrophoresis electrospray ionization-mass spectrometry method using a borate background electrolyte for the fingerprinting analysis of flavonoids in *Ginkgo biloba* herbal supplements." *Analytical Methods* 16 (8): 3325-3332. https://doi.org/10.1039/c6ay00463f.

[287] Wu, Tingni, Changzhu Yu, Rong Li, and Jun Li. 2018. "Minireview: Recent advances in the determination of flavonoids by capillary electrophoresis." *Instrumentation Science & Technology* 46 (4): 364-386. https://doi.org/10.1080/10739149.2017.1394877.

[288] Wang, Shu-Ping, and Kuo-Jun Huang. 2004. "Determination of flavonoids by high-performance liquid chromatography and capillary electrophoresis." *Journal of Chromatography A* 1032 (1): 273-279. https://doi.org/10.1016/j.chroma.2003.11.099.

[289] Saito, Muneo. 2013. "History of supercritical fluid chromatography: Instrumental development." *Journal of Bioscience and Bioengineering* 115 (6): 590-599. https://doi.org/10.1016/j.jbiosc.2012.12.008.

[290] Huang, Yang, Guangyun Tang, Tingting Zhang, Marianne Fillet, Jacques Crommen, and Zhengjin Jiang. 2018. "Supercritical fluid chromatography in traditional Chinese medicine analysis." *Journal of Pharmaceutical and Biomedical Analysis* 147: 65-80. https://doi.org/10.1016/j.jpba.2017.08.021.

[291] Jiang, Zheng-Meng, Lan-Jin wang, Wen-Jin Liu, Hui-Ying Wang, Ping-Ting Xiao, Ping Zhou, Zhi-Ming Bi, and E. Hu Liu. 2019. "Development and validation of a supercritical fluid chromatography method for fast analysis of six flavonoids in Citri

Reticulatae Pericarpium." *Journal of Chromatography B* 1133: 121845. https://doi.org/10.1016/j.jchromb.2019.121845.

[292] Wang, Bo, Xiao-Hua Liu, Zhi-Yuan Xue, Xiu-Yan Yang, Yao-Yao Fang, and Shi-Lan Feng. 2020. "Comparative study of ultra-high-performance supercritical fluid chromatography and ultra-high-performance liquid chromatography to simultaneous determination of ten components in Radix hedysari." *Pharmacognosy Magazine* 16 (67): 99-110. https://doi.org/10.4103/pm.pm_241_19.

[293] Gao, Wen, Xin Dong, Rui Wang, Xin-Guang Liu, Ping Li, and Hua Yang. 2016. "The use of ionic liquid as a mobile phase modifier in analytical supercritical fluid chromatography for the separation of flavonoids." *RSC Advances* 6 (66): 61418-61422. https://doi.org/10.1039/C6RA10975F.

[294] Wang, Bo, Xiao-hua Liu, Wei Zhou, Yan Hong, and Shi-lan Feng. 2017. "Fast separation of flavonoids by supercritical fluid chromatography using a column packed with a sub-2 μm particle stationary phase." *Journal of Separation Science* 40 (6): 1410-1420. https://doi.org/10.1002/jssc.201601021.

[295] Huang, Yang, Ying Feng, Guangyun Tang, Minyi Li, Tingting Zhang, Marianne Fillet, Jacques Crommen, and Zhengjin Jiang. 2017. "Development and validation of a fast SFC method for the analysis of flavonoids in plant extracts." *Journal of Pharmaceutical and Biomedical Analysis* 140: 384-391. https://doi.org/10.1016/j.jpba.2017.03.012.

[296] WHO, World Health Organization. 2002. *WHO Traditional Medicine Strategy 2002–2005.* (Geneva, Switzerland). https://www.who.int/medicines/publications/traditionalpolicy/en/.

[297] Siddiqui, Masoom Raza, Zeid A. AlOthman, and Nafisur Rahman. 2013. "Analytical techniques in pharmaceutical analysis: A review." *Arabian Journal of Chemistry* 10: S1409-S1421. https://doi.org/10.1016/j.arabjc.2013.04.016.

[298] Pharmacopeia, U. S. 2009. *U. S. Pharmacopeia*. Vol. 32. USA.

[299] Pharmacopoeia, The British. 2011. *The British Pharmacopoeia*. London UK.

[300] Pharmacopoeia, European. 2013. *European Pharmacopoeia 8.0*. Strasbourg.

[301] Galensa, Rudolf, and Karl Herrmann. 1980. "Analysis of flavonoids by high-performance liquid chromatography." *Journal of Chromatography A* 189 (2): 217-224. https://doi.org/10.1016/S0021-9673(00)81521-6.

[302] Jagota, Nirdosh K., and Steve F. Cheatham. 1992. "HPLC separation of flavonoids and flavonoid glycosides using a polystyrene/divinylbenzene column." *Journal of Liquid Chromatography* 15 (4): 603-615. https://doi.org/10.1080/10826079208018819.

[303] Robbins, Rebecca J., Jadwiga Leonczak, J. Christopher Johnson, Julia Li, Catherine Kwik-Uribe, Ronald L. Prior, and Liwei Gu. 2009. "Method performance and multi-laboratory assessment of a normal phase high pressure liquid chromatography–fluorescence detection method for the quantitation of flavanols and procyanidins in cocoa and chocolate containing samples." *Journal of Chromatography A* 1216: 4831-4840. https://doi.org/10.1016/j.chroma.2009.04.006.

[304] Piretti, Marco Vincenzo, and Paola Doghieri. 1990. "Separation of peracetylated flavanoid and flavonoid polyphenols by normal-phase high-performance liquid chromatography on a cyano-silica column and their determination." *Journal of Chromatography A* 514: 334-342. https://doi.org/10.1016/S0021-9673(01)89407-3.

[305] Piponski, Marjan, Tanja Bakovska Stoimenova, Marina Topkoska, Stefan Stefov, Magdalena Piponska, and Gordana Trendovska Serafimovska. 2018. "Development and validation of a fast and simple RP-HPLC method for the determination of diosmin and hesperidin in combined tablet dosage form." *Macedonian Journal of Chemistry and Chemical Engineering* 37 (2): 127-134. https://doi.org/10.20450/mjcce.2018.1448.

[306] Bloor, Stephen J. 2001. "Overview of methods for analysis and identification of flavonoids." In *Methods in Enzymology*, 3-14. Academic Press.

[307] Merken, Howard M., and Gary R. Beecher. 2000. "Measurement of food flavonoids by high-performance liquid chromatography: A review." *Journal of Agricultural and Food Chemistry* 48 (3): 577-599. https://doi.org/10.1021/jf990872o.

[308] Olszewska, Monika A. 2012. "New validated high-performance liquid chromatographic method for simultaneous analysis of ten flavonoid aglycones in plant extracts using a C18 fused-core column and acetonitrile–tetrahydrofuran gradient." *Journal of Separation Science* 35 (17): 2174-2183. https://doi.org/10.1002/jssc.201200287.

[309] Satinsky, Dalibor, Katerina Jagerova, Lucie Havlikova, and Petr Solich. 2013. "A new and fast HPLC method for determination of rutin, troxerutin, diosmin and hesperidin in food supplements using fused-core column technology." *Food Analytical Methods* 6: 1353-1360. https://doi.org/10.1007/s12161-012-9551-y.

[310] Biesaga, Magdalena, Urszula Ochnik, and Krystyna Pyrzynska. 2009. "Fast analysis of prominent flavonoids in tomato using a monolithic column and isocratic HPLC." *Journal of Separation Science* 32 (15-16): 2835-40. https://doi.org/10.1002/jssc.200800730.

[311] Repolles, Carme, Jose Manuel Herrero-Martinez, and Clara Rafols. 2006. "Analysis of prominent flavonoid aglycones by high-performance liquid chromatography using a monolithic type column." *Journal of Chromatography A* 1131 (1): 51-57. https://doi.org/10.1016/j.chroma.2006.07.012.

[312] Castellari, Massimo, Elisa Sartini, Alessandra Fabiani, Giuseppe Arfelli, and Aureliano Amati. 2002. "Analysis of wine phenolics by high-performance liquid chromatography using a monolithic type column." *Journal of Chromatography A* 973 (1): 221-227. https://doi.org/10.1016/S0021-9673(02)01195-0.

[313] Rahim, A. A., S. Nofrizal, and Bahruddin Saad. 2014. "Rapid tea catechins and caffeine determination by HPLC using microwave-assisted extraction and silica monolithic column." *Food Chemistry* 147: 262-268. https://doi.org/10.1016/j.foodchem.2013.09.131.

[314] Yuangang Zu, Chunying Li, Yujie Fu, and Chunjian Zhao. 2006. "Simultaneous determination of catechin, rutin, quercetin kaempferol and isorhamnetin in the extract of sea buckthorn (*Hippophae rhamnoides* L.) leaves by RP-HPLC with DAD." *Journal of Pharmaceutical and Biomedical Analysis* 41: 714-719. https://doi.org/10.1016/j.jpba.2005.04.052.

[315] Luo, Jiun-Lung, Fen-Ling Lu, Yi-Chu Liu, Yang-Chih Shih, and Chi-Fang Lo. 2013. "Fingerprint analysis of Ginkgo biloba extract and Ginkgo semen in preparations by LC-Q-TOF/MS." *Journal of Food and Drug Analysis* 21 (1): 27-39. https://doi.org/10.6227/jfda.2013210104.

[316] Haghi, Ghasem, and Alireza Hatami. 2010. "Simultaneous quantification of flavonoids and phenolic acids in plant materials by a newly developed isocratic high-performance liquid chromatography approach." *Journal of Agricultural and Food Chemistry* 58 (20): 10812-10816. https://doi.org/10.1021/jf102175x.

[317] Demirezer, L. Omur, Aslı Buyukkaya, Ebru Ucakturk, Ayse Kuruuzum-Uz, Zuhal Guvenalp, and Erhan Palaska. 2014. "Adulteration determining of pharmaceutical forms of Ginkgo biloba extracts from different international manufacturers." *Records of natural products* 8 (4): 394-400. http://hdl.handle.net/11655/19868.

[318] Sammani, Mohamad Subhi, Sabrina Clavijo, Alba Gonzalez, and Victor Cerda. 2019. "High-performance liquid chromatographic method for the simultaneous determination of four flavonols in food supplements and pharmaceutical formulations." *Analytical Letters* 52 (8): 1298-1314. https://doi.org/10.1080/00032719.2018.1536138.

[319] Wang, Chen, Minna Zhang, Limin Wu, Fang Wang, Lingxi Li, Shuting Zhang, and Baoshan Sun. 2021. "Qualitative and quantitative analysis of phenolic compounds in blueberries and protective effects on hydrogen peroxide-induced cell injury." *Journal of Separation Science*. https://doi.org/10.1002/jssc.202001264.

[320] L. C. Santos, M. Dachtler, F. D. P. Andrade, K. Albert, and W. Vilegas. 2000. "Application of HPLC-NMR coupling using C30 phase in the separation and identification of flavonoids of taxonomic relevance." *Fresenius j Anal Chem* 368: 540-542.

[321] Hollman, Peter C. H., John M. P. van Trijp, and Michel N. C. P. Buysman. 1996. "Fluorescence detection of flavonols in HPLC by postcolumn chelation with aluminum." *Analytical Chemistry* 68 (19): 3511-3515. https://doi.org/10.1021/ac960461w.

[322] Karoline Petrus, Heidi Schwartz, and Gerhard Sontag. 2011. "Analysis of flavonoids in honey by HPLC coupledwith coulometric electrode array detection and electrosprayionization mass spectrometry." *Anal Bioanal Chem* 400: 2555-2563. https://doi.org/10.1007/s00216-010-4614-7.

[323] da Silveira, Géssica Domingos, Monique Jung Motta, Larissa Sabo Müller, Osmar Lameira, Margareth L. Athayde, Mariana Piana, Marcelo Barcellos da Rosa, Carine Viana, and Leandro Machado de Carvalho. 2015. "Determination of phenolic antioxidants in Amazonian medicinal plants by HPLC with pulsed amperometric detection." *Journal of Liquid Chromatography & Related Technologies* 38 (13): 1259-1266. https://doi.org/10.1080/10826076.2015.1037450.

[324] Liang, Yan, Wei Cao, Wei-jun Chen, Xue-hong Xiao, and Jian-bin Zheng. 2009. "Simultaneous determination of four phenolic components in citrus honey by high performance liquid chromatography using electrochemical detection." *Food Chemistry* 114 (4): 1537-1541. https://doi.org/10.1016/j.foodchem.2008.11.024.

[325] Buszewski, Bogusław, and Sylwia Noga. 2012. "Hydrophilic interaction liquid chromatography (HILIC)--a powerful separation technique." *Analytical and bioanalytical chemistry* 402 (1): 231-247. https://doi.org/10.1007/s00216-011-5308-5.

[326] Sentkowska, Aleksandra, Magdalena Biesaga, and Krystyna Pyrzynska. 2015. "Retention study of flavonoids under different chromatographic modes." *Journal of Chromatographic Science* 54 (4): 516-522. https://doi.org/10.1093/chromsci/bmv174 %J Journal of Chromatographic Science.

[327] ____. 2016. "Application of hydrophilic interaction liquid chromatography for the quantification of flavonoids in Genista tinctoria extract." *Journal of Analytical Methods in Chemistry* 2016: 3789348. https://doi.org/10.1155/2016/3789348.

[328] ____. 2013. "Effects of the operation parameters on HILIC separation of flavonoids on zwitterionic column." *Talanta* 115: 284-290. https://doi.org/10.1016/j.talanta.2013.05.005.

[329] Kalili, Kathithileni M., and André de Villiers. 2010. "Off-line comprehensive two-dimensional hydrophilic interaction×reversed phase liquid chromatographic analysis of green tea phenolics." *Journal of Separation Science* 33 (6-7): 853-863. https://doi.org/10.1002/jssc.200900673.

[330] Muller, Magriet, Andreas G. J. Tredoux, and André de Villiers. 2019. "Application of kinetically optimised online HILIC × RP-LC methods hyphenated to high resolution MS for the analysis of natural phenolics." *Chromatographia* 82 (1): 181-196. https://doi.org/10.1007/s10337-018-3662-6.

[331] Zeng, Jing, Xiuli Zhang, Zhimou Guo, Jiatao Feng, Jianguo Zeng, Xingya Xue, and Xinmiao Liang. 2012. "Separation and identification of flavonoids from complex samples using off-line two-dimensional liquid chromatography tandem mass spectrometry." *Journal of Chromatography A* 1220: 50-56. https://doi.org/10.1016/j.chroma.2011.11.043.

[332] Russo, Marina, Francesco Cacciola, Ivana Bonaccorsi, Paola Dugo, and Luigi Mondello. 2011. "Determination of flavanones in

Citrus juices by means of one- and two-dimensional liquid chromatography." *Journal of Separation Science* 34 (6): 681-687. https://doi.org/10.1002/jssc.201000844.

[333] Willemse, Chandré M., Maria A. Stander, Jochen Vestner, Andreas G. J. Tredoux, and André de Villiers. 2015. "Comprehensive two-dimensional hydrophilic interaction chromatography (HILIC) × reversed-phase liquid chromatography coupled to high-resolution mass spectrometry (RP-LC-UV-MS) analysis of anthocyanins and derived Pigments in red wine." *Analytical Chemistry* 87 (24): 12006-12015. https://doi.org/10.1021/acs.analchem.5b03615.

[334] Beelders, Theresa, Kathithileni M. Kalili, Elizabeth Joubert, Dalene de Beer, and André de Villiers. 2012. "Comprehensive two-dimensional liquid chromatographic analysis of rooibos (*Aspalathus linearis*) phenolics." *Journal of Separation Science* 35 (14): 1808-1820. https://doi.org/10.1002/jssc.201200060.

[335] Ren, Qiang, Caisheng Wu, and Jinlan Zhang. 2013. "Use of on-line stop-flow heart-cutting two-dimensional high performance liquid chromatography for simultaneous determination of 12 major constituents in tartary buckwheat (*Fagopyrum tataricum* Gaertn)." *Journal of Chromatography A* 1304: 257-262. https://doi.org/10.1016/j.chroma.2013.07.008.

[336] Stoll, Dwight R., and Peter W. Carr. 2017. "Two-dimensional liquid chromatography: A state of the art tutorial." *Analytical Chemistry* 89 (1): 519-531. https://doi.org/10.1021/acs.analchem.6b03506.

[337] Wang, Jian, Xiaojuan Ren, Chunmei Wen, Yu Xu, and Yue Chen. 2020. "Separation and characterization of unknown impurities in rutin tablets using trap-free two-dimensional liquid chromatography coupled with ion trap/time-of-flight mass spectrometry." *Rapid Communications in Mass Spectrometry* 34 (10): e8739. https://doi.org/10.1002/rcm.8739.

[338] Mohamed, Dalia, and Shereen M. Tawakkol. 2013. "Fluorimetric determination of diosmin and hesperidin in combined dosage

forms and in plasma through complex formation with terbium." *Bulletin of Faculty of Pharmacy, Cairo University* 51: 81-88. https://doi.org/10.1016/j.bfopcu.2012.12.001.

[339] Sanli, Senem, Onur Guneser, Sinem Kilicarslan, and Nurullah Sanlı. 2020. "Screening of eighteen polyphenolic compounds in different carob pekmez by green capillary electrophoresis method." *SN Applied Sciences* 2 (4): 576. https://doi.org/10.1007/s42452-020-2387-y.

[340] Karabat, Raad Radi, Ashraf Saad Rasheed, and Mohammed Jasim M. Hassan. 2020. "A new method in some German grape wines using ZIC-HILIC technology with UV detection to separate and identify quercetin." *Plant Archives* 20 (1): 2692-2696. http://plantarchives.org/20-1/2692-2696%20(6210).pdf.

In: The Book of Flavonoids
Editors: Chao-Hui Feng et al.
ISBN: 978-1-68507-376-3

Chapter 2

DETERMINATION OF TOTAL PHENOL AND FLAVONOID CONTENT AND THE EVALUATION OF LARVICIDAL AND ANTIMICROBIAL ACTIVITIES OF *BIDENS SUBALTERNANS* DC

Iasmin Fernandes Emediato[1],
Thaís Paula Rodrigues Gonçalves[1,2]
and Luciana Alves Rodrigues dos Santos Lima[1,2,3,*]
[1]Phytochemistry Laboratory,
Universidade Federal de São João Del Rei,
Campus Centro-Oeste Dona Lindu, Divinópolis, MG, Brazil
[2]Programa de Pós-Graduação em Biotecnologia,
Universidade Federal de São João Del Rei,
Campus Centro-Oeste Dona Lindu, Divinópolis, MG, Brazil
[3]Programa de Pós-Graduação em Ciências Farmacêuticas,
Universidade Federal de São João Del Rei,
Campus Centro-Oeste Dona Lindu, Divinópolis, MG, Brazil

[*] Corresponding Author's E-mail: luarsantos@ufsj.edu.br.

ABSTRACT

The species *Bidens subalternans* DC, belonging to the Asteraceae family, is a typical plant from the Midwest, Southeast, and South regions of Brazil, and is popularly known as 'carrapicho-de-pontas,' 'coambi,' 'erva-picão,' 'fura-capa,' 'goambu,' 'picão,' 'picão-do-campo,' and 'picão-preto.' This species develops in areas occupied by annual or perennial crops and areas with fruit growing, where it becomes undesirable because it is a host to certain pests, as well as by the spinning edges of the fruits, which adhere to animals' bodies. Continuing the study of this species, the objective of this work was phytochemical screening, the determination of phenol and total flavonoid content, and to evaluate the larvicidal and antimicrobial activities of the ethanol extract from the leaves of *B. subalternans* DC. The extract was obtained by turbo-extraction with ethanol 70% (v:v). The total phenol content was quantified with the Folin-Ciocalteu reagent and the flavonoids with aluminum chloride. Larvicidal activity with *Artemia salina* larvae was determined by their mortality when exposed to different concentrations of the extract (1000, 500, 250, and 125 μg/mL). Antimicrobial activity by means of the minimum inhibitory concentration (MIC) and minimum bactericidal and fungicidal concentration (MBC and MFC, respectively) were evaluated by the broth microdilution method. Phytochemical screening revealed the presence of steroids, triterpenoids, flavonoids, alkaloids, and coumarins. The extract had a total phenolic compound content of 6.46 ± 1.13 μg/mL and total flavonoid content of 33.13 ± 0.48 μg/mL. The results showed that there was no mortality of the larvae of *A. salina*, indicating that the extract was not toxic. The ethanol extract showed bactericidal and bacteriostatic action on *Enterococcus faecalis*, *Aeromonas hydrophila, Escherichia coli* EHEC, and *Enterobacter cloaceae*, with MIC and MBC values of 0.5 mg/mL and 1 mg/mL for *E. faecalis*, 0.5 mg/mL and 1 mg/mL for *A. hydrophila*, and 0.25 mg/mL and 0.5 mg/mL for *E. cloaceae,* respectively. There was no activity against *Candida glabrata* and *C. infanticola*. Therefore, the ethanol extract of *B. subalternans* leaves demonstrated an antibacterial potential on the tested species.

Keywords: *Bidens subalternans* DC, larvicide, antibacterial

1. Introduction

The prolonged and extensive use of antibiotics has resulted in the development of antimicrobial resistance. Since the beginning of the 21st century, resistance has become a global health problem requiring intervention [1]. Some factors that contributed to the increase in antimicrobial resistance were: inadequate prescription methods, misuse of antibiotics by doctors and patients, and inadequate diagnosis. Therefore, with this continuation of scenarios, common infections became the main cause of death [2].

The loss of the effectiveness of antimicrobials poses a threat not only to health but also to the economy, security, and development [3]. The search for new antimicrobial agents is urgent, and the production of drugs from natural products has been encouraged in several countries [4]. This is due to the presence of chemical constituents in plant extracts that serve as prototypes for the discovery of new drugs, and the development of herbal medicines [5].

The chemical constituents, including secondary metabolites, have biological functions, such as defending against microorganisms, among other ecological roles, that can be used for various purposes, such as medicines, insecticides, and herbicides [6]. Millions of people across the planet use medicinal plants for the treatment of various diseases, which can be justified by the ease of access and low cost. Because these are natural products, the population believes they have fewer side effects and no toxicity [5].

However, several phytochemical compounds used or ingested on a daily basis can be toxic, producing harmful effects to the organism; thus, toxicological tests are developed that determine the effects of chemical substances on the biological system [7]. One of these tests is the bioassay on *Artemia salina*, which consists of assessing the acute toxicity of the extract. The assay has demonstrated good acceptability due to the fact that it is fast and easy to perform, in addition to having a low cost, and high sensitivity. Lethality on this organism has been used to identify biological responses, where variables such as death or life are the only ones involved [8].

In this context, the species *Bidens subalternans* is a good candidate. *B. subalternans* DC. ('carrapicho-de-pontas,' 'coambi,' 'picão-preto') is a weed native to South America, and infests annual and perennial crops. This species has a high similarity with the species *B. pilosa* L., which also belongs to the genus *Bidens* with wide incidence in agricultural fields. *B. subalternans* is used in traditional medicine to treat hepatitis, jaundice, fever, throat disorders, and coughs. However, the biological activities of *B. subalternans* have not been evaluated, and the traditional uses of this species need to be investigated [9].

2. Materials and Methods

2.1. Plant Material and Extraction

Bidens subalternans DC fresh leaves were collected in Gafanhoto Park, Divinópolis City, Minas Gerais State, Brazil (20º 6' 38.459"S, and 44º 50' 47.184"W). Fertile samples were collected, and the vouchers were identified by Andréia Fonseca Silva, and deposited in the PAMG Herbarium (PAMG 57686) at the Agricultural Research Company of Minas Gerais (EPAMIG). This study has access permission to the components of plant genetic heritage registered in the SisGen Platform (Register A6704FE), according to Brazilian Biodiversity Law (13.123/2015).

The fresh vegetable material was dried in an oven at 40°C for seven days. After drying, the material was crushed in a knife mill, and extracted with 70% (v.v) ethanol (9:1 alcohol/plant material), using turbo-extraction. After this period, the material was filtered, and solvent removed using rotatory evaporator.

2.2. Phytochemical Trial

The presence of secondary metabolites was evaluated by specific experimental in assays tubes. Steroids and triterpenoids were evaluated by Lieberman-Burchard reaction. Flavonoids were evaluated with change the color to red, yellow or orange after addition of 13 sulfuric acid (H_2SO_4) drops. Saponins presence was evaluated with observation of persistent foam after 30 minutes of intense shake with water. Tannins was evaluated with precipitate formation after addiction of 3 Iron (II) chloride ($FeCl_2$) drops. Alkaloids were evaluated by Dragendorff reaction. Coumarins and anthraquinones were evaluated with change the color to yellow or purple after addiction of 5 sodium hydroxide (NaOH) (1 mol/L) drops, respectively [10].

2.3. Total Phenolic Content (TPC)

The total phenolic content was estimated using the Folin-Ciocalteu test [11] with modifications. The Folin-Ciocalteu aqueous solution (2250 μL; 1:4 v:v) was added to samples (250 μL) and, subsequently, sodium carbonate (Na_2CO_3) solution (250 μL) was added. After vigorous shaking, these solutions were kept at rest for 30 min at room temperature. The absorbance was determined by spectrophotometry at 750 nm (Thermo Scientific Genesys 10S, USA), after 30 minutes of incubation at room temperature, with a blank sample as well as a standard solution and samples. Gallic acid was used as a reference compound, and the total phenolic content were expressed as micrograms of gallic acid equivalents per milliliter (mL). All assays were performed in triplicate.

2.4. Total Flavonoid Content (TFC)

The total flavonoid content was estimated according to the aluminum chloride ($AlCl_3$) method [12]. Exactly 1900 μL of ethanol 50% (v:v) were added 100 μL of extracts and fractions, and 500 μL of solution of aluminum chloride. The absorbance was read at 425 nm using a spectrophotometer (Thermo Scientific Genesys 10S, USA) after 30 min. Quercetin was used as a reference compound to produce a standard curve, and total flavonoid content were expressed as micrograms of quercetin equivalents per milliliter (mL). All assays were performed in triplicate.

2.5. *Artemia salina* Larvicide Bioassay

Artemia salina eggs were incubated in 400 mL of seawater (0.5 mg/mL) under artificial light at 28°C, pH 7-8. After incubation for 48 h, 10 nauplii were transferred to assay tubes with 10 mL of the samples (in triplicate) in 1000, 500, 250, and 125 μg/mL concentrations. Negative controls containing seawater, and 1, 0.5, 0.25, and 0.125 percent of DMSO were included in each experiment [13]. The number of survivors was counted after 24 h and recorded.

2.6. Antimicrobial Effect

The microorganisms used in this study originated from American Type Culture Collection (ATCC), and were provided by the microorganism reference laboratory of the Fundação Oswaldo Cruz (FIOCRUZ, Rio de Janeiro, Brasil). Five species were used: Gram-positive bacteria (*Enterococcus faecalis* ATCC 14506, and *Bacillus cereus* ATCC 11778), and Gram-negative bacteria (*Escherichia coli* EHEC ATCC 43895, *Enterobacter cloacae* ATCC 23355, and *Aeromonas hydrophila* ATCC 11036). In antifungal tests, *Candida infanticola* (UFSJ6A), and *Candida glabrata* (ATCC 2001) were used.

The bacterial susceptibility test was performed according to the CLSI [14] document guidelines M100-S25. The bacterial inoculum colonies isolated from a 24-hour growth streak were transferred to tubes containing saline solution 0.85% (p:v). Inoculum was prepared by dilution of a suspension corresponding to 0.5 on the McFarland density scale (DO_{530nm} = 0.19 - 0.21) in Mueller-Hinton broth (MH) (1:10) to obtain a final inoculum with 7.5 x 10^5 UFC/mL, which was used to assay. The fungal suspension was prepared in a similar way [15], to obtain a final inoculum with 1 x 10^3 UFC/mL, which was used to assay. Stretch marks of 48 h of growth were used to obtain the suspension in saline, and later adjusted to 0.5 on the McFarland scale (DO_{530nm} = 0.19 - 0.21), representing a cell density of approximately 1 x 10^6 UFC/mL. To obtain the final inoculum, two dilutions were performed, produced by transferring 20 µL of the suspension to 980 µL of Sabouraud dextrose (SD) broth and, subsequently, from 1 mL of the resulting suspension to 19 mL of SD broth. The samples were dissolved in water:dimethylsulfoxide (DMSO) (9:1 v:v), obtaining the initial concentration of 40 mg/mL. The MIC was carried out in 96-well polystyrene microplates containing 100 µL of the liquid medium, 20 µL of the samples solution (ranging from 2000 to 2 µg/mL, dissolved in 9:1 water:DMSO), and 100 µL of the different microbial inoculum in each well. After microdilution, they were incubated in a bacteriological incubator (Nova Instruments) at 37°C for 24 h for bacteria, and at 28°C for 48 h for yeasts, and the minimum inhibitory concentration (MIC) was defined as the lowest concentration of samples without visible growth of microorganisms. Subsequently, the minimal bactericidal concentration (MBC) was defined as the lowest concentration of samples that inhibit bacterial growth, as described by Lyu et al. [16]. Cephalexin, ciprofloxacin, and ketoconazole were used as positive controls. Water:DMSO (9:1 v:v) was used as solvent control. The tests were carried out in triplicate wells, with three independent repetitions between them.

3. Results and Discussion

Phytochemical screening revealed the presence of steroids/triterpenes, flavonoids, coumarins, and alkaloids in the ethanol extract of *B. subalternans*. Phytochemical studies carried out with *B. pilosa*, another species of the *Bidens* genus, also detected the presence of terpenes and flavonoids in the plant extracts [17].

The total phenolic and flavonoid contents present in the ethanol extract were 6.46 ± 1.13 μg of gallic acid equivalents/mL, and 33.13 ± 0.48 μg of quercetin equivalents/mL, respectively.

Liu et al. [18] determined the total phenolic compound content in honey from *B. pilosa* flowers and *B. subalternans* leaves (0.822 ± 0.03 mg/g, and 3.87 ± 0.47 mg/g, respectively). Comparing the results, the leaves of *B. subalternans* exhibited a higher content of phenolic compounds than the honey of *B. pilosa* flowers. Borella et al. [19] quantified an average flavonoid content of 0.080 ± 0.015% (m:m) in samples of *B. pilosa* under different forms of fertilization. Comparing the results, ethanol extract of *B. subalternans* presented a high content of total phenolic and flavonoid compounds.

The toxicity of the extract was evaluated by the larvicidal assay on *A. salina*. In all concentrations evaluated (125, 250, 500, and 1000 μg/mL), there was no mortality of *A. salina*, indicating that the extract of *B. subalternans* leaves is not toxic. No reports of the larvicidal activity of *B. subalternans* on *A. salina* were found in the literature to compare to these results.

The antimicrobial effect was evaluated on seven microorganisms. The results showed that the ethanol extract had no antifungal activity on *C. glabrata* and *C. infanticola*. Fabri et al. [20] demonstrated low activity of the ethanol extract of the leaves of *B. segetum* on *Candida albicans*. However, no reports of antifungal activity by *B. subalternans* on *C. glabrata* and *C. infanticola* were found for comparison with our results.

Antibacterial activity was tested on five microorganisms, and the results are described in Table 1.

Table 1. Result of antimicrobial activity of ethanol extract of *B. subalternans* leaves

Microorganisms	MBC (mg/mL)	MIC (mg/mL)
B. cereus	-	-
E. cloaceae	0.50	0.25
A. hydrophila	1.00	0.50
E. faecalis[*]	1.00	0.50
E. coli EHEC	2.00	1.00

[*]*Enterococcus faecalis* used to be classified as *Streptococcus faecalis* due to its morphology.

The ethanol extract demonstrated bacteriostatic and bactericidal activity on *E. cloaceae*, *A. hydrophila*, *E. faecalis,* and *E. coli* EHEC. The best results were on *E. cloaceae*, *A. hydrophila,* and *E. faecalis*, with MIC values of 0.25, 0.50, and 0.50 mg/mL, respectively. Khan et al. [21], reported little antibacterial activity of *B. pilosa* on *E. coli,* and no effect on *B. cereus*, corroborating the results observed in this work. Unlike the study of Kahn et al. [21], which showed no activity on *Streptococcus faecalis*, the *B. subalternans* extract exhibited moderate activity. No reports of antibacterial activity of *B. subalternans* on *E. cloaceae* and *A. hydrophila* were found in the literature for comparison with our results.

CONCLUSION

The results were promising in relation to the antibacterial activity of *B. subalternans*. The biological activity of this plant species is still poorly known, making the results found in this study of great relevance. In addition, the *B. subalternans* extract was not toxic to *A. salina*. In this sense, it is necessary to perform futher research to evaluate the effectiveness of this species, and its use may be promising.

ACKNOWLEDGMENTS

The authors are grateful to the Universidade Federal de São João Del Rei (UFSJ), Conselho Nacional de Desenvolvimento Científico e Tecnológico (CNPq), Coordenação de aperfeiçoamento de Pessoal de Nível Superior (CAPES), and Fundação de Amparo à Pesquisa do Estado de Minas Gerais (FAPEMIG) for the financial support. "This study was financed in part by the Coordenação de Aperfeiçoamento de Pessoal de Nível Superior - Brasil (CAPES) - Finance code 001."

REFERENCES

[1] Abushaheen, M. A., Muzaheed, Fatani, A. J., Alosaimi, M., Mansy, W., George, M., Acharya, S., Rathod, S., Divakar, D. D., Jhugroo, C., Vellappally, S., Khan, A. A., Shaik, J., Jhugroo, P. (2020). Antimicrobial resistance, mechanisms and its clinical significance. *Dis. Mon.*, 66(6): 100971.

[2] Alam, A., Majumdar, R. S., Alam, P. (2019). Comparative study of metabolites and antimicrobial activities of essential oils extracted from three *Amomum subulatum* cultivars. *Asian J. Pharmaceut. Clin. Res.*, 12: 219-223.

[3] Chandler, C. I. R. (2019). Current accounts of antimicrobial resistance: stabilisation, individualisation and antibiotics as infrastructure. *Palgrave Commun.*, 5: 53.

[4] Gonçalves, T. P. R. (2020). Estudo fitoquímico e avaliação do potencial antimicrobiano, sinérgico e antioxidante das flores de *Tecoma stans* (L.) Juss. ex Kunth (Bignoniaceae) [Phytochemical study and evaluation of the antimicrobial, synergistic and antioxidant potential of *Tecoma stans* (L.) Juss. ex Kunth (Bignoniaceae)]. *Dissertation* (Pharmaceutical Sciences), Divinópolis, 92 p.

[5] Farahbod, F., Soureshjani, S. H. (2018). A systematic review of medicinal plants and plant derivatives affecting increase in endometrial thickness. *Int. J. Pharm. Sci. Res.*, 9(1): 53-57.

[6] Ullah, A., Bano, A., Janjua, H. T. (2020). Microbial secondary metabolites and defense of plant stress. In: Singh, J. S., Vimal, S. R. *Microbial Services in Restoration Ecology.* Amsterdam: Elsevier, 37-46.

[7] Araújo, M. G. F., Galeane, M. C., Castro, A. D., Salgado, H. R. N., Almeida, A. E., Cunha, W. R., Veneziani, R. C. S., Moreira, R. R. D. (2010). Pharmacognostical evaluation of fruits of *Solanum lycocarpum* A. St. -Hill. (Solanaceae). *Pharmacogn. J.*, 2(9): 248-253.

[8] Carvalho, C., Matta, S., Melo, F., Andrade, D., Carvalho, L., Nascimento, P., Silva, M., Rosa, M. (2009). Cipó-cravo (*Tynnanthus fasciculatus* Miers - Bignoniaceae): estudo fitoquímico e toxicológico envolvendo *Artemia salina* [Stem herb (*Tynnanthus fasciculatus* Miers - Bignoniaceae): phytochemical and toxicological study involving *Artemia salina*]. *Rev. Eletrônica Farm.*, 6(1): 51-57.

[9] Gonçalves, T. P. R., Silva, I. C. A., Lima, W. G., Coimbra, M. C., Parreira, A. G., Lima, L. A. R. S. (2018). Efeito antioxidante e análise fitoquímica de extratos etanólicos de *Bidens subalternans* DC. In: Sousa, B. J. M., Alves, B. C., Duarte-Almeida, J. M. (Eds). *Plantas Medicinais: Os Avanços das Pesquisas no Século XXI.* Divinópolis: Gulliver Editora, 109-118. [Antioxidant effect and phytochemical analysis of ethanol extracts from Bidens subalternans DC. In: Sousa, B.J.M., Alves, B.C., Duarte-Almeida, J.M. (Eds). *Medicinal Plants: Research Advances in the 21st Century.*]

[10] Simões, C. M. O., Schenkel, E. P., Gosmann, G., De Melo, J. C. P., Mentz, L. A., Petrovick, P. R. (2007). *Farmacognosia: da Planta ao Medicamento.* Porto Alegre: UFGRS, 1104 p.

[11] Zielinski, A. A. F., Haminiuk, C. W. I., Alberti, A., Nogueira, A., Demiate, I. M., Granato, D. (2014). A comparative study of the

phenolic compounds and the *in vitro* antioxidant activity of different Brazilian teas using multivariate statistical techniques. *Food. Res. Int.*, 60: 246-254.

[12] Brasil. (2010). Ministério da Saúde. Agência Nacional de Vigilância Sanitária. *Farmacopeia Brasileira.* 5ª ed. Anvisa, Brasília, v. 2, 889 p. [Ministry of Health. National Health Surveillance Agency. *Brazilian Pharmacopoeia*]

[13] Meyer, B. N., Ferrigni, N. R., Putnam, J. E., Jacobsen, L. B., Nichols, D. E., McLaughlin, J. L. (1982). Brine shrimp: A convenient general bioassay for active plant constituents. *Planta. Med.*, 45(5): 31-34.

[14] Clinical and Laboratory Standards Institute. (2015). Performance Standards for Antimicrobial Susceptibility Testing; Twenty-Fifth Informational Supplement. *CLSI document* M100-S25, 240 p.

[15] Clinical and Laboratory Standards Institute. (2017). Reference Method for Broth Dilution Antifungal Susceptibility Testing of Yeasts; Fourth Edition. *CLSI Document* M27, 46 p.

[16] Lyu, Y., Yang, Y., Lyu, X., Dong, N., Shan, A. (2016). Antimicrobial activity, improved cell selectivity, and mode of action of short PMAP-36-derived peptides against bacteria and *Candida*. *Sci. Rep.*, 6: 27258.

[17] Lastra Valdés, H. A., Ponce de León Rego, H. (2001). *Bidens pilosa* Linné/28. *Rev. Cuba. Plant. Med.*, 6(1): 28-33.

[18] Liu, J. R., Ye, Y. L., Lin, T. Y., Wang, Y. W., Peng, C. C. (2013). Effect of floral sources on the antioxidant, antimicrobial, and anti-inflammatory activities of honeys in Taiwan. *Food Chem.*, 139(1-4): 938-943.

[19] Borella, J. C., Borella, P. H., Gastaldi, M. D., Miranda, C. E. (2019). *Bidens pilosa* - picão preto: influência da adubação orgânica e da luminosidade na produtividade e no teor de flavonoides [*Bidens pilosa* - hairy beggartick: influence of organic fertilization and luminosity on productivity and flavonoid content]. *Rev. Fitos*, 13(4): 261-269.

[20] Fabri, R. L., Nogueira, M. S., Dutra, L. B., Bouzada, M. L. M., Scio, E. (2011). Potencial antioxidante e antimicrobiano de espécies da família Asteraceae [Antioxidant and antimicrobial potential of Asteraceae species]. *Rev. Bras. Pl. Med.*, 13(2):183-189.

[21] Khan, M. R., Kihara, M., Omoloso, A. D. (2001). Anti-microbial activity of *Bidens pilosa, Bischofia javanica, Elmerillia papuana* and *Sigesbekia orientalis*. *Fitoterapia*, 72(6): 662-665.

In: The Book of Flavonoids
Editors: Chao-Hui Feng et al.
ISBN: 978-1-68507-376-3

Chapter 3

EVALUATION OF THE ANTIOXIDANT AND LARVICIDAL POTENTIAL OF PHENOLIC AND FLAVONOID COMPOUNDS FROM DIFFERENT EXTRACTS OF THE FLOWERS OF *MATRICARIA RECUTITA* L.

***Brayan Jonas Mano-Sousa*[1], *Thaís Paula Rodrigues Gonçalves*[1], *Bruna Cristina Alves*[1], *Perla Fonseca Amorim de Araújo*[2], *Luciana Alves Rodrigues dos Santos Lima*[1,3,*] and *Joaquim Maurício Duarte-Almeida*[1,3]**

[1]Programa de Pós-Graduação em Ciências Farmacêuticas da Universidade Federal de São João Del Rei, Divinópolis, MG, Brazil
[2]Programa de Pós-Graduação em Ciências da Saúde da Universidade Federal de São João Del Rei, Divinópolis, MG, Brazil
[3]Programa de Pós-Graduação em Biotecnologia da Universidade Federal de São João Del Rei, Divinópolis, MG, Brazil

[*] Corresponding Author's E-mail: luarsantos@ufsj.edu.br.

ABSTRACT

Matricaria recutita L. (chamomile) is widely used by the population due to its therapeutic potential, which is attributed to the various phytochemicals present in this species. The aim of this study was to evaluate the extraction efficiency of phenolic compounds and flavonoids obtained in the flowers of *M. recutita* by two different extraction methods (soxhlet [SXT] and ultrasound assisted extraction [UAE]), as well as to evaluate the antioxidant and larvicidal potential of the extracts and their fractions. The extracts were obtained by extraction in a soxhlet apparatus (SXT) and extraction assisted by ultrasound (UAE) with ethanol. The fractions were obtained by partitioning the extracts with solvents of increasing polarity: hexane, dichloromethane, and ethyl acetate. The extract yield was directly influenced by the extraction method, which was more efficient when obtained by the SXT method. The ethanol extracts, hexane and dichloromethane fractions, showed traces of essential oils, suggesting the presence of α-bisabolol A/B oxide, bisabolol, and farnesene. In the samples analyzed by high performance liquid chromatography with diode array detector (HPLC-DAD), it was possible to suggest the presence of catechin gallate, isoflavone, chlorogenic acid, undefined derivative of luteolin, apigenin, undefined derivative of apigenin, luteolin-3,' 7'-di-*O*-glycoside, apigenin-8-*C*-glycoside, and galangin. Greatest Pearson's correlations between antioxidant activity and flavonoids were found in the hydroethanol (SXT) and ethyl acetate (UAE) fractions, while the ethyl acetate (SXT), dichloromethane (SXT), hydroethanol (UAE) and dichloromethane (UAE) fractions showed greater correlations between antioxidant activity and phenolic compounds. However, there was no significant influence of the extraction method on the antioxidant activity using 1,1-diphenyl-2-picrylhydrazyl (DPPH) radicals and on the larvicidal evaluation using lethality tests in *Artemia salina*. The ethyl acetate and hydroethanol fractions exhibited antioxidant activity and showed no larvicidal potential against *A. salina*, which may indicate low toxicity of these samples.

Keywords: antioxidant, extractive methods, chamomile, larvicidal, phenolic compounds

1. Introduction

Matricaria recutita L. (German chamomile) is an herbaceous plant belonging to the Asteraceae family, which is native to southwestern Europe and western Asia, and is found throughout Europe, North America, Australia, and, most recently, in South America [1]. This species is one of the most popular and widely used plant in traditional medicine [2]. The flowers are mainly used for their anti-inflammatory and spasmolytic properties [3]. This species contains therapeutically active substances [2] that show antioxidant [4], anti-allergic [5], antimicrobial [6], anti-inflammatory, and anticancer activities [7]. Sesquiterpenes, flavonoids, phenolic compounds, and coumarins are the main bioactive compounds of this plant [8]. The sesquiterpenes found in chamomile have been correlated to various biological activities exhibited by the species [9].

The various biological activities presented by natural compounds have increased interest in the extraction, production, and use of these substances, leading to a greater need for the development of better and more efficient extractive methods for obtaining these compounds in a shorter time and at a lower cost. Therefore, it is important to evaluate the effectiveness of the method for extracting the bioactive compounds from the plant material [10]

Traditional methods used for extraction of plant materials include steam distillation (indicated for processing large quantities of plant material on an industrial scale) and extraction with organic solvents by percolation, maceration, or soxhlet techniques [11]. The phenolic compounds extracted from chamomile by different extraction methods have been investigated extensively in the last 10 years [12]. The main methodologies use ethanol, methanol, chloroform, or other organic solvents for extraction; however, the most efficient extractive method for chamomile compounds has not yet been defined [13, 14].

Thus, the objective of this study was to explore the use of two extraction techniques [soxhlet (SXT) and ultrasound assisted extraction (UAE)] to compare the yield of extracts from flowers of *M. recutita* and to evaluate the antioxidant and larvicidal potential of the ethanol extracts and fractions obtained.

2. Materials and Methods

2.1. Drugs and Reagents

Folin-Ciocalteu reagent, 2,6-di-tert-butyl-4-methylphenol (BHT), 1,1-diphenyl-2-picrylhydrazyl (DPPH), rutin, quercetin, dimethylsulfoxide (DMSO), gallic acid, and thin-layer chromatography (TLC) on Merck silica gel 60 F_{254} aluminium plates were purchased from Sigma-Aldrich Inc. (USA). All other reagents and solvents were of analytical grade were purchased from Vetec (Brazil). For the highperformance liquid chromatography with diode array detector (HPLC-DAD) analyses, methanol, and formic acid were purchased from J. T. Baker, and the ultrapure water (Milli-Q, Millipore®) was also used.

2.2. Plant Material and Extraction

Dried flowers of *M. recutita* were commercially purchased at *Naturallis Suplementos* in the Divinópolis City, Minas Gerais State, Brazil, in August 2018. This study has access permission to the components of plant genetic heritage registered in the SisGen Platform (Register AF2518F), according to Brazilian Biodiversity Law (13.123/2015).

The material was extracted by two methods, UAE and SXT, using 10 g of dried flowers and 200 mL of ethanol 96 °GL in both methods. The cold method (UAE) was performed at room temperature (27 ± 2ºC), being extracted in ultrasound equipment for 30 minutes followed by rest for 30 minutes, until totaling 6 hours. The hot extraction (SXT) was carried out for 6 hours on heating at 65°C without intervals. After filtration, the solvent was removed using rotatory evaporator. The yield obtained in the extraction processes was calculated from the equation:

$$Yield = \frac{Dry\ extract\ weight_{(g)}}{Dry\ plant\ weight_{(g)}} \times 100 \quad (1)$$

Part of the extracts was solubilized in ethanol:water (7:3 v:v) and partitioned with hexane, dichloromethane and ethyl acetate to obtain hexane (Hex), dichloromethane (Dcm), ethyl acetate (Ac) and the hydroethanol (Hid) fractions, respectively. The fractions were filtered and solvents removed using rotatory evaporator.

2.3. Extraction of Essential Oil

The essential oil was extracted from 10 g of dried flowers of *M. recutita* with 200 mL of water by steam distillation in a Clevenger apparatus for 6 hours on heating at 65°C without intervals.

2.4. Thin-Layer Chromatography (TLC) Analysis

The evaluation of the presence of essential oil in *M. recutita* in ethanol extracts and Hex and Dcm fractions was performed by thin-layer chromatography (TLC) on Merck silica gel 60 F_{254} aluminium plates, using toluene:ethyl acetate (93:7 v:v) as eluent and vanillin and sulfuric acid followed by heating at 100°C for 10 minutes as developer [15]. The essential oil of *M. recutita* was used as a standard for comparison between the samples applied in TLC plates. Compound determination was performed comparing the retention factor (R_F) of samples and standards, as well as with data from literature [15, 16]. The R_F was calculated by equation:

$$R_F = \frac{\text{distance spot moved}}{\text{distance solvent moved}} \quad (2)$$

2.5. Color Determination

The color naming was confirmed by the Sci-Chromus® (BR512019002743-4, UFSJ-CCO, Minas Gerais, Brazil) software, as proposed by Mano Sousa et al. [17]. This software was developed to standardize color names and eliminating human subjectivity in these names.

2.6. High Performance Liquid Chromatography Analysis

The chromatographic analysis was conducted using UFLC Proeminence chromatographic system (Shimadzu, Kyoto, Japan), composed of a binary pump system (LC-20AD), autosampler (SIL-20AHT), communicator (CBM-20A) and degasser, controlled by the LabSolutions software (version 1.25, Shimadzu, Kyoto, Japan). The experiments were carried out in an air-conditioned room (14 ± 2ºC). The chromatographic column was a Kinetex C_{18} (5 μm, 100 × 2.1 mm, Phenomenex), eluted with methanol (solvent B) and ultrapure water added formic acid 0.1% (v:v). The elution profile applied was: 0-8.5 min: 10 to 90% of B; 9-10 min: 90% of B and; 10-11 min: 10% of B. All samples (rutin, extracts and fractions) were solubilized (1 mg/mL) and filtered through a 0.45 mm PTFE filter syringe. The injection volume was 10 μL of samples (rutin, extracts and fractions), and the flow rate was 0.5 mL/min. The spectra were recorded in the range of 200 to 600 nm. Compound determination was performed comparing the retention times and spectrum in the UV region of the standards, and correlating with data found in the literature, according to Costa et al. [18].

2.7. Total Phenolic Compounds (TPC)

The total phenolic content in the extracts and fractions was estimated using the Folin-Ciocalteu test described by Zielinski et al. [19] with modifications. The Folin-Ciocalteu aqueous solution (2.250 μL; 1:4 v:v) was added to samples (250 μL) and, subsequently, sodium carbonate solution (250 μL) was added. After vigorous shaking, these solutions were kept at rest for 30 min at room temperature. The absorbance was determined by spectrophotometry at 750 nm (Thermo Scientific Genesys 10S, USA) after 30 min of incubation at room temperature with a blank sample as well as a standard solution and samples. Gallic acid was used as a reference compound, and the total phenolic contents were expressed as micrograms of gallic acid equivalents (GAE) per milliliter (mL). All assays were performed in triplicate.

2.8. Total Flavonoid Content (TFC)

The total flavonoid content was estimated according to the aluminum chloride method [20]. Exactly 1900 μL of ethanol 50% (v:v) was added 100 μL of extracts or fractions and 500 μL of solution of aluminum chloride. The absorbance was read at 425 nm using a spectrophotometer (Thermo Scientific Genesys 10S, USA) after 30 min. Quercetin was used as a reference compound to produce a standard curve, and total flavonoid contents were expressed as micrograms of quercetin equivalents (QE) per milliliter (mL). All assays were performed in triplicate.

2.9. DPPH Radical Scavenging Assay

The radical scavenging abilities of extracts and fractions were analyzed compared to standard, BHT based on the reaction with DPPH. The determination of antioxidant activity using the DPPH method was adapted for use with microplates [21, 22]. Briefly, a DPPH solution (0.002% w:v) was prepared in ethanol. Exactly 75 μL of the samples (1, 10, 100, 250 and 500 μg/mL) was added to the wells in a 96-well flat-bottom plate containing 150 μL of DPPH solution. The plate was then covered and left in the dark at room temperature (25°C). After 30 min, the absorbance at 517 nm was measured with a spectrophotometer (Biotek Power Wave XS2, USA), and ethanol was used for the baseline correction. Scavenging ability was expressed as the inhibition percentage and was calculated using Burda and Oleszek [23] equation:

$$\text{Scavenging ability} = \frac{\text{Abs}_{\text{Control}} - \text{Abs}_{\text{sample}}}{\text{Abs}_{\text{control}}} \times 100 \quad (3)$$

where Abs control = absorbance of DPPH radical in ethanol and Abs sample = absorbance of the extract or fractions in ethanol + DPPH. The antioxidant activity of all of the samples was expressed as IC_{50}, which was defined as the concentration (in μg/mL) of sample required to inhibit the formation of the DPPH radicals by 50%. All assays were performed in triplicate.

2.10. *Artemia salina* larvicide bioassay

A. salina encysted eggs (200 mg) were incubated in 400 mL of seawater under artificial light at 28°C, pH 7-8. After incubation for 48 h, metanauplii were collected with a Pasteur pipette. The samples (in triplicate) to be assayed were dissolved in DMSO and diluted serially (25, 50, 100 and 200 μg/mL) in seawater. To each set of tubes containing the samples, 10 nauplii were added. Negative controls containing seawater and 0.025, 0.050, 0.1 and 0.2% DMSO were included in each experiment [24]. The number of survivors was counted after 24 h and recorded.

2.11. Statistical Analysis

The lethal dose (LD_{50}) was calculated by simple linear regression using Probit analysis [25]. The data were subjected to an analysis of variance (ANOVA) followed by Tukey Test at a 5% probability level.

All statistical parameters were calculated using Sisvar software 5.4 [26]. The criterion for statistical significance was set at $p < 0.05$. The correlations of antioxidant test and phenol and flavonoid contents were calculated using Pearson's correlation coefficient in the Microsoft Excel 19 (Microsoft Office 2019, Redmont, USA).

3. RESULTS AND DISCUSSION

The extraction methods used in this study showed different yields for the ethanol extracts, with 11.6% by UAE and 24% by SXT. The results obtained suggest that the extraction method directly influences the yield. The SXT method demonstrated a higher extraction efficiency, similar to previously reported studies [27, 28]. Although the UAE method allows the compounds to dissolve more easily due to rupturing of the cell wall, causing an increase in the extraction yield in a shorter time [29], our results showed that the SXT technique was more efficient.

Many compounds are sensitive to heat, while others may have irreversible structural changes at high temperatures, which can make hot methods unviable [30]. Contrary to expectations, the method with the highest temperature obtained the highest yield. This fact can be explained by the increased solubility promoted by heat in addition to the constant renewal of the solvent in contact with the plant material promoted by the SXT technique.

The ethanol extracts and Hex and Dcm fractions were analyzed by TLC for the presence of essential oils. The extracts and fractions showed the same R_F and colors in spots corresponding to α-bisabolol oxide A/B, bisabolol, and farnesene (0.2 - yellow-green; 0.35 - violet; 0.99 - blue-violet, respectively) as described in literature [15]. From a qualitative point of view, it was observed that the chromatographic profiles obtained for the essential oil used as a standard, as well as for the samples (extracts, Hex and Dcm fractions) extracted by both methods, were similar to the profiles described by Wagner and Bladt [15]. The main compounds of *M. recutita* essential oil described are α-bisabolol oxide

A/B, α-bisabolol, camazulene, and α-farnesene [31, 32], confirming the presence of essential oil compounds in the extracts and fractions analyzed. This finding suggests that it is possible to detect traces of these compounds from *M. recutita* by these methods, even when heated for a long period (6 h) using organic solvents.

The extracts and fractions were analyzed by HPLC-DAD. The chromatograms of the extracts and their fractions are shown in Figures 1-4.

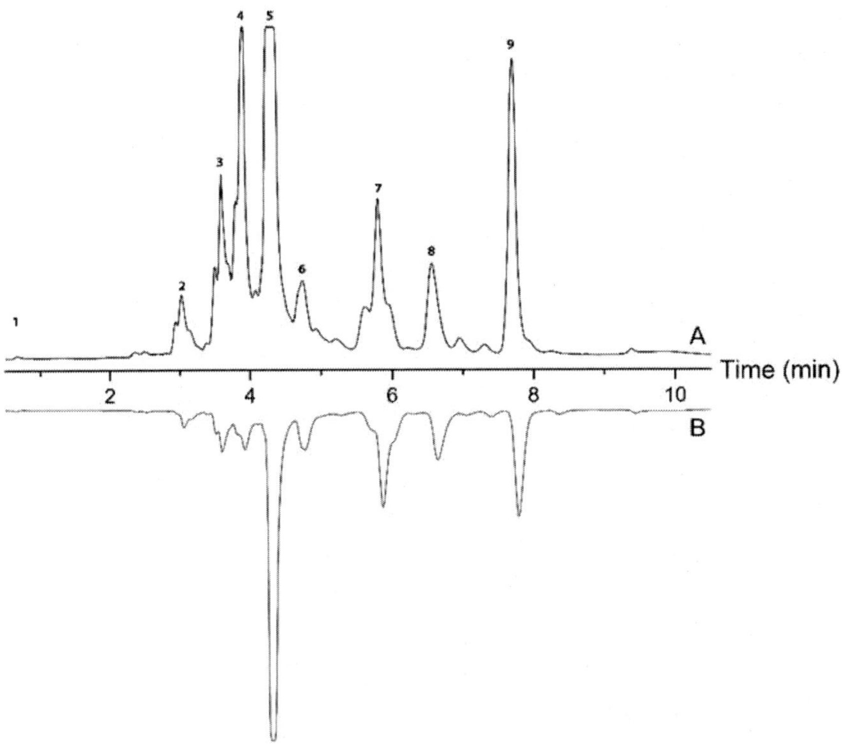

Figure 1. Chromatographic profile of *Matricaria recutita* ethanol extracts in high performance liquid chromatography (HPLC-DAD), 350 nm. (A) Soxhlet extract (SXT); (B) Ultrasound assisted extract (UAE). Suggested compounds: 1. Catechin gallate, 2. Isoflavone, 3. Chlorogenic acid, 4. Undefined luteolin derivative, 5. Apigenin, 6. Undefined apigenin derivative, 7. Luteolin-3,'7'-di-*O*-glycoside, 8. Apigenin-8-*C*-glycoside (Vitexin), 9. Galangin.

Based on the UV spectra obtained and comparison with published data, it was possible to suggest the presence of the following compounds: catechin gallate, isoflavone, chlorogenic acid, undefined luteolin derivative, apigenin, undefined apigenin derivative, luteolin-3,'7'-di-*O*-glycoside, apigenin-8-*C*-glycoside (vitexin), and galangin [33-36].

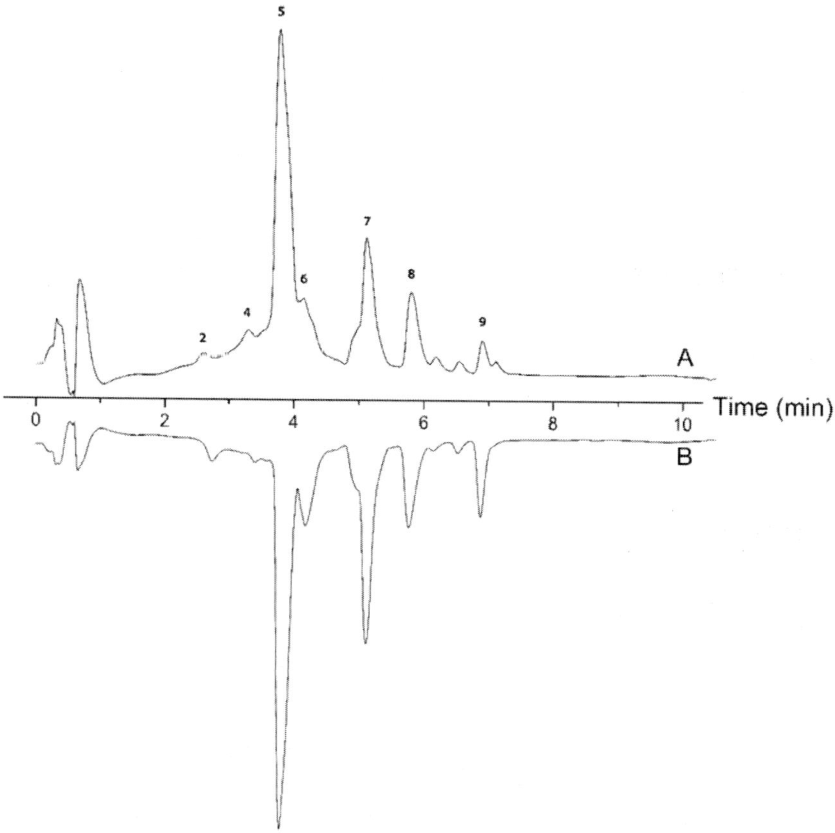

Figure 2. Chromatographic profile of dichloromethane (Dcm) fractions of *Matricaria recutita* ethanol extracts in high performance liquid chromatography (HPLC-DAD), 350 nm. (A) Soxhlet extract (SXT); (B) Ultrasound assisted extract (UAE). Suggested compounds: 2. Isoflavone, 4. Undefined luteolin derivative, 5. Apigenin, 6. Undefined apigenin derivative, 7. Luteolin-3,'7'-di-*O*-glycoside, 8. Apigenin-8-*C*-glycoside (Vitexin), 9. Galangin.

The presence of chlorogenic acid, apigenin, catechin gallate, and galangin have been reported in chamomile [13, 37-40]. The flavonoid luteolin-3,'7'-di-*O*-glycoside was also identified by Caleja et al. [33]; however, it is not a compound often found in *M. recutita*.

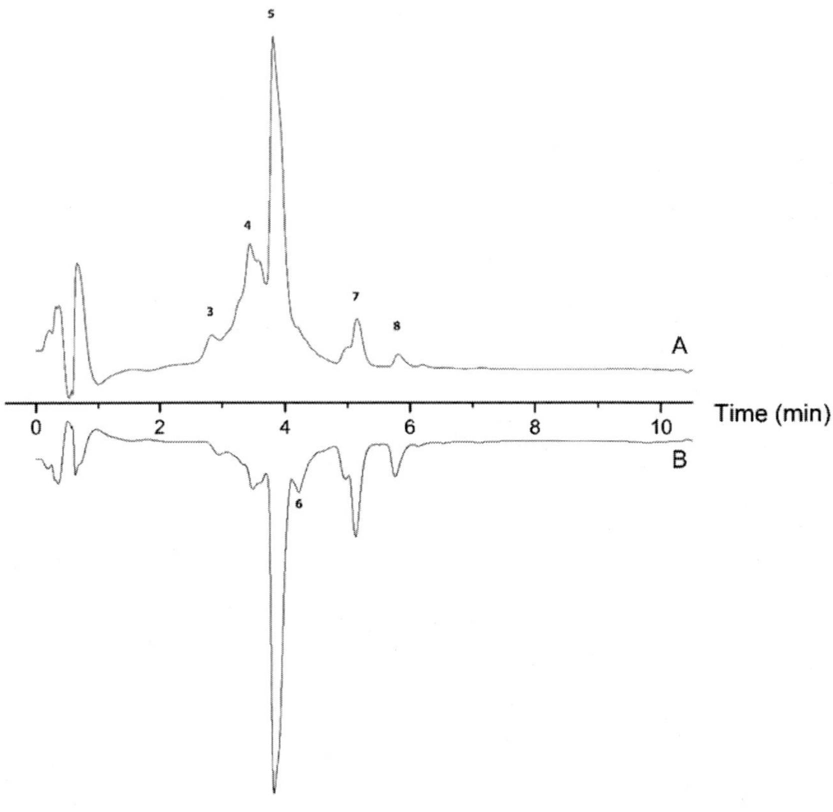

Figure 3. Chromatographic profile of ethyl acetate (Ac) fractions of *Matricaria recutita* ethanol extracts in high performance liquid chromatography (HPLC-DAD), 350 nm. (A) Soxhlet extract (SXT); (B) Ultrasound assisted extract (UAE). Suggested compounds: 3. Chlorogenic acid, 4. Undefined luteolin derivative, 5. Apigenin, 6. Undefined apigenin derivative, 7. Luteolin-3,'7'-di-*O*-glycoside, 8. Apigenin-8-*C*-glycoside (Vitexin).

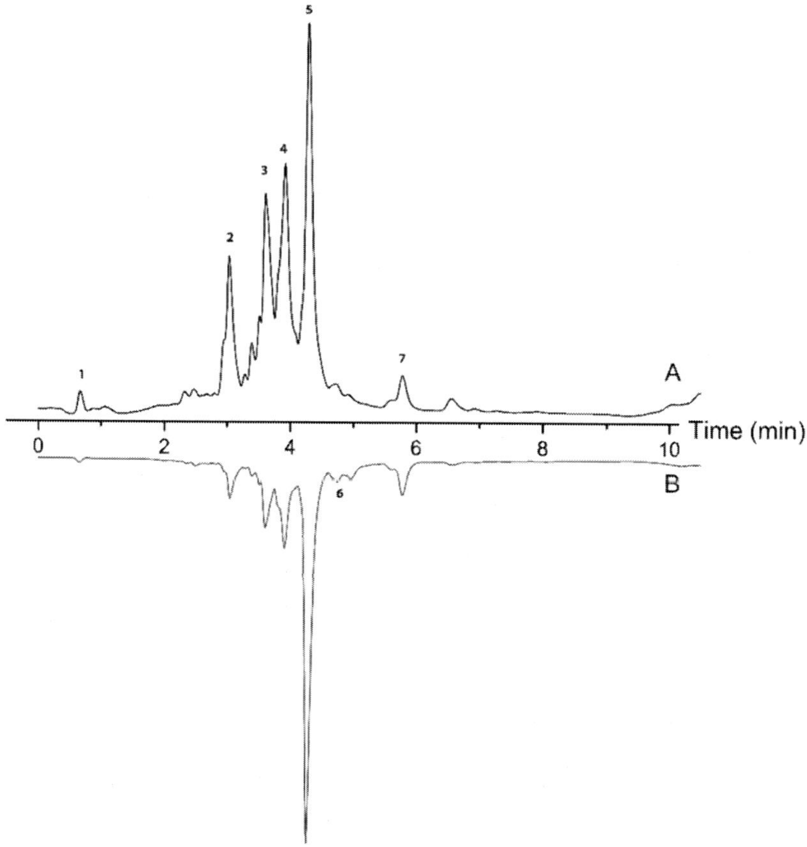

Figure 1. Chromatographic profile of hydroethanol (Hid) fractions of *Matricaria recutita* ethanol extracts in high performance liquid chromatography (HPLC-DAD), 350 nm. (A) Soxhlet extract (SXT); (B) Ultrasound assisted extract (UAE). Suggested compounds: 1. Catechin gallate, 2. Isoflavone, 3. Chlorogenic acid, 4. Undefined luteolin derivative, 5. Apigenin, 6. Undefined apigenin derivative, 7. Luteolin-3,'7'-di-*O*-glycoside.

The antioxidant potential of the extracts and fractions (Table 1) was not significantly influenced by the extraction method. Extracts, fractions, and BHT had an inhibitory effect that was concentration dependent. The ethanol extract, Ac and Hid fractions of SXT, and Dcm, Ac and Hid fractions of UAE showed more potent antioxidant activity with lower values of IC_{50} (0.12 - 4.76 μg/mL) and displayed a greater activity than BHT (IC_{50} = 16.36 μg/mL, $p < 0.001$).

Table 1. 1,1-diphenyl-2-picrylhydrazyl (DPPH) scavenging activity of the ethanol extracts and fractions of dried flowers of *M. recutita* L.

Sample	DPPH-scavenging activity (%)				
	1 μg/mL	10 μg/mL	100 μg/mL	250 μg/mL	500 μg/mL
	Ultrasound assisted extraction (UAE)				
EE	29.85 ± 0.29[a]	30.94 ± 0.21[a]	35.36 ± 0.29[a]	39.49 ± 1.06[a]	54.13 ± 1.64[a]
Hex	24.81 ± 0.73[a]	26.90 ± 0.62	30.80 ± 1.27[a]	42.91 ± 0.57[a]	50.85 ± 2.77[a]
Dcm	40.16 ± 0.93[a]	53.03 ± 0.57[a]	73.70 ± 1.89[a]	92.14 ± 0.85	97.08 ± 1.39[a]
Ac	43.53 ± 1.93[a]	86.25 ± 2.69[a]	97.32 ± 0.37[a]	98.84 ± 0.50[a]	99.60 ± 0.24[a]
Hid	42.77 ± 0.81[a]	53.13 ± 0.49[a]	70.99 ± 1.36[a]	94.09 ± 1.81[a]	97.79 ± 1.57[a]
BHT	18.50 ± 0.65	25.90 ± 0.64	86.00 ± 0.56	91.40 ± 0.28	94.02 ± 0.51
	Soxhlet extraction (SXT)				
EE	41.58 ± 0.89[a]	79.74 ± 0,92[a]	92.61 ± 0.37[a]	96.27 ± 0.59[a]	99.07 ± 0.29[a]
Hex	29.70 ± 0.21[a]	32.65 ± 1.16[a]	34.22 ± .53[a]	36.69 ± 0.67[a]	41.39 ± 0.87[a]
Dcm	42.91 ± 1.29[a]	46.81 ± 0.97[a]	49.71 ± 0.37[a]	69.71 ± 1.15[a]	95.13 ± 1.57
Ac	85.44 ± 1.53[a]	89.29 ± 0.21[a]	91.85 ± 0.41[a]	95.75 ± 1.02	99.36 ± 0.45[a]
Hid	41.44 ± 0.62[a]	52.13 ± 1.26[a]	75.60 ± 0.67[a]	96.27 ± 0.29[a]	98.79 ± 0.87[a]
BHT	18.50 ± 0.65	25.90 ± 0.64	86.00 ± 0.56	91.40 ± 0.28	94.02 ± 0.51

Each value in the table is the mean ± standard deviation ($n = 3$). [a]$p < 0.05$ compared with BHT. EE: ethanol extract; Hex: hexane fraction; Dcm: dichloromethane fraction; Ac: ethyl acetate fraction; Hid: hydroethanol fraction.

The samples with higher TPC had better antioxidant potential by DPPH inhibition (Table 2). The greatest correlations between antioxidant activity and flavonoids were found in the Hid (SXT) and Ac (UAE) fractions. Unlike flavonoids, there was an inversion of the methods in which the fractions of the phenolic compounds of chamomile obtained better correlations. In the SXT extraction method, the fraction Ac had a better correlation, followed by the Dcm fraction, while in the UAE method, this correlation was for Hid and Dcm fractions.

Phenolic compounds are capable of sequestering or inhibiting the reactive oxygen species, transferring electrons to free radicals, activating antioxidant enzymes, and inhibiting oxidase enzymes [41]. These mechanisms play an important role in the prevention of oxidative stress, which is associated with diseases such as arteriosclerosis, diabetes, cancer, and neurodegenerative diseases [42].

Table 2. Inhibitory concentration (IC_{50}) values for the antioxidant activity, total flavonoid and total phenolic content of the ethanol extract and fractions of dried flowers of *M. recutita* L.

Sample	IC_{50}[1]	TFC[2]	TPC[3]	IC_{50} vs TFC[4]	IC_{50} vs TPC[4]
Ultrasound assisted extraction (UAE)					
EE	443.40 ± 29.27[a]	124.84 ± 0.26	915.32 ± 0.74	-0.503	0.504
Hex	498.69 ± 33.95[a]	-	-	-	-
Dcm	4.42 ± 0.32[a]	134.35 ± 0.42	1002.91 ± 1.05	-0.816	0.816
Ac	1.30 ± 0.08[a]	78.69 ± 0.22	2851.45 ± 1.34	0.981	-0.201
Hid	4.76 ± 0.17[a]	46.69 ± 0.16	1643.43 ± 2.58	0.530	-0.836
BHT	16.36 ± 1.63	-	-	-	-
Soxhlet extraction (SXT)					
EE	1.53 ± 0.12[a]	176.56 ± 3.70	1398.90 ± 2.16	-0.147	0.147
Hex	632.55 ± 51.23[a]	-	-	-	-
Dcm	101.82 ± 4.95[a]	81.01 ± 1.21	220.00 ± 0.07	-0.875	0.872
Ac	0.12 ± 0.00[a]	53.18 ± 0.29	2134.30 ± 3.17	0.255	-0.972
Hid	4.14 ± 0.15[a]	29.53 ± 0.22	1172.62 ± 0.77	0.937	-0.334
BHT	16.36 ± 1.63	-	-	-	-

[1]IC_{50}: concentration (in μg/mL) of sample required to inhibit 50% the formation of DPPH radicals. [2]Total flavonoids content (TFC): results expressed as μg of quercetin equivalents/mL of extract or fraction. [3]Total phenolic content (TPC): results expressed as μg of gallic acid equivalents/mL of extract or fraction. Each value in the table is the mean ± standard deviation (n = 3). [a]$p < 0.05$ compared with BHT. [4]Pearson correlation coefficients. EE: ethanol extract; Hex: hexane fraction; Dcm: dichloromethane fraction; Ac: ethyl acetate fraction; Hid: hydroethanol fraction.

The antioxidant activity of *M. recutita* by the DPPH method has been reported previously. Ethanolic extracts obtained by maceration of flowers exhibited an IC_{50} value of 26.7 μg/mL [43]. The decoction extract showed an IC_{50} value of 165 μg/mL [40]. Other studies have also demonstrated antioxidant potential with different extracts of *M. recutita* [44-46]. Our results showed a greater antioxidant effect than those reported in the literature.

The brine shrimp (*Artemia salina*) lethality bioassay is an efficient, rapid, and inexpensive test that correlates well with toxic activity in some human solid tumours and with pesticidal activity [24]. The results presented in Table 3 show the mortality rate of *A. salina* larvae exposed to samples of *M. recutita*. The larvae of *A. salina* were highly susceptible to the Hex (SXT) fraction at all concentrations tested, with a mortality rate between 96.66 - 100%. The Dcm (UAE) fraction and ethanol extract (SXT), at a concentration of 50 µg/mL, and ethanol extract (UAE), Hex (UAE) and Dcm (SXT) fractions, at a concentration of 100 µg/mL, had mortality rates higher than 50%.

The lethal concentration required to kill 50% of *A. salina* larvae (LD_{50}) was determined for each sample (Table 3). The samples of *M. recutita* exhibited LD_{50} values of 82.85, 77.98, 30.27, 48.21, 5.7, and 86.72 µg/mL for the ethanol extract (UAE), Hex (UAE) fraction, Dcm (UAE) fraction, ethanol extract (SXT), Hex (SXT) fraction and Dcm (SXT) fraction, respectively. The results showed that lethal doses of the Ac and Hid fractions occurred above 1000 µg/mL, which indicate that these substances may not be toxic [24] (Table 3). The Dcm fractions (UAE and SXT) had a positive correlation with the TFC and LD_{50} (0.450 and 0.445, respectively), while the ethanol extracts (UAE and SXT) had a positive correlation with LD_{50} and TPC (0.644 and 0.500, respectively).

The Hex (UAE) fraction, Dcm (UAE) fraction, ethanol extract (SXT) and Hex (SXT) fraction showed high toxicity, and the ethanol extract (UAE) and Dcm (SXT) fraction exhibited moderate toxicity, according to Dolabela (1997). The following parameters should be considered: a sample with an LD_{50} of less than 80 µg/mL is highly toxic, between 80 µg/mL and 250 µg/mL is moderately toxic, and above 250 µg/mL has low toxicity or is non-toxic.

Few studies have shown the toxicity of metabolites of the *Matricaria* species on *A. salina*. The essential oil of the aerial parts of *M. chamomilla* exhibited toxicity on *A. salina* (LD_{50} = 31.7 µg/mL), which is a value similar to that found for the Dcm (UAE) fraction and higher than the Hex (SXT) fraction. The major compounds identified in the essential oil included (*E*)-*β*-farnesene (42.2%) and *α*-bisabolol oxide A (22.3%) [48]. Thus, the toxic effect of extracts and Hex and Dcm fractions on *A. salina* found in this study can be attributed, at least partially, to the presence of these substances.

Table 3. *A. salina* larvicidal bioassay of different concentrations of ethanol extracts and fractions of *M. recutita* L.

Sample	% Mortality Concentration (µg/mL)				LD_{50} (µg/mL)
	25 µg/mL	50 µg/mL	100 µg/mL	200 µg/mL	
			Ultrasound assisted extraction (UAE)		
EE	0.00[a]	13.33 ± 2.30[b]	63.33 ± 1.52[c]	100.00[d]	82.85
Hex	16.66 ± 0.57[a]	30.00 ± 2.64[b]	50.00 ± 4.35[c]	90.00 ± 1.73[d]	77.98
Dcm	33.33 ± 2.30[a]	86.66 ± 0.57[b]	100.00[c]	100.00[c]	30.27
Ac	0.00[a]	3.33 ± 0.57[b]	10.00 ± 1.73[c]	20.00 ± 0.00[d]	ND
Hid	0.00[a]	0.00[a]	3.33 ± 0.50[b]	16.66 ± 2.08[c]	ND
			Soxhlet extraction (SXT)		
EE	13.33 ± 2.30[a]	53.33 ± 0.57[b]	86.66 ± 1.52[c]	100.00[d]	18.21
Hex	96.66 ± 0.57[a]	100.00[b]	100.00[b]	100.00[b]	5.7
Dcm	0.00[a]	3.33 ± 0.57[b]	66.66 ± 1.15[c]	100.00[d]	86.72
Ac	0.00[a]	0.00[a]	0.00[a]	0.00[a]	ND
Hid	0.00[a]	0.00[a]	0.00[a]	10.00 ± 1.00[b]	ND

The results are means ± Standard Error (SE) ($n = 3$). Means followed by the same letter on the same line do not differ according to the Tukey test ($p < 0.05$). EE: ethanol extract; Hex: hexane fraction; Dcm: dichloromethane fraction; Ac: ethyl acetate fraction; Hid: hydroethanol fraction; ND: Not determined.

CONCLUSION

The results of our analysis showed that it is possible to differentiate the extraction potential of the employed methods, with the SXT method producing a higher yield. However, the extraction method did not influence the content of phenolic compounds and total flavonoids, as well as the biological tests. The samples of *M. recutita* showed antioxidant and larvicidal activity. The Ac and Hid fractions exhibited potential antioxidant effects and showed no larvicidal effect, which may indicate low toxicity of these samples. Finally, we emphasize that further studies are needed to evaluate the toxic effect and antioxidant activity for safe therapeutic use of the plant.

Acknowledgments

The authors are grateful to the Universidade Federal de São João Del Rei (UFSJ), Conselho Nacional de Desenvolvimento Científico e Tecnológico (CNPq - process number: 303177/2017-9), Coordenação de Aperfeiçoamento de Pessoal de Nível Superior (CAPES) and Fundação de Amparo à Pesquisa do Estado de Minas Gerais (FAPEMIG) for the financial support. "This study was financed in part by the Coordenação de Aperfeiçoamento de Pessoal de Nível Superior - Brasil (CAPES) - Finance Code 001."

References

[1] Srivastava, J. K., Shankar, E., Gupta, S. (2010). Chamomile: a herbal medicine of the past with a bright future (Review). *Mol. Med. Rep.,* 3(6): 895-901.

[2] McKay, D. L., Blumberg, J. B. (2006). A review of the bioactivity and potential health benefits of chamomile tea (*Matricaria recutita* L.). *Phytother. Res.,* 20(7): 519-530.

[3] Evans, W. C. (2004). *Trease and Evans Pharmacognosy*. Edinburg: Sauders/Elsevier, 600 p.

[4] Hernández-Ceruelos, A., Madrigal-Santillán, E., Morales-González, J. A., Chamorro-Cevallos, G. Cassani-Galindo, M., Madrigal-Bujaidar, E. (2010). Antigenotoxic effect of *Chamomilla recutita* (L.) Rauschert essential oil in mouse spermatogonial cells, and determination of its antioxidant capacity *in vitro*. *I. J. Mol. Sci.,* 11(10): 3793-3802.

[5] Chandrashekhar, V. M., Halagali, K. S., Nidavani, R. B., Shalavadi, M. H., Biradar, B. S., Biswas, D., Muchchandi, I. S. (2011). Anti-allergic activity of german chamomile (*Chamomilla recutita* L.) in mast cell mediated allergy model. *J. Ethnopharmacol.,* 137(1): 336-340.

[6] Silva, N. C. C., Barbosa, L., Seito, L. N., Fernandes-Junior, A. (2012). Antimicrobial activity and phytochemical analysis of crude extracts and essential oils from medicinal plants. *Nat. Prod. Res.,* 26(16): 1510-1514.

[7] Matić, I. Z., Juranić, Z., Šavikin, K., Zdunić, G., Nadvinski, N., Goddevac, D. (2013). Chamomile and marigold tea: chemical characterization and evaluation of anticancer activity. *Phytother. Res.,* 27(6): 852-858.

[8] Petronilho, S., Maraschin, M., Coimbra, M. A., Rocha, S. M. (2012). *In vitro* and *in vivo* studies of natural products: a challenge for their evaluation. The case study of chamomile (*Chamomilla recutita* L.). *Ind. Crop. Prod.,* 40(1): 1-12.

[9] Cavalieri, E., Mariotto, S., Fabrizi, C., De Prati, A. C., Gottardo, R., Leone, S., Berra, L. V., Lauro, G. M., Ciampa, A. R., Suzuki, H. (2004). Alpha-bisabolol, a nontoxic natural compound, strongly induces apoptosis in glioma cells. *Biochem. Biophys. Res. Commun.,* 315(3): 589-594.

[10] Heldmaier, M., Beyer-Koschitzke, J., Stahl-Biskup, E. (2009). Oil extracts of herbal drugs - optimisation of the extraction parameters. *Pharmazie,* 64(6): 403-406.

[11] Cvetanović, A., Švarc-Gajić, J., Zeković, Z., Savić, S., Vulić, J., Mašković, P., Ćetković, G. (2015). Comparative analysis of antioxidant, antimicrobiological and cytotoxic activities of native and fermented chamomile ligulate flower extracts. *Planta,* 242(3): 721-732.

[12] Haghi, G., Hatami, A., Safaei, A., Mehran, M. (2014). Analysis of phenolic compounds in *Matricaria chamomilla* and its extracts by UPLC-UV. *Res. Pharm. Sci.,* 9(1): 31-37.

[13] Singh, O., Khanam, Z., Misra, N., Srivastava, M. K. (2011). Chamomile (*Matricaria chamomilla* L.): an overview. *Pharmacog. Rev.,* 5(9): 82-95.

[14] Wagner, H., Bladt, S. (2001). *Plant drug analysis: a thin layer chromatography Atlas.* Berlin: Springer, 384 p.

[15] Sousa, B. J. M., De Souza, A. O., Dos Reis, L. F., Gomes, A. J. P. S., Pedrosa, A. M., Almeida, J. M. D. (2018). Avaliação da qualidade de medicamentos magistrais e fitoterápicos à base de *Ginkgo biloba* e *Maytenus ilicifolia*. In: *Plantas medicinais: os avanços das pesquisas no Século XXI*. Divinópolis: Gulliver Editora, 131-146. [Quality evaluation of magisterial and herbal medicines based on Ginkgo biloba and Maytenus ilicifolia. In: *Medicinal plants: research advances in the 21st century*.]

[16] Mano-Sousa, B. J., Maia, G. A. S., Lima, P. L., Campos, V. A., Negri, G., Chequer, F. M. D., Duarte-Almeida, J. M. (2020). Color determination method and evaluation of methods for the detection of cannabinoids by thin-layer chromatography (TLC). *J. Forensic Sci.,* 66(3): 854-865.

[17] Da Costa, G. A. F., Morais, M. G., Saldanha, A. A., Silva, I. C. A. A., Aleixo, A. A., Ferreira, J. M. S., Soares, A. C., Duarte-Almeida, J. M., Lima, L. A. R. Dos S. (2015). Antioxidant, antibacterial, cytotoxic, and anti-inflammatory potential of the leaves of *Solanum lycocarpum* A. St. Hil. (Solanaceae). *Evid. Based Complement. Alternat. Med.,* 1-8.

[18] Zielinski, A. A. F., Haminiuk, C. W. I., Alberti, A., Nogueira, A., Demiate, I. M., Granato, D. (2014). A comparative study of the phenolic compounds and the *in vitro* antioxidant activity of different brazilian teas using multivariate statistical techniques. *Food Res. Int.,* 60: 246-254.

[19] Brasil. (2010). Ministério da Saúde. Agência Nacional de Vigilância Sanitária. *Farmacopeia Brasileira*. 5ª ed. ANVISA, Brasília, v. 2, 889 p. [Ministry of Health. National Health Surveillance Agency. *Brazilian Pharmacopoeia*]

[20] Araújo, S. G., Pinto, M. E. A., Silva, N. L., Santos, F. J. L., Castro, A. H. F., Lima, L. A. R. Dos S. (2013). Antioxidant and allelopathic activities of extract and fractions from *Rosmarinus officinalis*. *BBR - Biochem. Biotechnol. Rep.,* 2(1): 35-43.

[21] Brand-Willians, W., Cevulier, M. E., Brest, C. (1995). Use of free radical method to evaluate antioxidant activity. *LWT - Food Sci. Technol.*, 28: 25-30.

[22] Burda, S., Oleszek, W. (2001). Antioxidant and antiradical activities of flavonoids. *J. Agric. Food Chem.*, 49(6): 2774-2779.

[23] Meyer, B. N., Ferrigni, N. R., Putnan, J. E., Jacobsen, L. B., Nichols, D. E., McLaughlin, J. L. (1982). Brine shrimp: a convenient general bioassay for active plant constituents. *Planta. Med.*, 45(5): 31-34.

[24] Finney, D. J. (1980). *Probit analysis: a statistical treatment of the sigmoid response curve.* Cambridge: Cambridge University Press, 318 p.

[25] Ferreira, D. F. (2011). Sisvar: a computer statistical analysis system. *Ciênc. Agrotec.*, 35(6): 1039-1042.

[26] Gnoatto, S. C. B., Bassani, V. L., Coelho, G. C., Schenkel, E. P. (2007). Influência do método de extração nos teores de metilxantinas em erva-mate (*Ilex paraguariensis* A. St. -Hil., Aquifoliaceae). *Quím. Nova.*, 30(2): 304-307. [Influence of the extraction method on the levels of methylxanthines in yerba mate (Ilex paraguariensis A. St. -Hil., Aquifoliaceae). *Chem. New*]

[27] Karabegović, I. T., Stojičević, S. S., Veličković, D. T., Todorović, Z. B., Nikolić, N. Č., Lazić, M. L. (2014). The effect of different extraction techniques on the composition and antioxidant activity of cherry laurel (*Prunus laurocerasus*) leaf and fruit extracts. *Ind. Crop. Prod.*, 54: 142-148.

[28] Shirsath, S. R., Sonawane, S. H., Gogate, P. R. (2012). Intensification of extraction of natural products using ultrasonic irradiations - A review of current status. *Chem. Eng. Process.*, 53: 10-23.

[29] Da Costa, J. C. F., Hoscheid, J. (2018). Perfil fitoquímico e avaliação da atividade antimicrobiana de extratos aquoso e etanólico de folhas de *Cecropia pachystachya*. *Rev. Fitos*, 12(2): 175-185.

[30] Cosmetic Ingredient Review Expert Panel. (2014). Safety assessment of *Chamomilla recutita* - derived ingredients as used in cosmetics. *Draft Tentative Report for Panel Review*, 1-45.

[31] Fajemiroye, J. O., Ferreira, N. L., De Oliveira, L. P., Elusiyan, C. A., Pedrino, G. R., Da Cunha, L. C., Da Conceição, E. C. (2016). *Chamomilla recutita* and its isolate-apigenin: economic value, ethnopharmacology and chemico-biological profiles in retrospect. *J. Pharmacog. Phytochem.,* 4(4): 17-31.

[32] Caleja, C., Barros, L., Oliveira, M. B. P. P., Santos-Buelga, C., Ferreira, I. C. F. R. (2017). Characterization of the phenolic profile of aqueous extracts of *Chamomilla recutita* L. obtained by decoction. *Rev. Ciên. Agr.,* 40: S136-S139.

[33] Dourado, R. S., Ladeira, A. M. L. (2008). Identification of flavonoids in *Hypericum cordatum* (Vell.) N. Robson (Clusiaceae). *Rev. Bras. Bot.,* 31(4): 611-620.

[34] De Mira, N. V. M., Barros, R. M. C., Schiocchet, M. A., Noldin, J. A., Lanfer-Marquez, U. M. (2008). Extraction, analysis and distribution of phenolic acids in pigmented and non-pigmented rice enotypes (*Oryza sativa* L.). *Ciên. Tecnol. Alim.,* 28(4): 994-1002.

[35] Sakakibara, H., Honda, Y., Nakagawa, S., Ashida, H., Kanazawa, K. (2003). Simultaneous determination of all polyphenols in vegetables, fruits, and teas. *J. Agric. Food Chem.,* 51(3): 571-581.

[36] Guimarães, R., Barros, L., Dueñas, M., Calhelha, R. C., Carvalho, A. M., Santos-Buelga, C., Queiroz, M. J. R. P., Ferreira, I. C. F. R. (2013). Infusion and decoction of wild german chamomile: bioactivity and characterization of organic acids and phenolic compounds. *Food Chem.,* 136(2): 947-954.

[37] Gupta, V., Mittal, P., Bansal, P., Khokra, S. K., Kaushik, D. (2010). Pharmacological potential of *Chamomilla recutita* - A Review. *I. J. Pharm. Sci. Drug Res.,* 2(1): 12-16.

[38] Mulinacci, N., Romani, A., Pinelli, E., Vincieri, F., Prucher, D. (2000). Characterization of *Chamomilla recutita* L. flower extracts by HPLC-MS and HPLC-DAD analysis. *Chromatographia,* 51(5): 301-307.

[39] Sebai, H., Jabri, M. A., Souli, A., Rtibi, K., Selmi, S., Tebourbi, O., Benna, J. E., Sakly, M. (2014). Antidiarrheal and antioxidant activities of chamomile (*Chamomilla recutita* L.) decoction extract in rats. *J. Ethnopharmacol.,* 152(2): 327-332.

[40] Dumitriu, D., Peinado, R. A., Peinado, J., De Lerma, N. L. (2015). Grape pomace extract improves the *in vitro* and *in vivo* antioxidant properties of wines from sun light dried Pedro Ximénez grapes. *J. Funct. Foods,* 17: 380-387.

[41] Asadi, S., Ahmadiani, A., Esmaeili, M. A., Sonboli, A., Ansari, N., Khodagholi, F. (2010). *In vitro* antioxidant activities and an investigation of neuroprotection by six *Salvia* species from Iran: a comparative study. *Food Chem. Toxicol.,* 48(5): 1341-1349.

[42] Al-Dabbagh, Bayan., Elthaty, I. A., Elhaw, M., Murali, C., Al-Mansoori, A., Awad, B., Amin, A. (2019). Antioxidant and anticancer activities of chamomile (*Chamomilla recutita* L.). *BMC Res. Notes,* 12(3).

[43] Roozbeh, F., Dong-Jin, L. (2017). Chemical constituents and antioxidant properties of *Chamomilla recutita* and *Chamaemelum nobile* essential oil growing in South West of Iran. *Free Rad. Bio. Med.,* 108: S24.

[44] Hajaji, S., Sifaoui, I., López-Arencibia, A., Reyes-Batlle, M., Jiménez, I. A., Bazzocchi, I. L., Valladares, Pinero, J. E., Lorenzo-Morales, J., Akkari, H. (2017). Correlation of radical-scavenging capacity and amoebicidal activity of *Chamomilla recutita* L. (Asteraceae). *Exp. Parasitol.,* 183: 212-217.

[45] Kolodziejczyk-Czepas, J., Bijak, M., Saluk, J., Ponczek, M. B., Zbikowska, H. M., Nowak, P., Tsirigotis-Maniecka, M., Pawlaczyk, I. (2015). Radical scavenging and antioxidant effects of *Matricaria chamomilla* polyphenolic-polysaccharide conjugates. *I. J. Biol. Macromol.,* 72: 1152-1158.

[46] Dolabela, M. F. (1997). *Triagem in vitro para atividade antitumoral e anti-Tripanossoma cruzi de extratos vegetais, produtos naturais e substâncias sintéticas.* Master's Thesis, Biology Sciences Institute, Federal University of Minas Gerais, 128

p. [*In vitro screening for antitumor and anti-Trypanosoma cruzi activity of plant extracts, natural products and synthetic substances*.]

[47] Satyal, P., Shrestha, S., Setzer, W. N. (2015). Composition and bioactivities of an (*E*)-*β*-Farnesene chemotype of chamomile (*Matricaria chamomilla*) essential oil from Nepal. *Nat. Prod. Commun.*, 10(8): 1453-1457.

In: The Book of Flavonoids
Editors: Chao-Hui Feng et al.
ISBN: 978-1-68507-376-3

Chapter 4

EVALUATION OF ANTIOXIDANT AND TOXICITY POTENTIAL OF PHENOLIC AND FLAVONOID COMPOUNDS FROM *PSIDIUM GUAJAVA* L. LEAF EXTRACTS

***Lucas Santos Azevedo*[1], *Carlos Ananias Aparecido Resende*[2], *Nathália Hiratsuka Camilo*[1], *Adriano Lopes da Silva*[2], *Paula Lamounier Lima*[2], *Brayan Jonas Mano-Sousa*[1], *Joaquim Maurício Duarte-Almeida*[1,2] and *Luciana Alves Rodrigues dos Santos Lima*[1,2,*]**

[1]Programa de Pós-Graduação em Biotecnologia da Universidade Federal de São João Del Rei, Campus Centro-Oeste Dona Lindu, Divinópolis, MG, Brazil

[2]Programa de Pós-Graduação em Ciências Farmacêuticas da Universidade Federal de São João Del Rei, Campus Centro-Oeste Dona Lindu, Divinópolis, MG, Brazil

[*] Corresponding Author's E-mail: luarsantos@ufsj.edu.br.

Abstract

Guava tree (*Psidium guajava* L.) is widely used by the population due to its therapeutic potential, which is attributed to the various phytochemical compounds present in this species. The aim of this work was to evaluate the extraction efficiency of phenolic compounds and flavonoids obtained from *P. guajava* fresh leaves by two different extraction methods (soxhlet [SXT] and ultrasound assisted extraction [UAE]), as well as to evaluate the antioxidant and larvicidal potential of the extracts and their fractions. The extracts were obtained by extraction in a soxhlet apparatus (SXT) and extraction assisted by ultrasound (UAE) with ethanol. The fractions were obtained by partitioning the extracts with solvents of increasing polarity: hexane, dichloromethane, and ethyl acetate. In the samples (SXT and UAE) analyzed by high performance liquid chromatography with diode array detector (HPLC-DAD), it was possible to suggest the presence of phenolic compounds and flavonoids, such as gallic acid isomers and quercetin derivatives. The extraction method directly influenced the extract yield, with the SXT method being more efficient. However, there was no significant influence of the extraction method on the antioxidant activity using 1,1-diphenyl-2-picryl-hydrazil (DPPH) radicals and toxicity on *Artemia salina* (brine shrimp). The antioxidant activity of extracts and fractions cannot be correlated with the total phenolic compounds and total flavonoids content. The ethanol extracts showed the highest levels of phenolic and flavonoid compounds, which corroborate the results obtained in the identification by HPLC-DAD analysis. The extracts of *P. guajava* fresh leaves did not show toxicity on *A. salina* except for the ethyl acetate fraction (UAE).

Keywords: guava leaves, extractive methods, medicinal plants, brine shrimp

1. Introduction

Psidium guajava Linn, popularly known as "goiabeira" in Brazil, is a small tree belonging to the Myrtaceae family, which is native to Central America and found mainly in tropical regions [1]. Roots, bark, leaves, and fruits from *P. guajava* are widely used in traditional medicine against anorexia, cholera, diarrhea, digestive problems, cutaneous inflammation, and sore throat, among others [2, 3].

Secondary metabolites of various classes have already been isolated from *P. guajava* leaves, such as quercetin, myricetin, guavinoside, and guavacoumaric acid, for example [2, 4]. Studies using its leaves showed antioxidant, anti-inflammatory, hepatoprotective, spasmolytic, and *in vitro* antitumor effects [2]. There are several methods of obtaining plant extract, which influence the extract yield and molecular profile. In addition, this difference can interfere with the chemical and biological effects of these extracts [5].

Thus, the objective of this work was to compare a hot [soxhlet extraction (SXT)] and a cold [ultrasound assisted extraction (UAE)] method, evaluating the yield, antioxidant and toxic potential of the ethanol extracts and fractions from *P. guajava* leaves.

2. Materials and Methods

2.1. Plant Material and Extraction

Psidium guajava L. fresh leaves were collected in Federal University of São João Del Rei, *Campus* Centro-Oeste Dona Lindu, in Divinópolis City, Minas Gerais State, Brazil, in August 2019. Fertile samples were collected and the vouchers were identified by Andréia Fonseca Silva, and deposited in the PAMG Herbarium (PAMG 58809) at the Agricultural Research Company of Minas Gerais (EPAMIG). This study has access permission to the components of plant genetic heritage registered in the SisGen Platform (Register A901EF7), according to Brazilian Biodiversity Law (13.123/2015).

The material was extracted by ultrasound assisted extraction (UAE) and soxhlet extraction (SXT) methods, using 38.5 g of fresh leaves and 400 mL of ethanol 96 °GL in both methods. UAE was performed at room temperature (27 + 2 °C), being extracted in ultrasound equipment for 30 minutes, followed by rest for 30 minutes, totaling 5 hours. SXT was carried out for 5 hours on heating at 65°C without intervals. After filtration, the solvent was removed using rotatory evaporator, obtained 2.78 g of EE (UAE) and 4.32 g of EE (SXT). The yield obtained in the extraction processes was calculated from the equation: Yield = (fresh leaves weight (g)/extract weight (g)) x 100.

Part of the ethanol extracts (EE) were solubilized in ethanol:water (7:3) and partitioned with hexane, dichloromethane and ethyl acetate to obtain hexane (Hex), dichloromethane (Dcm) and ethyl acetate (EAc) fractions. The fractions were filtered and solvents removed using rotatory evaporator.

2.2. Phytochemical Trial

The presence of secondary metabolites was evaluated by specific experimental in assays tubes. Steroids and triterpenoids were evaluated by Lieberman-Burchard reaction. Flavonoids were evaluated with change the color to red, yellow or orange after addition of 13 sulfuric acid (H_2SO_4) drops. Saponins presence was evaluated with observation of persistent foam after 30 min. of intense shake with water. Tannins presence was evaluated with precipitate formation after addiction of 3 Iron (II) chloride ($FeCl_2$) drops. Alkaloids were evaluated by Dragendorff reaction. Coumarins and anthraquinones were evaluated with change the color to yellow or purple after addiction of 5 sodium hydroxide (NaOH) (1 mol/L) drops, respectively [5].

2.3. Thin-Layer Chromatography

The evaluation of the presence of essential oils was performed by thin-layer chromatography (TLC) on Merck silica gel 60 F_{254} aluminium plates, using toluene/ethyl acetate (93:7 v:v) as eluent, vanillin and sulfuric acid followed by heating at 100°C for 10 minutes was used as developer. The evaluation of the presence of phenolic compounds was also performed by TLC using ethyl acetate/acetic acid/water (100:22:26 v:v:v) as eluent, diphenylboric acid-*β*-aminoethyl ester (NP)/ polyethylenoglycol-4000 (PEG) (NP/PEG) solution was used as developer with $\lambda = 254$ and 365 nm [6].

2.4. High Performance Liquid Chromatography Analysis

The chromatographic analysis was conducted using UFLC Proeminence chromatographic system (Shimadzu, Kyoto, Japan), composed of a binary pump system (LC-20AD), autosampler (SIL-20AHT), communicator (CBM-20A) and degasser, controlled by the LabSolutions software (version 1.25, Shimadzu, Kyoto, Japan). The experiments were carried out in an air-conditioned room (14 ± 2 °C). The chromatographic column was a Kinetex C_{18} (5 µm, 100 x 2.1 mm, Phenomenex), eluted with methanol (solvent B) and ultrapure water added formic acid 0.1% (v:v). The elution profile applied was: 0-8.5 min: 10 to 90% of B; 9-10 min: 90% of B and; 10-11 min: 10% of B. All samples (rutin, extracts and fractions) were solubilized (1 mg/mL) and filtered through a 0.45 mm PTFE filter syringe. The injection volume was 10 µL of samples (rutin, extracts and fractions), and the flow rate was 0.5 mL/min. The spectra were recorded in the range of 200 to 600 nm. Compound determination was performed comparing the retention times and spectrum in the UV region of the standards, and correlating with data found in the literature, according to Sousa et al. [7].

2.5. Total Phenolic Content (TPC)

The total phenolic content was estimated using the Folin-Ciocalteu test [8] with modifications. The Folin-Ciocalteu aqueous solution (2.250 µL; 1:4 v:v) was added to samples (250 µL) and, subsequently, sodium carbonate (Na_2CO_3) solution (250 µL) was added. After vigorous shaking, these solutions were kept at rest for 30 min at room temperature.

The absorbance was determined by spectrophotometry at 750 nm (Thermo Scientific Genesys 10S, USA) after 30 min of incubation at room temperature with a blank sample as well as a standard solution and samples. Gallic acid was used as a reference compound, and the total phenolic content were expressed as micrograms of gallic acid equivalents per milliliter (mL). All assays were performed in triplicate.

2.6. Total Flavonoid Content (TFC)

The total flavonoid content was estimated according to the aluminum chloride ($AlCl_3$) method [9]. Exactly 1900 μL of ethanol 50% (v:v) were added 100 μL of extracts and fractions and 500 μL of solution of aluminum chloride. The absorbance was read at 425 nm using a spectrophotometer (Thermo Scientific Genesys 10S, USA) after 30 min. Quercetin was used as a reference compound to produce a standard curve, and total flavonoid content were expressed as micrograms of quercetin equivalents per milliliter (mL). All assays were performed in triplicate.

2.7. DPPH Radical Scavenging Assay

The radical scavenging abilities of extracts and fractions were analyzed compared to standard, 2,6-di-tert-butyl-4-methylphenol (BHT) based on the reaction with 1,1-diphenyl-2-picrylhydrazyl radical (DPPH). The determination of antioxidant activity using the DPPH method was adapted for use with microplates [10, 11]. Briefly, a DPPH solution (0.002% w:v) was prepared in ethanol. Exactly 75 μL of the samples (1, 10, 100, 250 and 500 μg/mL) were added to the wells in a 96-well flat-bottom plate containing 150 μL of DPPH solution. The plate was then covered and left in the dark at room temperature (25°C).

After 30 min, the absorbance at 517 nm was measured with a spectrophotometer (Biotek Power Wave XS2, USA), and ethanol was used for the baseline correction. Scavenging ability was expressed as the inhibition percentage and was calculated using Burda and Oleszek's [12] equation: Scavenging ability = $\{(Abs_{control} - Abs_{sample})/Abs_{control}\}$ x 100, where $Abs_{control}$ = absorbance of DPPH radical in ethanol and Abs_{sample} = absorbance of the extract or fractions in ethanol + DPPH. The antioxidant activity of all of the samples was expressed as IC_{50}, which was defined as a concentration (μg/mL) of samples required to inhibit the formation of 50% DPPH radicals. All assays were performed in triplicate.

2.8. *Artemia salina* Larvicide Bioassay

Artemia salina eggs were incubated in 400 mL of seawater (0,5 mg/mL) under artificial light at 28°C, pH 7-8. After incubation for 48 h, 10 nauplii were transferred to assay tubes with 10 mL of the samples (in triplicate) in 1000, 500, 250 and 125 μg/mL concentrations. Negative controls containing seawater and 1, 0.5, 0.25 and 0.125 percent of DMSO were included in each experiment [13]. The number of survivors was counted after 24 h and recorded.

2.9. Statistical Analysis

The lethal dose (LD_{50}) was calculated by simple linear regression using Probit analysis [14].

The data were subjected to an analysis of variance (ANOVA) followed by Tukey Test at a 5% probability level, using ANOVA for balanced data. All statistical parameters were calculated using GraphPad Prism® 7.0 and only $p < 0.05$ were considered significant.

3. Results and Discussion

3.1. Extraction Methods and Phytochemical Trial

The extraction methods used in this work showed different yields for the ethanol extracts, with 11.19% by SXT and 7.22% by UAE. The results obtained suggest that the extraction method directly influences the yield. The SXT method demonstrated higher extraction efficiency, with similar results as in literature. However, a dried leaves re-maceration showed higher yield (36.91%) than both these methods [15, 16, 17].

Phytochemical screening (Table 1) showed the presence of alkaloids in all samples, tannins in the ethyl acetate (EAc) fraction, coumarins in the hexane (Hex), dichloromethane (Dcm), and EAc fractions from SXT extraction, while UAE extraction showed flavonoids in the EE and Dcm fraction, alkaloids in the EE and Hex fraction, steroids in the Hex fraction, and coumarins in the Dcm and EAc fractions.

Table 1. Phytochemical screening of ethanol extract and fractions from *P. guajava* L. leaves

Secondary metabolites groups	EE	Hex	Dcm	EAc
	Ultrasound assisted extraction			
Steroids	-	+	-	-
Triterpenoids	-	-	-	-
Flavonoids	+	-	+	-
Tannins	-	-	-	-
Alkaloids	+	+	-	-
Coumarins	+	-	+	+
Anthraquinones	-	-	-	-
Saponins	-	-	-	-
	Soxhlet extraction			
Steroids	-	-	-	-
Triterpenoids	-	-	-	-
Flavonoids	-	-	-	-
Tannins	-	-	-	+
Alkaloids	+	+	+	+
Coumarins	-	+	+	+
Anthraquinones	-	-	-	-
Saponins	-	-	-	-

Ethanol extract (EE), hexane (Hex), dichloromethane (Dcm) and ethyl acetate (EAc) fractions. (+) present, (-) absent.

3.2. Thin-Layer Chromatography and HPLC-DAD Analyses

Thin-layer chromatography showed the presence of flavonoid and phenolic compounds in Dcm and EAc fractions of both extraction methods. These results are corroborated with those obtained with high performance liquid chromatography with diode array detector (HPLC-DAD) (Figure 1) that suggest the presence of phenolic compounds and flavonoids (such as gallic acid isomers and quercetin derivatives) and literature that show a significant presence of quercetin derivatives and phenolic acids [2, 4]. Essential oils were not detected in the evaluated samples.

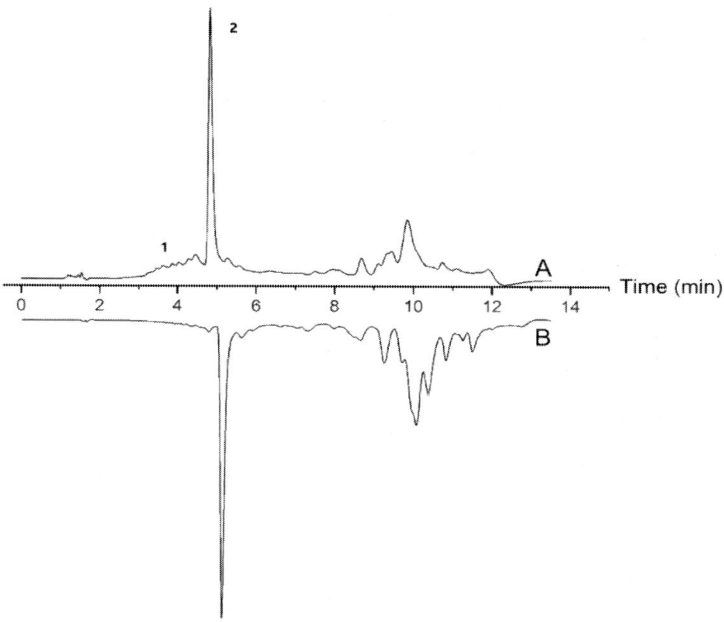

Figure 1. Chromatographic profile of *Psidium guajava* ethanol extracts in high performance liquid chromatography with diode array detector (HPLC-DAD), 350 nm. Soxhlet extract (SXT); (B) Ultrasound assisted extract (UAE). Suggested compounds: **1.** gallic acid isomers; **2.** Quercetin derivatives.

3.3. Total Flavonoid and Total Phenolic Contents

The assessments of total flavonoid (TFC) and total phenolic content (TPC) were performed using the reagents aluminum chloride and Folin Ciocalteu, respectively (Table 2). The results of these assays showed that the samples shared similar ranking (EE > Dcm > EAc), except for total flavonoids by samples obtained from UAE extraction.

Table 2. Total flavonoid and total phenolic content of ethanol extracts and fractions from *P. guajava* L. leaves

Sample	Ultrasound assisted extraction		Soxhlet extraction	
	TFC[1]	TPC[2]	TFC[1]	TPC[2]
EE	9.03 ± 1.17	20.17 ± 0.63	11.06 ± 0.84	16.05 ± 3.29
Hex	-	-	-	-
Dcm	0.68 ± 0.1	6.39 ± 1.68	3.85 ± 0.17	4.945 ± 0.44
EAc	3.45 ± 0.7	2.36 ± 0.11	0.35 ± 0.04	4.045 ± 0.44

Ethanol extract (EE), hexane (Hex), dichloromethane (Dcm) and ethyl acetate (EAc) fractions. [1]Total flavonoid content: results expressed as μg of quercetin equivalents/mL of extract or fraction. [2]Total phenolic content: results expressed as μg of gallic acid equivalents/mL of extract or fraction. Each value in the table is the mean ± standard deviation (n = 3), (-) not determined.

The highest concentration of TFC and TPC were exhibited by EE (SXT) and EE (UAE), respectively, with TPC = 11.06 μg/mL quercetin equivalent and TPC = 20.17 μg/mL gallic acid equivalent, respectively. In contrast, other studies showed high doses of TPC in the ethanol fraction (146.7 mg/g gallic acid equivalent) and ethyl acetate fraction (99.6 mg/g gallic acid equivalent) obtained from dried leaves [18]. Similarly, the dried leaves re-maceration ethanolic extract showed TFC of 118.90 mg/g quercetin equivalent [15]. With these results, it is observed that samples obtained from dry leaves have higher concentrations of phenolic compounds and flavonoids than those obtained with fresh leaves, as in the present study.

3.4. Evaluation of Antioxidant Activity

The antioxidant potential of the extracts and fractions (Table 3) was similar in both extraction methods. All samples showed IC_{50} (0.98 - 5.29 µg/mL), which were lower than the BHT (16.36 µg/mL). Among them, EAc showed the lowest IC_{50} (0.98 µg/mL for SXT and 0.99 µg/mL for UAE). No correlation was observed between the content of total phenolic compounds and total flavonoids with antioxidant activity.

Table 3. DPPH scavenging activity and IC_{50} values for the antioxidant activity of ethanol extracts and fractions from *P. guajava* L. leaves

Sample	DPPH-scavenging activity (%)					IC_{50}
	1 µg/mL	10 µg/mL	100 µg/mL	250 µg/mL	500 µg/mL	
	Ultrasound assisted extraction					
EE	49.31 ± 0.48 [a]	67.97 ± 1.55 [a]	96.29 ± 0.74 [a]	97.75 ± 0.56 [a]	97.90 ± 1.21 [a]	1.42 ± 0.07 [a]
Hex	49.65 ± 0.80 [a]	54.11 ± 0.53	81.68 ± 1.62	98.77 ± 0.70 [a]	99.07 ± 0.80 [a]	1.86 ± 0.25 [a]
Dcm	49.63 ± 1.00 [a]	62.66 ± 0.28 [a]	95.81 ± 1.47 [a]	98.23 ± 0.27 [a]	98.55 ± 1.27 [a]	1.57 ± 0.11 [a]
EAc	50.58 ± 0.00 [a]	61.51 ± 1.75 [a]	95.99 ±0.70 [a]	97.23 ± 0.92 [a]	98.77 ± 1.75 [a]	0.99 ± 0.09 [a]
BHT	18.50 ± 0.65	25.90 ± 0.64	86.00 ± 0.56	91.40 ±0.28	94.02 ± 0.51	16.36 ± 1.63
	Soxhlet extraction					
EE	47.37 ± 1.28 [a]	60.25 ± 2.01 [a]	96.14 ± 0.00 [a]	97.10 ± 1.28 [a]	98.39 ± 1.01 [a]	1.88 ± 0.17 [a]
Hex	48.27 ± 1.39 [a]	53.35 ± 1.60 [a]	73.21 ± 2.11 [a]	90.92 ± 1.48	96.15 ± 3.47	5.29 ± 0.39 [a]
Dcm	48.02 ± 1.22 [a]	60.57 ± 1.55 [a]	95.81 ± 1.00 [a]	96.94 ± 0.28 [a]	98.07 ± 0.96 [a]	1.80 ± 0.13 [a]
EAc	50.89 ± 0.53 [a]	65.82 ± 0.80 [a]	94.76 ± 1.62 [a]	97.38 ± 0.53 [a]	98.15 ± 1.66 [a]	0.98 + 0.08 [a]
BHT	18.50 ± 0.65	25.90 ± 0.64	86.00 ± 0.56	91.40 ±0.28	94.02 ± 0.51	16.36 ± 1.63

IC_{50}: concentration (µg/mL) of samples required to inhibit the formation of 50% DPPH radicals. Each value in the table is the mean ± standard deviation (n = 3). $^{a}p < 0.05$ compared with BHT. 2,6-di-tert-butyl-4-methylphenol (BHT), ethanol extract (EE), hexane (Hex), dichloromethane (Dcm) and ethyl acetate (EAc) fractions.

Studies of antioxidant activity with *P. guajava* have already been reported. Methanol, chloroform, and hexane dried leaves extracts presented IC_{50} of 89.82, 211.1, and 426.8 μg/mL, respectively [19]. The ethyl acetate fraction dried leaves showed an IC_{50} value between 40 - 80 μg/mL, while the ethanol fraction was between 20 - 40 μg/mL [18]. The dried leaves re-maceration ethanolic extract exhibited antioxidant activity of 87.65% at 13.3 mg/mL [15]. The essential oil from *P. guajava* bark demonstrated antioxidant activity of 19.33% and 71.83% at concentrations of 0.1 and 0.2 mg/mL, respectively [20]. Thus, the samples tested in this work demonstrated higher antioxidant activity when compared to other studies.

3.5. *Artemia salina* Toxicity

P. guajava leaves samples were evaluated on *Artemia salina* nauplii. Only the EAc fraction obtained from the UAE method showed lethality (Table 4). The EAc fraction exhibited mortality of 23.33% and 66.66% at concentrations of 500 and 1000 μg/mL, respectively, on *A. salina*. According to Meyer and collaborators [13], EAc exhibited toxicity.

Table 4. Larvicidal activity of different concentration of ethanol extracts and fractions from *P. guajava* L. leaves

Sample	% Mortality Concentration (μg/mL)			
	125 μg/mL	250 μg/mL	500 μg/mL	1000 μg/mL
	Ultrasound assisted extraction			
EE	0.0 ± 0.0	0.0 ± 0.0	0.0 ± 0.0	0.0 ± 0.0
Hex	0.0 ± 0.0	0.0 ± 0.0	0.0 ± 0.0	0.0 ± 0.0
Dcm	0.0 ± 0.0	0.0 ± 0.0	0.0 ± 0.0	0.0 ± 0.0
EAc	0.0 ± 0.0	0.0 ± 0.0	23.33 ± 1.15	66.66 ± 0.57
CTR	0.0 ± 0.0	0.0 ± 0.0	0.0 ± 0.0	0.0 ± 0.0
	Sohxlet extraction			
EE	0.0 ± 0.0	0.0 ± 0.0	0.0 ± 0.0	0.0 ± 0.0
Hex	0.0 ± 0.0	0.0 ± 0.0	0.0 ± 0.0	0.0 ± 0.0
Dcm	0.0 ± 0.0	0.0 ± 0.0	0.0 ± 0.0	0.0 ± 0.0
EAc	0.0 ± 0.0	0.0 ± 0.0	0.0 ± 0.0	0.0 ± 0.0
CTR	0.0 ± 0.0	0.0 ± 0.0	0.0 ± 0.0	0.0 ± 0.0

Control group (CTR), ethanol extract (EE), hexane (Hex), dichloromethane (Dcm) and ethyl acetate (EAc) fractions.

Few studies report the toxicity of *P. guajava* on *A. salina*. The essential oil from *P. guajava* barks showed high toxicity on *A. salina* (LC_{50} = 1.00 μg/mL) [20]. Aqueous extracts of dried leaves and barks showed LC_{50} values of 480.14 and 949.13 μg/mL, respectively, which indicates that the leaves are less toxic than the barks [21], which corroborates with our study. Methanol, chloroform, and hexane dried leaves extracts presented LC_{50} values of 63.81, 41.05, and 32.18 μg/mL, respectively [19]. Comparing the results, it is observed that the samples obtained from fresh leaves have less toxicity on *A. salina*.

CONCLUSION

The extraction method directly influenced the extract yield, with the SXT technique being more efficient. However, there was no significant influence of the extraction method on the antioxidant activity using DPPH radicals and toxicity on *Artemia salina* (brine shrimp). The antioxidant activity of extracts and fractions cannot be correlated with the total phenolic compounds and total flavonoids content. The ethanol extracts showed the highest levels of phenolic and flavonoid compounds, which corroborate the results obtained in the identification by HPLC-DAD analysis. The samples of *P. guajava* fresh leaves did not show toxicity on *A. salina* except for the ethyl acetate fraction (UAE).

ACKNOWLEDGMENTS

The authors are grateful to the Universidade Federal de São João Del Rei (UFSJ), Conselho Nacional de Desenvolvimento Científico e Tecnológico (CNPq), Coordenação de aperfeiçoamento de Pessoal de Nível Superior (CAPES) and Fundação de Amparo à Pesquisa do Estado de Minas Gerais (FAPEMIG) for the financial support. This study was financed in part by the Coordenação de Aperfeiçoamento de Pessoal de Nível Superior – Brasil (CAPES) – Finance code 001".

References

[1] Ravi, K., Divyashree, P. (2014). *Psidium guajava*: A review on its potential as na adjunct in treating periodontal disease. *Phcog. Rev.*, 8(16): 96-100.

[2] Gutiérrez, R. M. P., Mitchell, S., Solis, R. V. (2008). *Psidium guajava*: A review of its traditional uses, phytochemistry and pharmacology. *J. Ethnopharmacol.*, 117: 1-27.

[3] Leonti, M., Vibrans, H., Sticher, O., Heinrich, M. (2001). Ethnopharmacology of the Popoluca, Mexico: an evaluation. *J. Pharm. Pharmacol.*, 53: 1653-1669.

[4] Feng, X. H., Wang, Z. H., Meng, D. L., Li, X. (2015). Cytotoxic and antioxidant constituents from the leaves of *Psidium guajava*. *Bioorg. Med. Chem. Lett.*, 25: 2109-2198.

[5] Simões, C. M. O., Schenkel, E. P., Gosmann, G., De Melo, J. C. P., Mentz, L. A., Petrovick, P. R. (2007). *Farmacognosia: da planta ao medicamento.* Porto Alegre: UFGRS, 1104 p.

[6] Wagner, H., Bladt, S. (2001). *Plant Drug Analyses: A Thin Layer Chromatography Atlas.* Berlin: Springer, 384 p.

[7] Sousa, B. J. M., Souza, A. O., Reis, L. F., Gomes, A. J. P. S., Pedrosa, A. M., Almeida, J. M. D. (2018). Avaliação da qualidade de medicamentos magistrais e fitoterápicos à base de *Ginkgo biloba* e *Maytenus ilicifolia*. In: *Plantas medicinais: os avanços das pesquisas no Século XXI.* Divinópolis: Gulliver Editora, 131-146. [Quality evaluation of magisterial and herbal medicines based on Ginkgo biloba and Maytenus ilicifolia. In: *Medicinal plants: research advances in the 21st century*.]

[8] Zielinski, A. A. F., Haminiuk, C. W. I., Alberti, A., Nogueira, A., Demiate, I. M., Granato, D. (2014). A comparative study of the phenolic compounds and the *in vitro* antioxidant activity of

different Brazilian teas using multivariate statistical techniques. *Food. Res. Int.*, 60: 246-254.

[9] Brasil. (2010). Ministério da Saúde. Agência Nacional de Vigilância Sanitária. *Farmacopeia Brasileira.* 5ª ed. ANVISA, Brasília, v. 2, 889 p. [Ministry of Health. National Health Surveillance Agency. *Brazilian Pharmacopoeia*]

[10] Brand-Williams, W., Cuvelier, M. E., Berset, C. (1995). Use of a free radical method to evaluate antioxidant activity. *LWT - Food. Sci. Technol.*, 28(1): 25-30.

[11] Araújo, S. G., Pinto, M. E. A., Silva, N. L., Santos, F. J. L., Castro, A. H. F., Lima, L. A. R. Dos S. (2013). Antioxidant and allelopathic activities of extract and fractions from *Rosmarinus officinalis. BBR - Biochem. Biotechnol. Rep.,* 2(1): 35-43.

[12] Burda, S., Oleszek, W. (2001). Antioxidant and antiradical activities of flavonoids. *J. Agric. Food. Chem.*, 49(6): 2774-2779.

[13] Meyer, B. N., Ferrigni, N. R., Putnam, J. E., Jacobsen, L. B., Nichols, D. E., McLaughlin, J. L. (1982). Brine shrimp: A convenient general bioassay for active plant constituents. *Planta. Med.*, 45(5): 31-34.

[14] Finney, D. J. (1980). *Probit Analysis. A Statistical Treatment of the Sigmoid Response Curve.* Cambridge: Cambridge University Press, 318 p.

[15] Braga, T. V., Dores, R. G. R., Ramos, C. S., Evangelista, C. G., Tinoco, L. M. S., Varotti, F. P., Carvalho, M. G., Sabino, A. P. (2014). Antioxidant, antibacterial and antitumor activity of ethanolic extract of the *Psidium guajava* leaves. *Amer. J. Plant. Sci.,* 5: 3492-3500.

[16] Karabegović, I. T., Stojičević, S. S., Veličković, D. T., Todorović, Z. B., Nikolić, N. Č., Lazić, M. L. (2014). The effect of different extraction techniques on the composition and antioxidant activity of cherry laurel (*Prunus laurocerasus*) leaf and fruit extracts. *Ind. Crop. Prod.*, 54: 142-148.

[17] Markom, M., Hasan, M., Daud, W. R. W., Singh, H., Jahin, J. M. (2007). Extraction of hydrolysable tannins from *Phyllanthus niruri*

Linn.: Effects of solvents and extraction methods. *Separat. Purif. Technol.*, 52: 487-496.

[18] Zahin, M., Ahmad, I., Aqil, F. (2017). Antioxidant and antimutagenic potential of *Psidium guajava* leaf extracts. *Drug. Chem. Toxicol.*, 40(2): 146-153.

[19] Ashraf, A., Sarfraz, R. A., Rashid, M. A., Mahmood, A., Shahid, M., Noor, N. (2016). Chemical composition, antioxidant, antitumor, anticancer and cytotoxic effects of *Psidium guajava* leaf extracts. *Pharm. Biol.,* 54(10): 1971-1981.

[20] Fasola, T. R., Oloyede, G. K., Aponjolosun, B. S. (2011). Chemical composition, toxicity and antioxidant activities of essenctial oils of stem bark of nigerian species of guava (*Psidium guajava* Linn.). *Excli J.*, 10: 34-43.

[21] Bautista, M. L., Pomer, P., Salonga, J. J. (2018). Comparison of toxic effects of *Psidium guajava* leaf and bark extracts against Brine Shrimp (*Artemia salina*). *Int. J. Agr. Technol.*, 14(7): 1033-1038.

In: The Book of Flavonoids
Editors: Chao-Hui Feng et al.
ISBN: 978-1-68507-376-3

Chapter 5

BIOLOGICAL ACTIVITY OF PHENOLIC COMPOUNDS FOUND IN *TECOMA STANS* (L.) JUSS. EX KUNTH (BIGNONIACEAE)

***Thaís Paula Rodrigues Gonçalves*[1,2,*], *Adriano Lopes da Silva*[1,3,*], *Iasmin Fernandes Emediato*[1,*], *Jéssica Elvira Siqueira Medeiros*[1], *Iara Cotta Mendes*[1], *Paula Avelar Amado*[1,2], *Izabela Caputo Assis Silva*[1,2], *Lucas Santos Azevedo*[1,2] and *Luciana Alves Rodrigues dos Santos Lima*[1,2,3,†]**

[1]Phytochemistry Laboratory,
Universidade Federal de São João Del Rei, Divinópolis, Brazil
[2]Programa de Pós-Graduação em Biotecnologia da Universidade Federal de São João Del Rei, Divinópolis, Brazil
[3]Programa de Pós-Graduação em Ciências Farmacêuticas da Universidade Federal de São João Del Rei, Divinópolis, Brazil

---

[*] These authors contributed equally to this work.
[†] Corresponding Author's E-mail: luarsantos@ufsj.edu.br.

Abstract

Tecoma stans (L.) Juss. ex Kunth (Bignoniaceae) is a species that has arboreal or shrubby bearing, reaching up to 12 meters in height, popularly known as ipê-mirim or yellow bell. It is originally from Mexico and the southern United States, and can be found in North, Central and South America. In some countries, such as Mexico and India, the aerial parts of *T. stans* are used in traditional medicine. Research into the identification of phenolic compounds in plants has been growing worldwide because of their important biological potential. Currently, the therapeutic use of plants requires isolation, chemical characterization and bioactivity analysis of the pharmacologically active compounds present in plants. In this sense, the objective of this review is to report flavonoids and phenolic compounds already identified in *T. stans* and their biological potential. Data in the literature revealed that extracts of *T. stans* have anti-hyperglycemic, antimicrobial, antioxidant, anti-inflammatory, analgesic, anti-cancer, protective hepatorenal and wound healing properties. The studies related the biological activities with the phenolic compounds identified in the phytochemical analysis, highlighting the presence of flavonoids, flavones, glycosides, tannins, saponins, coumarins and anthraquinones in different parts of the plant. Thus, the *T. stans* species is promising in the production of new drugs, requiring the isolation of active substances, clinical analysis and toxicity of plant extracts and fractions.

Keywords ipê-mirim, Bignoniaceae, flavonoids, phenolic compounds, biological potential.

1. Introduction

The species *Tecoma stans* belongs to the family Bignoniaceae, popularly known as "ipê-de-jardim," "yellows," "ipê-mirim," or "bells-yellow"; it has the arboreal or shrub bearing with a lot of ramifications. *T. stans* is originally from Mexico and the southern United States, and can be found in North America (southern Florida, southern Arizona and Mexico), Central America (including the Antilles) and South America (in the Andean region, in addition to Argentina and Brazil) [1]. In some countries, such as Mexico and India, the aerial parts of *T. stans* are used in traditional medicine to reduce blood glucose, in the control of fungal infections, and as a diuretic, vermifuge and tonic. The infusion of flowers and leaves of *T. stans* is used to treat diabetes and digestive problems, and the decoction of flowers is used to improve stomach pain [2].

The plants produce various chemical compounds, such as secondary metabolites, the health benefits of which and their importance in preventing chronic diseases have been extensively investigated [3]. Among the phenolic compounds, the flavonoids that are associated with health promotion for humans and plant defense systems stand out [4]. In plants, flavonoids and other phenylpropanoids accumulate in the vacuoles of epidermal cells, preventing different biotic and abiotic stresses, in addition to providing immunity against various pathogens. The flavonoids present in the human diet are naturally occurring antioxidants, considered to be nutritionally valuable compounds in the prevention of chronic diseases [5, 6].

Phenolic compounds provide several medicinal properties, such as anti-allergic, antioxidant, antimicrobial, anti-inflammatory, hepatoprotective, antiviral, anti-proliferative and anti-carcinogenic [7, 8]. The protective action of plant phenolic compounds has been well studied and has aroused the interest of scientists from all over the world for their medicinal activities. According to the WHO, among the countless plants used for medicinal purposes worldwide, few have been scientifically examined, so studies are needed to assess the effectiveness of medicinal plants and ensure their safe use [9, 10]. In this context, this work aims to review the phenolic compounds found in the species *T. stans* and the biological activities related to these metabolites.

2. Phenolic Compounds

The secondary metabolites of higher plants can be divided into three main groups according to their biosynthetic origins: terpenes, alkaloids and phenolic compounds [11]. The phenolic compounds are mainly produced through the L-phenylalanine and L-tyrosine chiquimate pathway, and structurally contain one or more hydroxyl groups directly linked to the aromatic ring [12].

The main classes of phenolic compounds are: simple phenols, benzoquinones (C_6), hydroxybenzoic acids (C_6-C_1), hydroxycinnamic acids, phenilpropanoids (C_6-C_1), acetophenones, phenylacetic acids (C_6-C_3), xanthones (C_6-C_1-C_6), stilbenes, anthraquinones (C_6-C_2-C_6), flavonoids, isoflavones (C_6-C_3-C_6), lignans, neolignans $(C_6\text{-}C_3)_2$, lignins $(C_6\text{-}C_3)_n$, condensed tannins $(C_6\text{-}C_3\text{-}C_6)_n$, and hydrolyzable tannins (C_6-C_1), $(C_6\text{-}C_1)_2$ and $(C_6\text{-}C_1)_2$. In the representation of structures, C is a carbon atom and the subscribed numbers are the carbon numbers [11].

These compounds have a defense function in plants against pests and diseases and are the most abundant secondary metabolites [12]. The search for new biologically active compounds for medical use is the reason why phytochemical research for the isolation of compounds is important [13].

Some phenolic compounds obtained by different extractions methods and different parts of the plant have already been identified in *T. stans*. Flavonoids, tannins, saponins, and coumarins have already been reported in leaf extracts of *T. stans* [14-20]; coumarins, flavonoids, anthraquinones, saponins, and tannins in flower extracts [21-23]; and saponins, flavonoids, and tannins in the roots and stem bark extracts [24].

In the review by Anand and Basavaraju [25], various phenolic compounds were listed, such as phenolic acids (chlorogenic acid, cinnamic acid, ferulic acid, isoferulic acid, gallic acid, caffeic acid, vanillic acid, *o*-coumaric acid, *p*-coumaric acid, rosmarinic acid, protocatechuic acid, ellagic acid, *p*-hydroxybenzoic acid, 3,4,5-trimethoxy cinnamic acid, and sinapic acid), verbascoside, pyrogallol, catechin, and flavonoids (flavonone, apigenin, chrysoeriol, kaempferol, luteolin, quercetin, rutin, naringin, quercitrin, hesperetin, 7-hydroxy-flavone, 7,8-dihydroxy-4,6-dimethoxy flavone, rutin, luteolin 7-*O*-β-D-neohespridoside, luteolin 7-*O*-β-D-glucopyranoside, luteolin 7-*O*-β-D-glucuronopyranoside, diosmetin 7-*O*-β-D-glucopyranoside, diosmetin luteolin 7-*O*-β-D-glucopyranoside, and diosmetin 7-*O*-β-D-glucuronopyranoside methyl ester).

3. Biological Activities

Bioassays allow analysis of the activity of natural compounds and indicate which are the most promising phytochemical compounds in plants. In addition, they guarantee proof of the potential of plants used in traditional medicine, and the investigation of the toxicity of extracts for the safe use of these compounds [26]. Studies have been found with the following biological activities for *T. stans*.

3.1. Anti-Hyperglycemic Potential

A study carried out in Egypt by Taher et al. [19] showed that methanol extract from *T. stans* leaves had an anti diabetic effect. The results indicated that the crude methanol extract had the highest potential, followed by the methylene chloride fraction rich in alkaloid. In conclusion, the study suggested that alkaloids may act synergistically, as an anti-diabetic agent, with other bioactive compounds of *T. stans,* especially flavonoids. Flavonoids, glycosides, tannins and saponins were found in the extracts and fractions of *T. stans*.

The fractions obtained from the ethanol extract leaves of *T. stans* showed a high content of flavonoids and a considerable inhibitory effect of the enzyme pancreatic lipase. The flavones identified (luteolin and chrysoeriol) exhibited the greater inhibition of this enzyme in a concentration-dependent manner. These results allow a new mechanism of action to be described for *T. stans* and may help to prevent the development of type 2 diabetes [14].

3.2. Antimicrobial Activity

In Brazil, Gonçalves et al. [22] evaluated the antimicrobial potential of the ethanol extract and fractions obtained from flowers of *T. stans* by the broth microdilution assay on 10 isolates of clinical interest. The samples demonstrated antibacterial activity and fungistatic potential on *Proteus mirabilis, Proteus vulgaris, Staphylococcus aureus, Streptococcus mutans* and *Candida infanticola*, in particular the ethanol extract and dichloromethane and ethyl acetate fractions, with values of minimum inhibitory concentration (MIC) between 500 and 2000 μg/mL. In addition, this study showed the antimicrobial potential of combinations of *T. stans* samples with commercial drugs, with synergistic and additive effects. In the samples, phenolic compounds were found, such as coumarins, flavonoids and tannins, which may be related to the antimicrobial activity.

A study carried out in India investigated the antimicrobial potential of methanol and ethanol extracts using the paper diffusion disk method, determining the zone of inhibition by the halo formed. The best effects were against *Escherichia coli, Xanthomonas axanopodis pv. malvacearum, Clavibacter michiganensis sub sp. michiganensis, Staphylococcus aureus, Klebsiella pneumoniae, Pseudomonas aeruginosa, Pseudomonas fluorescens, Aspergillus niger, Aspergillus flavus, Alternaria carthami* and *Alternaria helianthi.* The phytochemical screening revealed the presence of tannins, flavonoids, phenols and anthraquinones in the samples of *T. stans* [27].

Salem et al. [16] determined the antimicrobial activity of *T. stans* leaves by the MIC and disk diffusion methods. The methanol, ethyl acetate and chloroform extracts of *T. stans* showed a significant effect on the bacteria tested (*Bacillus subtilis, Micrococcus luteus, Sarcina lutea, Staphylococcus aureus, Escherichia coli, Serratia marcescens, Salmonella typhi, Proteus vulgaris and Pseudomonas aeruginosa*). Flavonoids and tannins were identified in this study. Thus, the extracts from *T. stans* indicated promising potential for the development of new antimicrobial agents.

3.3. Antioxidant Effect

The search for compounds with antioxidant potential becomes increasingly important, since they can slow or prevent the oxidation of substrates and the formation of free radicals. In addition, they can prevent premature aging and the genesis of various diseases, such as cancer, diabetes and cardiovascular problems [28]. The analysis of the antioxidant activity of *T. stans* was evaluated *in vitro* by the ability to capture 2,2-diphenyl-1-picrylhydrazyl (DPPH) free radicals and the Ferric Reducing Antioxidant Power (FRAP) method. The ethanol, methanol and aqueous extracts demonstrated strong activities in these assays [27]. Extracts and fractions of the leaves and branches of *T. stans* demonstrated antioxidant activity by the sequestering capacity of DPPH [19]. These studies indicated that *T. stans* is a promising candidate to be used as a natural antioxidant.

3.4. Anti-Inflammatory and Antinociceptive Potential

New potent analgesics and anti-inflammatories from natural sources without considerable side effects are being evaluated [29]. The anti-inflammatory activity and anti-nociceptive potential of *T. stans* were assessed.

Dash et al. [18] investigated the ethanol extract of *T. stans* leaves, highlighting the concentration of 400 mg/kg of sample, which inhibited 49.38% of paw edema carrageenan-induced *in vivo* tests, and presented mild to moderate analgesic activity in the acetic acid-induced abdominal contortions. In the study of Prasanna et al. [20], the aqueous and alcoholic extracts of *T. stans* leaves were analyzed, and showed that the alcoholic extract at a concentration of 500 mg/kg showed the greatest anti-inflammatory effect (76.92%, after 24 hours), with inhibition of the nociceptive response. The study suggested that the presence of phenolic compounds and flavonoids may be responsible for the bioactive potential of *T. stans* [18, 20].

The methanol extract of the flowers of *T. stans* was also evaluated and demonstrated anti-nociceptive and anti-inflammatory potential [23]. The study suggested that inhibition of the inflammation induced by carrageenan could be due to inhibition of the enzyme cyclooxygenase, and the subsequent inhibition of prostaglandin synthesis, while an analgesic effect could be due to inhibition of the prostaglandin pathway [23].

The anti-inflammatory effect of *T. stans* can be attributed the presence of flavonoids, which are known to target prostaglandins that are involved in the late phase of acute inflammation and in the perception of pain [20].

3.5. Anticancer Activity

Cancer is characterized by the uncontrolled division of cells, being represented by more than a hundred different types. It is one of the biggest causes of death in the world. [30]. Breast cancer is an important cause of human suffering and premature mortality among women worldwide [31]. There is an ongoing effort in the search for naturally occurring anticancer agents, which are responsible for preventing, combining or reversing the development of cancer [30]. A large number of plant-derived chemotherapeutic agents, such as vimblastine, taxol, camptothecin and podophyllotoxin, are used as anti-cancer agents [32].

Thirumal et al. [33] analyzed the anti-cancer activity of *T. stans* leaf ethanolic extract against the breast cancer cell line (MCF-7). Different concentrations of extract were tested in MCF-7 by 3-(4,5-dimethylthiazol-2-yl) -2,5-diphenyltetrazolium (MTT) bromide assay. *T. stans* showed significant cancer cell anti-proliferative activity with a minimum inhibition of 14.6%, at a concentration of 7.8 μg/mL, and the maximum inhibition (95.9%) was observed at 1000 μg/mL. Phytochemical screening revealed the presence of saponins, flavonoids, tannins and other phenolic compounds.

The same authors [24] evaluated *in vitro* the anti-proliferative activity of crude ethanolic extract of the root (ERETS), stem bark (ESETS) and flowers (EFETS) of *T. stans* on cell lines of human breast cancer (MCF-7) in different concentrations to identify bioactive compounds for direct use as drugs or which can be used in the formulation of semi-synthetic drugs. Preliminary results attributed anticancer activity to the roots, stem bark and flowers of *T. stans*. Although all extracts have shown significant activity, the stem bark (ESETS) exhibited potential anti-cancer activity *in vitro* on the MCF-7 cell line at increasing concentrations when compared to other extracts. The effective concentration (IC_{50}) of the root extracts (ERETS), stem bark (ESETS) and flowers (EFETS) was 46.0 μg/mL, 42.0 μg/mL and 70 μg/mL, respectively. Thus, the stem bark (ESETS) of *T. stans* may be a potential alternative agent for the therapy of human breast cancer [24].

Therefore, it is suggested that *T. stans* is a plant with potential anti-cancer activity, which can be considered for further studies.

3.6. Wound Healing Activity

Wound healing is a natural process that involves cellular, biochemical and molecular mechanisms to regenerate injured tissue. The research into pharmaceutical products seeks to improve the wound healing process with the objective of immunological protection and against infections. In addition, it is necessary to promote healing with minimal pain, discomfort, and scarring, in less time. Thus, the search for new agents with fewer side effects has fostered research in the field of natural products in recent decades [34].

The healing activity of *T. stans* wounds has been studied. The analysis was performed with methanol extracts obtained from the bark of *T. stans* to verify its healing activity in albino rats. The results suggested that the administration of methanol extract from the bark has shown more significant healing activity in wound excision and incision models and supports the popular use of wound plants in folk medicine.

Phytochemicals compounds, such as triterpene, phytosterol, glycosides, phenols, flavonoids, saponins and tannins, individually or in combination, can exhibit synergistic effects for wound healing [35].

3.7. Protective Hepatorenal

The methanol extract and fractions of the leaves from *T. stans* were examined for their effects in subchronic doses on liver functions. In this study, it was observed that the levels of glutathione (GSH), which protect the body's cells against the toxic effects of lipid peroxidation, were significantly depleted in the liver tissue of the control group (diabetic animals) compared to healthy rats. Treatments with different preparations of *T. stans* samples for 28 days exhibited an increased GSH of the liver tissue, showing positive effects. Different phenolic compounds were identified, such as flavonoids, tannins and saponins, in this work. The high polyphenolic contents of *T. stans* preparations were correlated with this activity [19].

Raju et al. [36] investigated the nephroprotective activity of the ethyl acetate fraction of *T. stans* dried flowers on gentamicin-induced nephrotoxicity in albino rats. Nephrotoxicity was induced by the intraperitoneal administration of gentamicin (80 mg/kg/day, for 8 days), concomitant with the ethyl acetate fraction (100, 200 and 300 mg/kg/day, orally). The results indicated that there was no toxicity and that the fraction significantly protected the kidneys of rats from the histopathological changes induced by gentamicin. Preliminary phytochemical studies revealed the presence of flavonoids, saponins, tannins and glycosides in the fraction [36].

Conclusion

The pharmacological potential of *T. stans* is notable, which may be related to its phytochemical constituents, in particular phenolic compounds. The species showed anti-hyperglycemic, antimicrobial, antioxidant, anti-inflammatory, analgesic, anti-cancer, hepatorenal protective and wound healing properties, being a source of compounds for the production of new drugs in the future. However, there is a need for further clinical studies and toxicity analyses of the extracts obtained from this species to ensure their safe use.

ACKNOWLEDGMENTS

The authors are grateful to the Universidade Federal de São João Del Rei (UFSJ), Conselho Nacional de Desenvolvimento Científico e Tecnológico (CNPq), Coordenação de aperfeiçoamento de Pessoal de Nível Superior (CAPES) and Fundação de Amparo à Pesquisa do Estado de Minas Gerais (FAPEMIG) for the financial support. "This study was financed in part by the Coordenação de Aperfeiçoamento de Pessoal de Nível Superior - Brasil (CAPES) - Finance code 001."

REFERENCES

[1] Gentry, A. H. (1992). Bignoniaceae - Part II (Tribe Tecomeae). *Flora Neotropica Monograph*, v. 25, 370 p.

[2] Singh, A., Nagori, B. P., Mathur, K. (2013). *Tecoma stans*: An Important Medicinal Plant. *Int. J. Pharm. Erud.*, 3(2): 13-21.

[3] Belhaoues, S., Amri, S., Bensouilah, M. (2020). Major phenolic compounds, antioxidant and antibacterial activities of *Anthemis praecox* Link aerial parts. *S. Afr. J. Bot.*, 131: 200-205.

[4] Wang, T. Y., Li, Q., Bi, K. S. (2018). Bioactive flavonoids in medicinal plants: structure, activity and biological fate. *Asian J. Pharm. Sci.*, 13(1): 12-23.

[5] Petrussa, E., Braidot, E., Zancani, M., Peresson, C., Bertolini, A., Patui, S., Vianello, A. (2013). Plant flavonoids-biosynthesis, transport and involvement in stress responses. *Int. J. Mol. Sci.*, 14(7): 14950-14973.

[6] Piasecka, A., Jedrzejczak-Rey, N., Bednarek, P. (2015). Secondary metabolites in plant innate immunity: conserved function of divergent chemicals. *New Phytol.*, 206(3): 948-964.

[7] Jucá, M. M., Cysne Filho, F. M. S., de Almeida, J. C., Mesquita, D. D. S., Barriga, J. R. M., Dias, K. C. F., Barbosa, T. M., Vasconcelos, L. C., Leal, L. K. A. M., Ribeiro, J. E., Vasconcelos, S. M. M. (2020). Flavonoids: biological activities and therapeutic potential. *Nat. Prod. Res.*, 34(5): 692-705.

[8] Lalani, S., Poh, C. L. (2020). Flavonoids as antiviral agents for *Enterovirus A71 (EV-A71). Viruses*, 12(2): 184.

[9] Zhang, J., Onakpoya, I. J., Posadzki, P., Eddouks, M. (2015). The safety of herbal medicine: from prejudice to evidence. *Evid. -Based Complement. Alternat. Med.*, 2015: 316706.

[10] Larbie, C., Owusu, K. P., Torkornoo, D., Asibey, O. (2016). Acute and sub-chronic toxicity of aqueous ethanolic extract of *Ficus pumila* leaves in rats. *European J. Biomed. Pharm. Sci.*, 3(8): 22-27.

[11] Vuolo, M. M., Lima, V. S., Junior, M. R. M. (2019). Phenolic compounds: structure, classification, and antioxidant power. In: Campos, M. R. S. (Ed.). *Bioactive Compounds.* Cambridge: Woodhead Publishing, 33-50.

[12] Kumar, N., Goel, N. (2019). Phenolic acids: natural versatile molecules with promising therapeutic applications. *Biotechnol. Rep.*, 24: e00370.

[13] Sharma, D., Singh, V. P., Singh, R. K., Joshi, C. S., Sharma, V. (2021). Isolation and characterization of bioactive compounds from natural resources: metabolomics and molecular approaches. In: Srivastava, A. K., Kannaujiya, V. K., Singh, R. K., Singh, D. (Ed.). *Evolutionary Diversity as a Source for Anticancer Molecules.* London: Academic Press, 77-101.

[14] Ramirez, G., Zamilpa, A., Zavala, M., Perez, J., Morales, D., Tortoriello, J. (2016). Chrysoeriol and other polyphenols from *Tecoma stans* with lipase inhibitory activity. *J. Ethnopharmacol.,* 185:1-8.

[15] Larbie, C., Nyarkoh, C. O., Adjei, C. O. (2019). Phytochemical and safety evaluation of hydroethanolic leaf extract of *Tecoma stans*

(L.) Juss. ex Kunth. *Evid. -Based Complement. Alternat. Med.*, 2019: 7417624.

[16] Salem, M. Z. M., Gohar, Y. M., Diaz, L. M. C., El-Shanhorey, N. A., Salem, A. Z. M. (2013). Antioxidant and antibacterial activities of leaves and branches extracts of *Tecoma stans* (L.) Juss. ex Kunth against nine species of pathogenic bacteria. *Afr. J. Microbiol. Res.*, 7(5): 418-426.

[17] Thein, A. M., Oo, K. C. C. (2019). *Phytochemical investigation of Tecoma stans (L.) H. B. K. leaves and its brine shrimp lethality test.* 2nd. Myanmar Korea Conference, Dagon University, 1, 1-11.

[18] Dash, S., Das, C., Sahoo, D. C., Sahoo, A. C. (2011). Phytochemical composition, anti-inflammatory and analgesic activities of *Tecoma stans* Linn. (Bignoniaceae). *Nat. Pharm. Technol.*, 1(2): 5-8.

[19] Taher, M. A. H., Dawood, D. H., Sanad, M. I., Hassan, R. A. (2016). Searching for anti-hyperglycemic phytomolecules of *Tecoma stans. Eur. J. Chem.*, 7(4): 397-404.

[20] Prasanna, V. L., Lakshman, K., Hedge, M. M., Bhat, V. (2013). Antinociceptive and anti-inflammatory activity of *Tecoma stans* leaf extracts. *Indian Journal of Research in Pharmacy and Biotechnology*, 1(2): 156-160.

[21] Anburaj, G., Marimuthu, M., Rajasudha, V., Manikandan, R. (2016). Phytochemical screening and GC-MS analysis of ethanolic extract of *Tecoma stans* (Family: Bignoniaceae) Yellow Bell Flowers. *J. Pharmacogn. Phytochem.*, 5(6): 172-175

[22] Gonçalves, T. P. R., Parreira, A. G., Lima, L. A. R. S. (2020). Study of the antimicrobial activity of *Tecoma stans* (L.) ex Kunth (Bignoniaceae). *Period. Tchê Quím.*, 18(35): 1037-1043.

[23] Kameshwaran, S., Suresh, V., Arunachalam, G., Frank P. R., Manikandan, V. (2012). Evaluation of antinociceptive and anti-inflammatory potential of flower extract *Tecoma stans. Indian J. Pharmacol.*, 44(4): 543-544.

[24] Thirumal, M., Kishore, G., Srimanthula, S. (2013). Anti proliferative activity of various parts of *Tecoma stans* (L.) against human breast cancer cells *in vitro. Res. J. Pharm. Biol. Chem. Sci.*, 4(2): 305-313.

[25] Anand, M., Basavaraju, R. (2021). A review on phytochemistry and pharmacological uses of *Tecoma stans* (L.) Juss. ex Kunth. *J. Ethnopharmacol.*, 265: 113270.

[26] Seca, A. M. L., Moujir, L. M. (2020). Natural Compounds: A Dynamic Field of Applications. *Appl. Sci.*, *10*(11): 4025.

[27] Govindappa, M., Sadananda, T. S., Channabasava, R., Jeevitha, M. K., Pooja, K. S., Vinay, R. B. (2011). Antimicrobial, antioxidant activity and phytochemical screening of *Tecoma stans* L. Juss. Ex Kunth. *J. Phytol.*, 3: 68-76.

[28] Gonçalves, T. P. R. (2020). Estudo fitoquímico e avaliação do potencial antimicrobiano, sinérgico e antioxidante das flores de *Tecoma stans* (L.) Juss. ex Kunth (Bignoniaceae) [Phytochemical study and evaluation of the antimicrobial, synergistic and antioxidant potential of *Tecoma stans* (L.) Juss. ex Kunth (Bignoniaceae)]. *Dissertation* (Pharmaceutical Sciences), Divinópolis, 92 p.

[29] Adnan, M., Nazim Uddin Chy, M., Mostafa Kamal, A. T. M., Barlow, J. W., Faruque, M. O., Yang, X., Uddin, S. B. (2019). Evaluation of anti-nociceptive and anti-inflammatory activities of the methanol extract of *Holigarna caustica* (Dennst.) Oken leaves. *J. Ethnopharmacol.*, 236: 401-411.

[30] Akhtar, M. F., Saleem, A., Rasul, A., Faran Ashraf Baig, M. M., Bin-Jumah, M., Abdel Daim, M. M. (2020). Anticancer natural medicines: an overview of cell signaling and other targets of anticancer phytochemicals. *Eur. J. Pharmacol.*, 888: 173488.

[31] Coughlin, S. S. (2019). Epidemiology of breast cancer in women. In: Ahmad, A. (Ed.). *Breast Cancer Metastasis and Drug Resistance: Challenges and Progress.* Springer Nature Switzerland AG, 9-29.

[32] Raskin, I., Ribnicky, D. M., Komarnytsky, S., Ilic, N., Poulev, A., Borisjuk, N., Brinker, A., Moreno, D. A., Ripoll, C., Yakoby, N., O'Neal J. M., Cornwell, T., Pastor, I., Fridlender, B. (2002). Plants and human health in the twenty-first century. *Trends Biotechnol.*, 20(12): 522-531.

[33] Thirumal, M., Kishore, G., Prithika, R., Das, S., Nithya, G. (2012). *In vitro* anticancer activity of *Tecoma stans* (L.) ethanolic leaf

extract on human breast cancer cell line (MCF-7). *Intl. J. Pharma. and Bio. Sci*., 2(4): 488-493.

[34] Zimmermann-Klemd, A. M., Konradi, V., Steinborn, C., Ücker, A., Falanga, C. M., Woelfle, U., Huber, R., Jürgenliemk, G., Rajbhandari, M., Gründemann, C. (2019). Influence of traditionally used Nepalese plants on wound healing and immunological properties using primary human cells *in vitro*. *J. Ethnopharmacol.*, 235: 415-423.

[35] Das, C., Dash, S., Sahoo, D. C., Mohanty, A. (2010). Evaluation of methanolic bark extract of *Tecoma stans* Linn., for wound healing in albino rats. *Int. J. Pharm. Technol.*, 2(3): 735-742.

[36] Raju, S., Kavimani, S., Uma Maheshwara Rao, V., Sreeramulu Reddy, K., Vasanth Kumar, G. (2011). Floral extract of *Tecoma stans*: a potent inhibitor of gentamicin-induced nephrotoxicity *in vivo*. *Asian Pac. J. Trop. Med.*, 4(9): 680-685.

In: The Book of Flavonoids
Editors: Chao-Hui Feng et al.
ISBN: 978-1-68507-376-3

Chapter 6

FLAVONOIDS, PHENOLIC COMPOUNDS, AND POTENTIAL BIOLOGICAL ACTIONS FROM *SMILAX BRASILIENSIS* SPRENGEL AND *SMILAX FLUMINENSIS* STEUD

***Izabela Caputo Assis Silva*[1,2,*],**
***Lucas Santos Azevedo*[1,2,*],**
***Iasmin Fernandes Emediato*[1],**
***Adriano Lopes da Silva*[1,3],**
***Thaís Paula Rodrigues Gonçalves*[1,2],**
***Jéssica Elvira Siqueira Medeiros*[1], *Iara Cotta Mendes*[1],**
***Paula Avelar Amado*[1,2],**
***Ana Hortência Fonsêca Castro*[2,3]**
***and Luciana Alves Rodrigues dos Santos Lima*[1,2,3,†]**

---

* These authors contributed equally to this work.
† Corresponding Author's E-mail: luarsantos@ufsj.edu.br.

[1]Phytochemistry Laboratory,
Universidade Federal de São João Del Rei,
Campus Centro-Oeste Dona Lindu, Divinópolis, MG, Brazil,
[2]Programa de Pós-Graduação em Biotecnologia da Universidade Federal de São João Del Rei, Campus Centro-Oeste Dona Lindu, Divinópolis, MG, Brazil,
[3]Programa de Pós Graduação em Ciências Farmacêuticas da Universidade Federal de São João Del Rei,
Campus Centro-Oeste Dona Lindu, Divinópolis, MG, Brazil

ABSTRACT

The Smilacaceae family is constituted of approximately 350 species, represented by five genera, *Anikenton*, *Coprosmanthus*, *Heterosmilax*, *Nemexia,* and *Smilax*. They are found in hot and humid climates in tropical and subtropical regions and are located in both hemispheres. Plants of the genus *Smilax* are popularly known as 'japecanga' and/or 'sarsaparilla' in South American countries. The roots and aerial parts of these species are widely used in folk medicine to treat sexually transmitted infections, skin conditions, and inflammatory diseases. Within this genus, *Smilax brasiliensis* Sprengel and *S. fluminensis* Steud. are endemic in Brazil, native to the Cerrado, and poorly studied. The objective of this work was to verify the flavonoids and phenolic compounds present in *S. brasiliensis* Sprengel and *S. fluminensis* Steud. with those reported in the literature and to verify which biological activities are related to them. In the literature, it was reported that *S. brasiliensis* produces several phenolic compounds, such as flavonoids (rutin, 3-*O*-β-galactopyranosyl quercetin, 3-*O*-β-glicopyranosyl quercetin, quercetin, naringenin, and kaempferol), tannins, and phenolic acids (chlorogenic acid, caffeic acid, *p*-coumaric acid, ferulic acid, *trans*-cinnamic acid, quinic acid, 3-*O*-*E*-caffeoyl quinic acid, *O*-feruloyl quinic acid, and *O*-caffeoyl shikimic acid). These phenolic compounds showed antioxidant, allelopathic, antimicrobial, anticancer, antihyperglycemic, and antihyperlipidemic biological activities. Phenolic compounds were also found in *S. fluminensis*. The obtained extracts allowed the identification of chlorogenic acid, caffeic acid, and ferulic acid in the methanol extract of roots, leaves, and rhizophores, rutin in the methanol extract of roots and leaves, *p*-cinnamic acid and *trans*-cumaric acid in the methanol extract of roots, and taxifoline, quercitrin, rutin, quercetin, kaempferol, apigenin, agatisflavone, amentoflavone, quercetin-3-*O*-β-D-galactopyranoside (hiperin), quercetin-3-*O*-α-L-ramnopyranosyl (1'''→6'')-*O*-β-D-glucopyranoside, and two unpublished phenolic metabolites in the ethanol extract of leaves. The biological effects observed were toxicological activity in *Artemia salina* with an LD_{50} = 33.43 μg/mL and antioxidant activity quantified by the DPPH method with EC_{50} ranging from 4 to 50 μg/mL.

Keywords: *Smilax*, flavonoids, phenolic acids, biological potential, Smilacaceae

1. INTRODUCTION

The Smilacaceae family is distributed in both hemispheres in tropical and subtropical regions and is rarely observed in temperate regions. It is represented by five genera, the *Smilax* genus being the most important and largest of the family, with approximately 310 species [1]. Among these species, 33 are distributed in Brazil [2], and 19 are endemic in the Brazilian territory [1].

The uses of plants in the genus *Smilax* are not only medicinal, but throughout the world, it has been used as a food from rhizomes, stems, and leaves and in the civil industry as fibers for construction. In popular medicine, it has been used since antiquity for presenting depurative, diaphoretic, diuretic, emollient, expectorant, aperitive, eupeptic, anti-prose, myotonic, and sweat properties [3] and is widely used to treat sexually transmitted infections (STIs), skin conditions, vesicular and renal inflammatory diseases, and rheumatic diseases [4, 5].

Species of the genus *Smilax* are known by the same popular name, sarsaparilla, japecanga, or ijuapeca guasu [2, 6]. Sarsaparilla has several biological actions described in the literature, such as antioxidant, allelopathic, larvicidal, antimicrobial, analgesic, antidiabetic, cytotoxic, antimutagenic, cytoprotective, hepatoprotective, anti-inflammatory, immunomodulatory, antitumor, pesticide [4, 5, 7], hemolytic, antisyphilitic, and anti-rheumatic properties [8]. This diversity of biological actions, presented by species of the genus *Smilax*, is associated with a variety of chemical constituents, with saponins being the main class synthesized by these plants [7].

Among the endemic species, *S. brasiliensis* Sprengel and *S. fluminensis* Steud. stand out, with few studies presenting the phytochemical profile and biological actions.

2. *Smilax brasiliensis* Sprengel

S. brasiliensis Sprengel is a plant native to the Brazilian Cerrado. This species belongs to the family Smilacaceae, of the order Liliales. It is monocotyledonous and dioecious, characterized by presenting oval leaves with a trio of central ribs, spine stems, a sclerenchymatic ring in continuous form throughout the length of the stem, and the presence of spherical and simple starch grains around the vascular bundle. It produces greenish flowers, globose berry fruits, and reddish seeds [4, 7].

2.1. Phenolic and Flavonoid Compounds of *S. brasiliensis*

Throughout evolution, plants have developed defense and attraction mechanisms. The secondary metabolites are highly important for plant growth and development, being divided into three chemical groups, terpenes or terpenoids, phenolics or phenols, and nitrogen compounds [9, 10].

Phenolic compounds have an aromatic ring and at least one hydroxyl group. Among these constituents, there are several structures of phenolic acids, derived from coumarin, flavonoids, lignins, and tannins [9, 10]. Flavonoids are a class of polyphenols widely distributed among secondary metabolites in plants and are biosynthesized by the phenylpropanoid route and have 15 carbon atoms in its fundamental nucleus, in addition to the phenyl groups linked to a three-carbon chain [9].

In works with *S. brasiliensis,* the presence of phenolic and flavonoid components were related. A study on the stems identified the presence of coumarins, saponins, flavonoids, tannins condensed, rutin, quercetin, and chlorogenic acid in the ethanol extracts and fractions by thin layer chromatography (TLC) [4]. Rutin, chlorogenic acid, coumarins, flavonoids, saponins, and tannins were identified in the leaves [11, 12].

Analysis by liquid chromatography coupled with a diode array detector and mass spectrometry (LC-DAD-MS) identified phenolic and flavonoid compounds, such as quinic acid, galocatechin, 3-*O*-*E*-caffeoyl quinic acid, 5-*O*-*E*-caffeoyl quinic acid, *O*-feruloyl quinic acid, ferulic acid, *O*-methyl quercetin, *O*-methyl kaempferol, *O*-deoxyhexosyl-hexosyl luteolin, *O*-deoxyhexosyl-hexosyl quercetin, 3-*O*-β-galactopyranosyl quercetin, 3-*O*-β-glucopyranosyl quercetin, quercetin, naringenin, and kaempferol [13, 14].

Analysis using LC-DAD and LC-MS techniques demonstrated the presence of phenolic substances, such as gallic acid, chlorogenic acid, caffeic acid, and flavonoids, including rutin and quercetin [15]. Martins et al. [16] studied the chemical composition of *Smilax* spp., including *S. brasiliensis,* by high performance liquid chromatography coupled with a diode array detector (HPLC-UV-PDA) and identified chlorogenic acid, caffeic acid, rutin, *p*-coumaric acid, ferulic acid, and *trans*-cinnamic acid in methanol extracts of the roots, rhizophores, and leaves.

2.2. Biological Activities Associated with the Phenolic and Flavonoid Compounds of *S. brasiliensis*

S. brasiliensis is a poorly studied species; however, there are reports that it has antioxidant, larvicidal [13, 17], anticancer [18], antimicrobial [19], anti-hyperglycemic, anti-hyperlipidemic [15], and allelopathic activities, cytotoxic, and antigenotoxic effects [14].

The effects of phenolic and flavonoid compounds in relation to antioxidant activity are known. *S. brasiliensis* was evaluated by a 1,1-diphenyl-2-picryl-hydrazyl (DPPH) radical assay. The ethanol extracts and fractions of leaves showed high antioxidant potential, with values of the effective concentration to decolorize 50% of DPPH radicals (EC_{50}) between 2.73 and 11.75 μg/mL [11, 12]. Amado et al. [13, 17, 20] also demonstrated the antioxidant potential of methanol extracts and fractions from leaves with EC_{50} values of 0.70-14.01 μg/mL. These results can be correlated to the high content of phenolic and flavonoid compounds present in the samples and also to compounds identified by TLC and LC-DAD-MS [11-13, 20].

Ethanol extracts and fractions obtained from the stems showed high antioxidant potential, with EC_{50} values of 1.71, 1.91, and 2.77 μg/mL for ethyl acetate and dichloromethane fractions and ethanol extract, respectively. The antioxidant action was correlated with the presence of phenolic compounds, such as quercetin, rutin, chlorogenic acid, coumarins, flavonoids and tannins, in the extracts and fractions [4].

The antioxidant action of 26 species, including *S. brasiliensis*, were evaluated by three methods, namely DPPH, ferric reducing antioxidant power (FRAP), and nitric oxide (NO) assays, with EC_{50} values of 171.83, > 800, and > 800 μg/mL, respectively. The data were related to the total phenolic compounds, flavonoids, and coumarins of 1.47 ± 0.09, 0.32 ± 0.00 and 1.14 ± 0.00 mg/g, respectively, and expressed as mg of tannic acid, rutin, and coumarin, respectively [19].

S. brasiliensis was also evaluated for antimicrobial potential. The species showed moderate activity on *Enterococcus faecalis* and a weak effect on *Aspergillus fumigatus*, with a minimal inhibitory concentration (MIC) of 400 and 800 μg/mL, respectively [7, 19]. Methanol, dichloromethane, and aqueous extracts of roots, leaves, and flowers were evaluated on strains of *Candida albicans, C. glabrata,* and *C. parapsilosis*. The extracts showed antifungal activity, and the best effects were observed for dichloromethane extracts obtained from the roots of *C. albicans* and *C. parapsilosis*, with an MIC value of 0.06 mg/mL [21].

The methanol extract of *S. brasiliensis* was evaluated for its inhibition of tumor cell line proliferation (murine skin, human leukemia, human breast, and human colon), with greater inhibition for human breast (77.9%), leukemia (56.3%) and human colon tumor cells (53.6%) [18].

The allelopathic and cytotoxic activities were evaluated under the action of the methanol extracts and fractions of leaves, and quercetin and rutin were used as a positive control and glyphosate as a negative control. The extracts and fractions inhibited or promoted lettuce hypocotyl and radicle growth. The samples inhibited onion hypocotyl growth at all concentrations and inhibited or promoted radicle growth. The extracts and fractions did not cause a cytotoxic effect and showed antigenotoxic potential at all tested concentrations; however, they demonstrated genotoxic action [14]. Ethanol extracts and fractions from leaves of *S. brasiliensis* inhibited the growth of roots and hypocotyls of onion seeds at concentrations of 125, 250, and 500 μg/mL. For lettuce, the effect was diversified, showing inhibition and growth stimulus for roots and hypocotyls [11].

A study with the ethanol extracts and fractions of the stems showed an inhibitory effect on seed growth, especially onion seeds. During cytogenetic analysis, there was a cytotoxic effect. Genotoxic changes were not observed in meristematic cells of onions submitted to samples of *S. brasiliensis*, as the extracts and fractions were considered non-toxic [1].

S. brasiliensis and *Herreria salsaparilha* are known as sarsaparilla and were evaluated for metabolic activities. The species demonstrated high hypolipidemic potential, and the effects were attributed to the identified phenolic substances and saponins [7, 15].

Methanol extracts and fractions of the leaves from *S. brasiliensis* were evaluated for their larvicidal effect on *Culex quinquefasciatus* and were observed until adulthood. The dichloromethane fraction was the most efficient, killing the larvae in the highest concentrations evaluated, 500 and 1000 μg/mL, with lethal concentration (LC_{50}) values of 469.78 μg/mL. Larvicidal activity on *Artemia salina* larvae was also evaluated, with a low mortality for dichloromethane and ethyl acetate at the highest concentration tested (1000 μg/mL), while the other samples did not show larval mortality, as the samples of *S. brasiliensis* were considered non-toxic [17].

3. *Smilax Fluminensis* Steud

S. fluminensis Steud. is a plant distributed in Central and South America, in Argentina, Brazil, Bolivia, Costa Rica, Colombia, Ecuador, Panama, Peru, and Venezuela. It has two synonyms - *S. chinesa* Vell. and *S. syringoides* Griseb - that belong to the Smilacaceae family and Liliales order. They are popularly known as japicanga, yuapeca guasu, salsa, salshina, zarzaparrilla, ijuapeca guasu, and vena [6, 22].

It is monocotyledonous and dioecious and characterized by having lanceolate, glabrous, coriaceous, or membranous oval leaves. Its inflorescences are umbelliferous and arranged in racemes with petals of 3 to 5 mm; fruits are yellow-orange berries (8 to 10 mm in diameter) when ripe [23-26].

Although it is a poorly studied plant, there are reports that it has antioxidant, toxicological, and antineoplastic activity [27-29].

3.1. Phenolic and Flavonoid Compounds of *S. fluminensis*

Some studies have observed the presence of flavonoids in *S. fluminensis* Steud. One study evaluated the ethanol extracts obtained by turbo extraction of leaves and roots using TLC, indicating the presence of flavonoids in the root extract [26]. Phytochemical screening revealed the presence of phenolic and flavonoid compounds in the ethanol extract of stems and leaves obtained by maceration. The dosage of total phenolic compounds and flavonoids were 145.91 mg/g of gallic acid and 49.23 mg/g of quercetin, respectively. Analysis by TLC identified the presence of quercetin and two quercetin glycoflavonoids [28].

The flavonoid profile of the ethanol extract from leaf maceration was performed by HPLC coupled to the dual wavelength ultraviolet detector (UV) and were identified as agatisflavone, amentoflavone, apigenin, kaempferol, quercetin, quercitrin, rutin, and taxifoline. The highest concentration was amentoflavone (44.1 μg/mL), followed by agatisflavone and apigenin, with 40.9 μg/mL and 38.5 μg/mL, respectively [27]. Another study with methanol extracts obtained by macerating the roots, rhizophores, and leaves of *S. fluminensis* Steud. by HPLC-DAD revealed the presence of caffeic, chlorogenic, and ferulic acids in all extracts, *p*-cinnamic and *trans*-cinnamic acids in root extracts, and rutin in the root and leaf extracts [30].

Petrica et al. [31], using HPLC coupled to an electro-spray ionization mass spectrometer (ESI-MS) and nuclear magnetic resonance (NMR) for ^{1}H and ^{13}C, identified two flavonoids: quercetin-3-*O*-α-L-ramnopyranosyl(1'''→6'')-*O*-β-D-glucopyranoside and quercetin-3-*O*-β-D-galactopyranoside [31, 32].

3.2. Biological Activities Associated with the Phenolic and Flavonoid Components of *S. fluminensis* Steud

Studies evaluating the biological activity of *S. fluminensis* Steud are scarce, which encourages further research with this species. The antioxidant activity of ethanol extracts obtained by maceration from leaves and its twenty-two fractions were evaluated by the DPPH assay. The samples presented EC_{50} values ranging from 4 to 50 μg/mL. Among them, an extract and eight fractions were the best samples (EC_{50} = 4 μg/mL) [29].

Activity of the hydroethanol extract on murine fibroblast (3T3) and murine melanoma (B16F10) cell lines was preliminarily evaluated by the 3-(4,5-dimethylthiazol-2-yl)-2,5-diphenyltetrazolium (MTT) bromide assay, with cytotoxic activity in both lines, being more toxic for B16F10 at inhibitory concentrations (IC_{50}) of 250 μg/mL, followed by IC_{50} = 500 μg/mL for the 3T3 cell [29].

Artemia salina toxicity was assessed with ethanol extracts by maceration from leaves and stems, which showed a LC_{50} value of 33.43 μg/mL, a concentration considered to be very toxic [28].

CONCLUSION

S. brasiliensis Sprengel showed antioxidant, allelopathic, antimicrobial, anticancer, antihyperglycemic, and antihyperlipidemic activities, and *S. fluminensis* Steud. exhibited toxic, antineoplastic, and antioxidant activities, which were correlated with the phenolic compounds and flavonoids present in these species.

The results reported in this review show that both *Smilax* spp. have potential as sources for new biological effects and, consequently, new drugs.

ACKNOWLEDGMENTS

The authors are grateful to the Universidade Federal de São João del-Rei (UFSJ), Conselho Nacional de Desenvolvimento Científico e Tecnológico (CNPq), Coordenação de aperfeiçoamento de Pessoal de Nível Superior (CAPES), and Fundação de Amparo à Pesquisa do Estado de Minas Gerais (FAPEMIG) for the financial support. "This study was financed in part by the Coordenação de Aperfeiçoamento de Pessoal de Nível Superior - Brasil (CAPES) - Finance code 001".

REFERENCES

[1] Andreata, R. H. P., Watanabe, M. T. C. (2018). Flora of the *canga* of the Serra dos Carajás, Pará, Brazil: Smilacaceae. *Rodriguésia*, 69(1), 245-250.

[2] Soares, M. K. M., Martins, A. R., Montanari Júnior, I., Figueira, G. M., Zucchi, M. I., Bajay, M. M., Appezzato-da-Glória, B. (2014). The sarsaparilla market in the state of São Paulo (Brazil) and the challenges of cultivation. *Rev. Bras. Farmacogn.*, 24(1), 73-79.

[1] Medeiros, M. F. T., Senna-Valle, L., Andreata, R. H. P. (2007). Histórico e o uso de "salsa parrilha" (*Smilax* spp.) pelos boticários no Mosteiro de São Bento. *Rev. Bras. Biociênc.*, 5(S1), 27-29. [History and the use of "parrilha parrilha" (*Smilax* spp.) by apothecaries in the São Bento Monastery.]

[2] Silva, I. C. A. (2019). Avaliação das atividades antioxidante, alelopática e dos efeitos citotóxico e genotóxico dos caules de *Smilax brasiliensis* Sprengel (Smilacaceae) [Evaluation of the antioxidant, allelopathic and cytotoxic and genotoxic effects of *Smilax brasiliensis* stems Sprengel (Smilacaceae)]. *Dissertation* (Biotechnology), Divinópolis, 82 p.

[3] Vila Verde, G. M., Paula, J. R., Carneiro, D. M. (2003). Levantamento etnobotânico das plantas medicinais do cerrado utilizadas pela população de Mossâmedes (GO). *Rev. Bras. Farmacogn.*, 13(S1), 64-66. [Ethnobotanical survey of medicinal plants from the cerrado used by the population of Mossâmedes (GO).]

[4] Ferrufino-Acosta, L. (2010). Taxonomic revision of the genus *Smilax* (Smilacaceae) in Central America and the Caribbean Islands. *Willdenowia,* 40(2), 227-280.

[5] Silva, I. C. A., Fonseca, J. C., Medeiros, J. E. S., Castro, A. H. F., Lima, L. A. R. S. (2019). Phytochemical aspects and biological properties of *Smilax brasiliensis* Sprengel. In: Ryan, I. (Ed.). *Phytochemicals: Plant Sources and Potential Health Benefits*. New York: Nova Science Publishers, 175-189.

[6] Soares, M. K. M. (2012). *Anatomia e perfil químico de salsaparrilha comercializada no Estado de São Paulo*. Dissertation (Sciences), São Paulo, 70 p. [*Anatomy and chemical profile of sarsaparilla sold in the State of São Paulo*]

[7] Simões, C. M. O., Schenkel, E. P., Gosmann, G., Mello, J. C. P., Mentz, L. A., Petrovick, P. R. (2007). *Farmacognosia da planta ao medicamento.* Porto Alegre/Florianópolis: Editora da UFRGS/Editora da UFSC, 6ª ed., 1104 p.

[8] Sampaio, B. L. (2010). *Influência dos fatores ambientais sobre a concentração de compostos fenólicos nas folhas e na casca do caule de Lafoensia pacari A. St. -Hil (Lythraceae)* [*Influence of environmental factors on the concentration of phenolic compounds in leaves and stem bark of Lafoensia pacari A. St. -Hil. (Lythraceae)*]. Dissertation (Pharmaceutical Sciences). Goiânia, 62 p.

[9] Fonseca, J. C., Barbosa, M. A., Silva, I. C. A., Duarte-Almeida, J. M., Castro, A. H. F., dos Santos Lima, L. A. R. (2017). Antioxidant and allelopathic activities of *Smilax brasiliensis* Sprengel (Smilacaceae). *S. Afr. J. Bot.*, 111, 336-340.

[10] Barbosa, M. A. (2015). *Anatomia foliar e avaliação das atividades antioxidante e alelopática de folhas de Smilax brasiliensis Spreng. (Smilacaceae).* [*Leaf anatomy and evaluation of the antioxidant and allelopathic activities of leaves of Smilax brasiliensis Spreng. (Smilacaceae)*]. Dissertation (Pharmaceutical Sciences), Divinópolis, 103 p.

[11] Amado, P. A., Castro, A. H. F., Alves, S. N., Silva, D. B., Carollo, C. A., Lima, L. A. R. S. (2020). Phenolic compounds: antioxidant and larvicidal potential of *Smilax brasiliensis* Sprengel leaves. *Nat. Prod. Res.,* 34(17): 2545-2553.

[12] Amado, P. A., Castro, A. H. F., Zanuncio, V. S. S., Stein, V. C., Silva, D. B., Lima, L. A. R. S. (2020). Assessment of allelopathic, cytotoxic, genotoxic and antigenotoxic potential of *Smilax brasiliensis* Sprengel leaves. *Ecotoxicol. Environ. Saf.*, 192, 110310.

[13] Pereira, F. L., Oliveira, V. B., Viana, C. T. R., Campos, P. P., Silva, M. A. N., Brandão, M. G. L. (2015). Antihyperlipidemic and antihyperglycemic effects of the Brazilian salsaparrilhas *Smilax brasiliensis* Spreng. (Smilacaceae) and *Herreria salsaparrilha*

Mart. (Agavaceae) in mice treated with a high-refined-carbohydrate containing diet. *Food Res. Int.*, 76(Pt 3), 366-372.

[14] Martins, A. R. (2009). *Morfoanatomia dos órgãos vegetativos e perfil químico de espécies do gênero Smilax L. (Smilacaceae).* [*Morphoanatomy of vegetative organs and chemical profile of species of the genus Smilax L. (Smilacaceae)*]. Thesis (PhD in Plant Biology), Campinas, 160 p.

[15] Amado, P. A., Ferraz, V., Silva, D. B., Carollo, C. A., Castro, A. H. F., Lima, L. A. R. S. (2018). Chemical composition, antioxidant and cytotoxic activities of extracts from the leaves of *Smilax brasiliensis* Sprengel (Smilacaceae). *Nat. Prod. Res.*, 32(5), 610-615.

[16] dos Santos Júnior, H. M., Oliveira, D. F., de Carvalho, D. A., Pinto, J. M. A., Campos, V. A. C., Mourão, A. R. B., Pessoa, C., de Moraes, M. O., Costa-Lotufo, L. V. (2010). Evaluation of native and exotic Brazilian plants for anticancer activity. *J. Nat. Med.*, 64(2), 231-238.

[17] Silva, L. I., Karuppusamy, A., Miyajima, F., Violante, I. M. P., Bieski, I. G. C., Balogun, S. O., Martins, D. T. O. (2017). Antimicrobial and antioxidant activities of selected plants used by populations from Juruena valley, legal Amazon, Brazil. *Int. J. Pharm. Pharm. Sci.*, 9(5), 179 191.

[18] Amado, P. A. (2016). *Avaliação das atividades antioxidante, alelopática, larvicida e dos efeitos citotóxico e genotóxico das folhas de Smilax brasiliensis Spreng. (Smilacaceae)* [*Evaluation of antioxidant, allelopathic, larvicide activities and cytotoxic and genotoxic effects of Smilax brasiliensis Spreng. (Smilacaceae) leaves*]. Dissertation (Pharmaceutical Sciences), Divinópolis, 101 p.

[19] Rodrigues, K., Ramos, D. F., Carrion, L. L., Cursino, L. M. C., Jefreys, M. F., Pedroza, L. S., Osorio, M. I. C., Oliveira, J. L., Andrade, J. I. A., Fernandes, C. C., Nunez, C. V., Silva, P. E. A. (2014). Antifungal activity of Brazilian Amazon plants extracts against some species of *Candida* spp. *Int. J. Phytopharmacol.*, 5(6), 445-453.

[20] Martins, A. R., Bombo, A. B., Soares, A. N., Appezzato-da-Glória, B. (2013). Aerial stem and leaf morphoanatomy of some species of *Smilax*. *Rev. Bras. Farmacogn.*, 23(4), 576-584.

[21] Andreata, R. H. P. (2009). A new species of *Smilax* and a key to all species from Minas Gerais, Brazil. *Syst. Bot.*, 34(1), 28-31.

[22] Appezzato-da-Glória, B., Silva, J. M., Soares, M. K. M., Soares, A. N., Martins, A. R. (2015). Axillary bud and pericycle involved in the thickening process of the rhizophore nodes in *Smilax* species. *Braz. J. Biol.*, 75(3), 718-725.

[23] Martins. A. R., Abreu, A. G., Bajay, M. M., Villela, P. M. S., Batista, C. E. A., Monteiro, M., Alves-Pereira, A., Figueira, G. M., Pinheiro, J. B., Appezzato-da-Glória, B., Zucchi, M. I. (2013). Development and characterization of microssatélite markers for the medicinal plant *Smilax brasiliensis* (Smilacaceae) and related species. *Appl. Plant Sci.*, 1(6), 1-3.

[24] Martins, A. R., Soares, M. K. M., Redher, V. L. G., Bajay, M. M., Villela, P. M. S., Zucchi, M. I., Appezzato-da-Glória, B. (2014). Use of anatomical, chemical, and molecular genetic characteristics in the quality control of medicinal species: a case study of Sarsaparilla (*Smilax* spp.). *Econ. Bot.*, 68(4), 410-425.

[25] Petrica, E. E. A. (2012). Análise de flavonoides por cromatografia líquida de alta eficiência (CLAE), com ensaios *in vitro* do potencial antioxidante em folhas de *Smilax fluminensis* Steud. (Smilacaceae). *Dissertation* (Chemistry), Cuiabá, 100 p.

[26] Mendonça, L. A. B. M., Matias, R., Zanella, D. F. P., Porto, K. R. A., Guilhermino, J. F., Moreira, D. L., Roel, A. R., Pott, A., Carvalho, C. M. E. (2020). Toxicity and phytochemistry of eight species used in the traditional medicine of sul-mato-grossense, Brazil. *Braz. J. Biol.*, 80(3), 574-581.

[27] Lopes, M. E., Charro, G. J., Souza, I. L., Fernandes, H. M., Santos, W. J. S., Lima, C. A. V., Giese, P. I., Melão, J., Farias, T., Oliveira, R. J., Dourado, D. M., Matias, R., Mauro, M. O. (2019). Avaliação do potencial citotóxico de extrato alcoólico de *Smilax fluminensis* em fibroblastos e melanoma murinho. In: *71ª Reunião Anual da*

SBPC, UFMS, Campo Grande/MG, Brazil, 1-3. [Evaluation of the cytotoxic potential of alcoholic extract of Smilax fluminensis in murine fibroblasts and melanoma. In: *71st Annual Meeting of the SBPC*]

[28] Soares, A. N. (2010). *Morfoanatomia, perfil químico e propagação de Smilax fluminensis Steud. (Smilacaceae)* [*Morphoanatomy, chemical profile and propagation of Smilax fluminensis Steud. (Smilacaceae)*]. Dissertation (Science), Piracicaba, 75 p.

[29] Petrica, E. E. A., Sinhorin, A. P., Sinhorin, V. D. G., Júnior, G. M. V. (2014). First phytochemical studies of japecanga (*Smilax fluminensis*) leaves: flavonoids analysis. *Rev. Bras. Farmacogn.*, 24(4), 443-445.

[30] Salas-Coronado, R., Hernández-Carlos, B., Llaguno-Sánchez, J., Santos-Sánchez, N. F. (2017). Phenolic compounds in genus *Smilax* (Sarsaparilla). In: Soto-Hernandez, M., Palma-Tenango, M., Garcia-Mateos, R. (Ed.). *Phenolic Compounds - Natural Sources, Importance and Applications.* London: IntechOpen, 233-260.

In: The Book of Flavonoids
Editors: Chao-Hui Feng et al.
ISBN: 978-1-68507-376-3

Chapter 7

PHENOLIC COMPOUNDS IDENTIFIED IN *SOLANUM LYCOCARPUM* AND THEIR BIOLOGICAL ACTIVITIES

Paula Avelar Amado[1,2,*],
Jéssica Elvira Siqueira Medeiros[1,*],
Iara Cotta Mendes[1,*], ***Izabela Caputo Assis Silva[1,2]***,
Lucas Santos Azevedo[1,2], ***Iasmin Fernandes Emediato[1]***,
Thaís Paula Rodrigues Gonçalves[1,2],
Adriano Lopes da Silva[1,3]
and Luciana Alves Rodrigues dos Santos Lima[1,2,3,†]
[1]Phytochemistry Laboratory,
Universidade Federal de São João Del Rei, Divinópolis, MG, Brazil
[2]Programa de Pós-Graduação em Biotecnologia da Universidade Federal de São João Del Rei, Divinópolis, MG, Brazil
[3]Programa de Pós-Graduação em Ciências Farmacêuticas da Universidade Federal de São João Del Rei, Divinópolis, MG, Brazil

* These authors contributed equally to this work.
† Corresponding Author's E-mail: luarsantos@ufsj.edu.br.

Abstract

Plants have the ability to synthesize a wide variety of chemical compounds that perform or modify important biological functions. Many extracts prepared from plants have biological activities in assays *in vitro* and *in vivo*. Among the compounds found in nature are phenolic compounds, which can be present in vegetables in free forms or complexed with sugars. *Solanum lycocarpum* (Solanaceae), popularly known as "lobeira," it is a shrub of about 2 to 5 meters, which blooms and bears fruit throughout the year; it is widely distributed in the Brazilian Cerrado and is used in folk medicine and for the manufacture of sweets and jellies. Phenolic compounds such as flavonoids (for example, apigenin, and kaempferol), which have the capacity to act as antioxidants, and capture free radicals and reactive oxygen species, and phenolic acids (for example, chlorogenic acid, caffeic acid, 3,5-di-*O*-*E*-caffeoylquinic acid, 4,5-di-*O*-*E*-caffeoylquinic acid, *O*-coumaroyl caffeoylquinic acid, and 3,4,5-tri-*O*-*E*-caffeoylquinic acid), which are associated with the prevention of chronic diseases such as diabetes mellitus, cardiovascular disease and other diseases related to oxidative stress, in addition to other derivatives, have already been identified in the leaves and fruits of S. *lycocarpum.* Antioxidant, antibacterial, anti-inflammatory, antitumor, antinociceptive, allelopathic and larvicidal activities have been reported for this species, suggesting that the presence of phenolic compounds in *S. lycocarpum* is responsible for the biological effects exhibited. This review comprehensively highlights aspects related to the extraction, characterization and biological activities of phenolic compounds from *S. lycocarpum*.

Keywords: flavonoid, phenolic acid, Solanaceae, Brazilian Cerrado, Lobeira

1. Introduction

The species *Solanum lycocarpum*, found in the Brazilian Cerrado, is a shrub that develops in unfavorable conditions, such as acidic soil and ground poor in nutrients, and is resistant to long periods of drought and arid climates, in addition to burning. Its fruiting ranges from May to August, with approximately 40 to 100 fruits. The fruit is 13 cm in diameter and green in color, but when ripening has brown tones, and presents several biological activities described in the literature [1].

The ripe fruits can be consumed in natura, as well as in the manufacture of jellies and jams. The fruits have high amounts of vitamin C, total soluble sugars (AST), sucrose, phosphorus and iron when compared to other fruits such as pineapple, banana, orange, and mango, among others [2]. The powder form of green fruits is applied in the treatment of diabetes, and their components include the alkaloids, solasodine and solamargine, which contain a chemical structure similar to steroidal hormones [3].

Several secondary metabolites are commonly described in this species, making them important for the pharmaceutical industry. These substances, for the most part, are alkaloids such as atropine, scopolamine, hyoscyamine and solasodine, and are used in medicine for the treatment of various clinical conditions such as respiratory diseases, Parkinson's disease and anti-inflammatory disease [4].

The antioxidant activity for *S. lycocarpum* has already been described in the literature, and several compounds that promote this activity have been identified, such as chlorogenic acid and caffeic acid; these are associated with the prevention of chronic diseases such as diabetes mellitus, and cardiovascular diseases, as well as others related to oxidative stress [4].

This review comprehensively highlights aspects related to phenolic compounds found in this species and their biological activities.

2. Phenolic Compounds

Phenolic compounds, such as phenolic acids and flavonoids, are usually consumed by the population, being found in foods such as beans and fruits, or present in herbal medicines, and are considered safe because they have low toxicological activity [5]. These compounds have been investigated as therapeutic alternatives for the treatment of a large number of pathologies [6]. Their chemical structure is formed by an aromatic ring with six carbons (C6) and may have one or more hydroxyl groups [7].

Phenolic compounds, such as flavonoids and phenolic acids, were previously described in *S. lycocarpum* species. The flavonoids apigenin and kaempferol have been identified in the leaves [4]. In the ripe fruits, phenolic acids such as chlorogenic acid, caffeic acid [2, 8], 3,5-di-*O-E*-caffeoylquinic acid, 4,5-di-*O-E*-caffeoylquinic acid, *O*-coumaroyl caffeoylquinic acid, 3,4,5-tri-*O-E*-caffeoylquinic acid, *O*-coumaroyl dicaffeoylquinic acid and *O*-caffeoyl dicoumaroylquinic acid were found [9].

Phytochemical screening in several studies revealed the presence of flavonoids, saponins, tannins and coumarins in the leaves [10], ripe fruits [1, 11, 12], unripe fruits [3, 13], and plant drugs from fruits [14] of *S. lycocarpum*, with these compounds being correlated to many biological activities described for this species.

3. Biological Activities

The species *S. lycocarpum* has been evaluated in relation to several biological activities. In part, the effects reported for this species have been attributed to the identified phenolic compounds.

Morais et al. [2] evaluated the antioxidant, antibacterial and cytotoxic potential of ripe fruits of *S. lycocarpum.* In the assessment of antioxidant activity using the 2,2-diphenyl-1-picrylhydrazyl (DPPH) method, dichloromethane and ethyl acetate fractions showed better antioxidant effects than the reference compound 2,6-di-tert-butyl-4-methylphenol (BHT). The extract and fractions exhibited an antibacterial effect on *Listeria monocytogenes* and *Bacillus cereus.* The samples were also active against *Streptococcus mutans* and *Klebsiella pneumoniae* (hexane fraction), *Streptococcus pyogenes*, *Staphylococcus aureus* and *Enterococcus faecalis* (dichloromethane fraction) and *S. pyogenes* (hydroethanol fraction). The extract and hydroethanol fraction showed moderate cytotoxicity, while dichloromethane and ethyl acetate fractions exhibited little cytotoxic activity against the LLC-MK2 cell line [2]. In another study, Morais et al. [8] investigated the action of ethanol extract and ethyl acetate fraction on the inhibition of the growth of solid Ehrlich tumors. The extract and fraction, at concentration of 100 mg/Kg, significantly reduced the thickness of the paw of mice when compared to the control group, inhibiting between 49.2% and 61.3% of tumor growth, respectively [8]. These activities may be due to the presence of cinnamic acids, such as caffeic and chlorogenic acids, which were identified in extracts and fractions of ripe fruits [2, 8].

Costa et al. [4] demonstrated the antioxidant, antibacterial, anti-inflammatory and cytotoxic potential of leaves of *S. lycocarpum*. The dichloromethane and ethyl acetate fractions were the most potent antioxidants. The hydroethanol fraction (75 and 150 mg/kg) and ethyl acetate fraction (150 mg/kg) exerted a significant anti-inflammatory action in the mice paw edema-induced carrageenan. The suppression of local edema by the highest dose of the hydroethanol and ethyl acetate fractions was 60% and 80%, respectively, 4h after the injection of carrageenan, and 67% and 75%, respectively, 6h after the inflammatory stimulus. The fractions showed antibacterial activity against Gram-positive bacteria *L. monocytogenes*. The dichloromethane and hydroethanol fractions were also active against *B. cereus*, hexane fraction against *S. aureus*, *S. mutans* and *E. coli*, dichloromethane fraction against *S. aureus* and *K. pneumoniae*, and ethyl acetate fraction against *P. aeruginosa*. The ethyl acetate fraction presented less of a cytotoxic effect on the LLC-MK2 cell line, dichloromethane fraction showed moderate cytotoxic activity, and other samples exhibited the highest cytotoxicity. These activities can be attributed to the flavonoids apigenin and caempferol which were identified in the dichloromethane and ethyl acetate fraction, respectively [4].

Morais et al. [9] also evaluated the anti-inflammatory and anti-nociceptive effect of ethanol extract obtained from ripe fruits. Ethanol extract showed significant anti-edematogenic activity in the model of paw edema induced by carrageenan. At 1, 2, 4 and 6 hours after the administration of carrageenan, with a dose of 300 mg/kg, the extract inhibited paw edema by 80.00%, 76.92%, 88.89% and 90.91%, respectively. In the fourth hour after the administration, extracts at doses of 30 and 100 mg/kg inhibited paw edema by 42.22% and 37.78%, respectively. The ethanol extract promoted the inhibition of nociception induced by acetic acid. Abdominal contortions in mice were reduced by 91.09%, 91.98% and 92.84% at doses of 30, 100 and 300 mg/kg, respectively, compared to the control group. In the formalin-induced nociception assay, the extract at doses of 100 and 300 mg/kg reduced the licking time by 53.85% and 81.04%, respectively, in the first phase. The extract also reduced the licking time in the second phase by 77.30%, 79.50% and 99.35%, at doses of 30, 100 and 300 mg/kg, respectively. In the hot plate test, ethanol extract at a dose of 300 mg/kg significantly increased the latency of the reaction 90 and 120 minutes after treatment. These activities can be attributed, in part, to the presence of phenolic acids identified in the ethanol extract of *S. lycocarpum* [9].

Bahia et al. [11] investigated the antioxidant, antibacterial and allelopathic activities of methanol extract and fractions from ripe fruits of *S. lycocarpum*. The results indicated that the antioxidant activity was more pronounced for the dichloromethane fraction, which showed a better activity than BHT. In antibacterial activity, *S. aureus* was more sensitive to hexane fraction, with a minimal inhibitory concentration (MIC) value of 250 μg/mL. The allelopathic potential on *Allium cepa* seeds showed promising results, with predominant effects of the inhibition of hypocotyls and radicles in the lower concentration tested (50 μg/mL). The best result for the inhibition of radicle and hypocotyl growth was found in the dichloromethane fraction at 100 μg/mL. Phytochemical tests revealed the presence of phenolic compounds, such as coumarins, flavonoids and tannins, which may have contributed in part to the detected biological activities [11].

Mendes et al. [1] verified the phytotoxic potential of ethanol extract from ripe fruits on *Lactuca sativa* and *Allium cepa* seeds. The extract completely inhibited the growth of the hypocotyl and radicle of *L. sativa* in all tested concentrations (125, 250 and 500 μg per plate). For *A. cepa* seeds, the concentration of 500 μg per plate was able to inhibit the growth of the hypocotyl and radicle by 100%. Phytotoxic activity can be correlated to the presence of coumarins, flavonoids and condensed tannins present in the extract [1]. Another study showed the allelopathic activity of hexane and methanol extracts obtained from the unripe fruits of *S. lycocarpum,* with the samples exhibiting heterogeneous effects on the growth of the hypocotyl and radicle of *A. cepa* seeds, as well as inhibitory effects on *L. sativa* seeds. This allelopathic effect may be correlated to presence of flavonoids observed in the phytochemical screening [3]. Allelopathic activity was also evaluated for ethanol extract and fractions of leaves from *S. lycocarpum* tested at concentrations of 250, 500 and 1000 μg/mL. The samples inhibited the radicle and hypocotyl growth of *L. sativa* and *A. cepa* seeds. These results can be attributed to presence of secondary metabolites, such as flavonoids and coumarins [10].

Pereira et al. [13] investigated the larvicidal activity of methanol extract and fractions of unripe fruits of *S. lycocarpum* on *Culex quinquefasciatus* larvae. The results demonstrated that the dichloromethane and ethyl acetate fractions exhibited the greatest larvicidal effect at a concentration of 200 mg/L (83.3 and 86.7%, respectively). The methanol extract and dichloromethane, ethyl acetate and hydromethanol fractions demonstrated an effect on *C. quinquefasciatus*, with lethal concentration (LC_{50}) values of 126.24, 75.13, 83.15 and 207.05 mg/L, respectively. The larvicidal effect was related to the presence of coumarins, flavonoids and tannins in the samples [13].

De Araújo et al. [15] evaluated the preliminary toxicity tests of ethanolic extract of *Solanum lycocarpum* fruits against brine shrimp larvae. The extract was fractionated with various solvents for toxicity testing against the *Artemia salina* larvae and the hydroalcoholic fraction exhibited considerable cytotoxicity (LC_{50} = 285.546 g/mL). The phytochemical assays showed the presence of phenols, tannins, saponins, alkaloids and free steroids.

Flavonoids, particularly apigenin and luteolin, have shown potent activity against oxidative stress [16]. Phenolic compounds, such as flavonoids, possess anti-inflammatory activity, which can be correlated to their antioxidant activity [17]. In addition, recent studies have shown anti-inflammatory, antimicrobial, and antioxidant properties for kaempferol and kaempferol rhamnosides [18, 19]. Cinnamic acids, such as caffeic and chlorogenic acids, are also well known for their antioxidant properties [20, 21]. Chlorogenic acid exhibits several important biological activities, including inhibition of the mutagenicity of carcinogenic compounds and the inhibition of HIV-1 integrase, as well as antioxidant and anti-inflammatory properties [22]. Caffeic acid presents important biological activities in the treatment of liver damage by inhibiting cyclooxygenases (COX-1 and COX-2) [23]. Phenolic compounds also exhibited allelopathic activity. They have the potential to affect the elasticity of the cell wall, in addition to blocking mitochondrial respiration [24]. Allelopathic effects are also mediated by coumarin compounds [25], which have previously been identified as potent inhibitors of both plant growth and seed germination, and have the ability to block mitosis [26].

CONCLUSION

The antioxidant, antibacterial, anti-inflammatory, anti-tumor, anti-nociceptive, allelopathic and larvicidal activities were confirmed, based on known information that is provided in different studies. Other groups suggest that the biological activities reported may be associated with the presence of phenolic compounds, especially flavonoids and phenolic acids, identified in the extract and fractions. The results reported in this review show that *S. lycocarpum* could be a promising source of new drugs.

ACKNOWLEDGMENTS

The authors are grateful to the Universidade Federal de São João Del Rei (UFSJ), Conselho Nacional de Desenvolvimento Científico e Tecnológico (CNPq), Coordenação de aperfeiçoamento de Pessoal de Nível Superior (CAPES) and Fundação de Amparo à Pesquisa do Estado de Minas Gerais (FAPEMIG) for the financial support. "This study was financed in part by the Coordenação de Aperfeiçoamento de Pessoal de Nível Superior - Brasil (CAPES) - Finance code 001."

REFERENCES

[1] Mendes, I. C., Amado, P. A., Morais, M. G., Lima, L. A. R. S. (2020). Phytochemical screening and evaluation of phytotoxic activity of *Solanum lycocarpum* (Solanaceae) ripe fruit. *Acta Brasiliensis*, 4(1): 45-48.

[2] Morais, M. G., Costa, G. A. F., Aleixo, A. A., Oliveira, G. T., Alves, L. F., Duarte-Almeida, J. M., Ferreira, J. M. S, Lima, L. A. R. S. (2015). Antioxidant, antibacterial and cytotoxic potential of the ripe fruits of *Solanum lycocarpum* A. St. Hil. (Solanaceae). *Nat. Prod. Res.*, 29(5): 480-483.

[3] Silva, I. C. A., Fonseca, J. C., Coimbra, M. C., Duarte-Almeida, J. M., Lima, L. A. R. S. (2019). Total phenol and total flavonoid content, antioxidant and allelopathic activities of extracts of the unripe fruits of *Solanum lycocarpum*. *Period. Tchê Quím.*, 16(31): 390-397.

[4] Costa, G. A. F., Morais, M. G., Saldanha, A. A., Silva, I. C. A., Aleixo, A. A., Ferreira, J. M. S., Soares, A. C., Duarte-Almeida, J. M., Lima, L. A. R. S. (2015). Antioxidant, antibacterial, cytotoxic, and anti-inflammatory potential of the leaves of *Solanum lycocarpum* A. St. Hil. (Solanaceae). *Evid. -Based Complement. Alternat. Med.,* 2015: 315987.

[5] Hollman, P. C. H., Katan, M. B. (1999). Health effects and bioavailability of dietary flavonols. *Free Radic. Res.*, 31(S1): 75-80.

[6] Graf, B. A., Milbury, P. E., Blumberg, J. B. (2005). Flavonols, flavones, flavanones, and human health: epidemiological evidence. *J. Med. Food*, 8(3): 281-290.

[7] Michalack, A. (2006). Phenolic compounds and their antioxidant activity in plants growing under heavy metal stress. *Pol. J. Environ. Stud.*, 15(4): 523-530.

[8] Morais, M. G., Costa, G. A. F., Silva, I. C. A., Silva, A. G., Ribeiro, R. I. M. A., Ferreira, J. M. S., Duarte-Almeida, J. M., Pinto, F. C. H., Lima, L. A. R. S. (2017). Inhibiting the development of solid Ehrlich tumor by extract of the ripe fruits of *Solanum lycocarpum*. In: Collins, H. R. (Ed.). *Caffeic Acid, Potential Uses and Health Benefits*. New York: Nova Science Publishers, 83-96.

[9] Morais, M. G., Saldanha, A. A., Rodrigues, J. P. C., Mendes, I. C., Ferreira, L. M., Amado, P. A., Farias, K. S., Zanuncio, V. S. S., Silva, D. B., Pinto, F. C. H., Soares, A. C., Lima, L. A. R. S. (2020). Chemical composition, antioxidant, anti-inflammatory and antinociceptive activities of the ethanol extract of ripe fruits of *Solanum lycocarpum* St. Hil. (Solanaeceae). *J. Ethnopharmacol.,* 262: 113125.

[10] Medeiros, J. E. S., Silva, I. C. A., Amado, P. A., Lima, L. A. R. S. (2020). Phytochemical screening and the evaluation of allelopathic activity of extract and fractions of *Solanum lycocarpum* A. St. Hil. (Solanaceae) leaves. In: Souza, V. A. (Ed.). *Solanum: An Overview*. New York: Nova Science Publishers, 181-198.

[11] Bahia, J. P., Silva, V. C. B., Morais, M. G., Costa, G. A. F., Ferreira, J. M. S., Duarte-Almeida, J. M., Lima, L. A. R. S. (2018). Total phenolic and total flavonoid content, antioxidant, antibacterial, and allelopathic activities of the methanol extract and fractions from *Solanum lycocarpum* A. St. Hil. (Solanaceae). In: Rocha, I. S. (Ed.). *Solanaceae: Cultivation, Nutrition and Health.* New York: Nova Science Publishers, 63-78.

[12] Morais, M. G., Mendes, I. C., Amado, P. A., Silva, I. C. A., Lima, L. A. R. S. (2018). Ripe fruits of *Solanum lycocarpum* A. St. Hil. (Solanaceae): composition and biological activities. In: Rocha, I. S. (Ed.). *Solanaceae: Cultivation, Nutrition and Health.* New York: Nova Science Publishers, 43-62.

[13] Pereira, T. M., Silva, V. C. B., Ribeiro Neto, J. A., Alves, S. N., Lima, L. A. R. S. (2014). Larvicidal activity of the methanol extract and fractions of the green fruits of *Solanum lycocarpum* (Solanaceae) against the vector *Culex quinquefasciatus* (Diptera: Culicidae). *Rev. Soc. Bras. Med. Trop.*, 47(5): 646-648.

[14] Araújo, M. G. F., Galeane, M. C., Castro, A. D., Salgado, H. R. N., Almeida, A. E., Cunha, W. R., Veneziani, R. C. S., Moreira, R. R. D. (2010). Pharmacognostical evaluation of fruits of *Solanum lycocarpum* A. St. -Hill. (Solanaceae). *Pharmacogn. J.*, 2(9): 248-253.

[15] De Araujo, M. G. F., Cunha, W. R., Veneziani, R. C. S. (2010). Estudo fitoquímico preliminar e bioensaio toxicológico frente a larvas de *Artemia salina* Leach. de extrato obtido de frutos de *Solanum lycocarpum* A. St.-Hill (Solanaceae) [Phytochemical screening and toxicological bioassay with brine shrimp larvae (*Artemia salina* Leach) of an extract obtained from fruits of *Solanum lycocarpum* A. St. -Hill.]. *Rev. Cienc. Farm. Basica Apl.*, 31(2): 205-209.

[16] Nielsen, S. E., Young, J. F., Daneshvar, B., Lauridsen, S. T., Knuthsen, P., Sandstrom, B., Dragsted, L. O. (1999). Effect of parsley (*Petroselinum crispum*) intake on urinary apigenin

excretion, blood antioxidant enzymes and biomarkers for oxidative stress in human subjects. *Br. J. Nutr.*, 81(6): 447-455.

[17] Loganayaki, N., Siddhuraju, P., Manian, S. (2012). Antioxidant, anti-inflammatory and anti-nociceptive effects of *Ammannia baccifera* L. (*Lythracceae*), a folklore medicinal plant. *J. Ethnopharmacol.*, 140(2): 230-233.

[18] Rho, H. S., Ghimeray, A. K., Yoo, D. S., Ahn, S. M., Kwon, S. S., Lee, K. H., Cho, D. H., Cho, J. Y. (2011). Kaempferol and kaempferol rhamnosides with depigmenting and anti-inflammatory properties. *Molecules*, 16(4): 3338-3344.

[19] Tatsimo, S. J. N., Tamokou, J. D., Havyarimana, L., Csupor, D., Forgo, P., Hohmann, J., Kuiate, J. R., Tane, P. (2012). Antimicrobial and antioxidant activity of kaempferol rhamnoside derivatives from *Bryophyllum pinnatum. BMC Res. Notes*, 5: 158.

[20] Duarte-Almeida, J. M., Salatino, A., Genovese, M. I., Lajolo, F. M. (2011). Phenolic composition and antioxidant activity of culms and sugarcane (*Saccharum officinarum L.)* products. *Food Chem.*, 125(2): 660-664.

[21] Sato, Y., Itagaki, S., Kurokawa, T., Ogura, J., Kobayashi, M., Hirano, T., Sugawara, M., Iseki, K. (2011). In vitro and in vivo antioxidant properties of chlorogenic acid and caffeic acid. *Int. J. Pharm.*, 403(1-2): 136-138.

[22] Sefkow, M. (2001). First efficient synthesis of clorogenic acid. *Eur. J. Org. Chem.*, 2001: 1137-1141.

[23] Buzzi, F. C., Franzoi, C. L., Antonini, G., Fracasso, M., Cechinel Filho, V., Yunes, R. A., Niero, R. (2009). Antinociceptive properties of caffeic acid derivates in mice. *Eur. J. Med. Chem.*, 44(11): 4596-4602.

[24] Weir, T. L., Park, S. W., Vivanco, J. M. (2004). Biochemical and physiological mechanisms mediated by allelochemicals. *Curr. Opin. Plant Biol.*, 7(4): 472-479.

[25] Baratto, L., Lang, K. L., Vanz, D. C., Reginatto, F. H., Oliveira, J. B., Falkenberg, M. (2008). Investigação das atividades alelopática e antimicrobiana de *Mikania laevigata* (Asteraceae) obtida de cultivos hidropônico e tradicional [Investigation of the allelopathic and antimicrobial activities of *Mikania laevigata* (Asteraceae) obtained in hydroponic and traditional crops]. *Braz. J. Pharmacogn.*, 18(4): 577-582.

[26] Abenavoli, M. R., Cacco, G., Sorgoná, A., Marabottini, R., Paolacci, A. R., Ciaffi, M., Badiami, M. (2006). The inhibitory effects of coumarin on the germination of durum wheat (*Triticum turgidum* ssp. durum, cv. Simeto) seeds. *J. Chem. Ecol.*, 32(2): 489-506.

In: The Book of Flavonoids
Editors: Chao-Hui Feng et al.
ISBN: 978-1-68507-376-3

Chapter 8

Flavonoids from *Bauhinia holophylla* (Bong.) Steud. (Fabaceae:Cercideae)

Thaiz Rodrigues Teixeira,
João Máximo de Siqueira,
Flávio José Leite dos Santos,
Carlos Alexandre Carollo
and Ana Hortência Fonseca Castro*
Universidade Federal de São João del-Rei,
Campus Centro Oeste, Divinópolis, MG, Brazil

Abstract

Bauhinia holophylla, also known as "pata-de-vaca" or "unha-de-vaca," is a woody species from the Brazilian Cerrado, which leaves are used in folk medicine to treat infections and diabetes. Preliminary phytochemical studies showed the presence of steroidal glycosides,

* Corresponding Author E-mail: acastro@ufsj.edu.br.

triterpenes, lactones and flavonoids in the leaves. This work aimed to evaluate and identify the flavonoids present in the leaves of *B. holophylla* and to relate them to the medicinal properties attributed to this species. For the phytochemical study, the crude hydroethanolic extract was obtained by percolation, which was subjected to liquid-liquid partition with *n*-hexane, chloroform, ethyl acetate and *n*-butanol. The isolation and identification of substances present in the ethyl acetate fraction were performed using the techniques of Classical Column Liquid Chromatography (CLC), monitored by Thin Layer Chromatography (TLC) and High Efficiency Liquid Chromatography coupled to DAD detectors and a mass spectrometer IES-Q-QTOF (HPLC-DAD-ESI-MS). Chromatographic analyses by HPLC-DAD-IES-EM suggested the presence of myricetin, kaempferol, and quercetin derivatives such as myricetin-*O*-pentoside, myricetin-*O*-deoxyhexoside, quercetin-*O*-hexoside, quercetin-*O*-xilopyranoside, kaempferol-3-*O*-glucoside, quercetin-*O*-pentoside, quercetin-*O*-deoxyhexoside, isorhamnetin-3-*O*-hexoside, kaempferol-*O*-pentoside, luteolin-deoxyhexose, and isorhamnetin in the leaves of *B. holophylla*. From a medicinal point of view, quercetin and its derivatives have high antioxidant and anti-inflammatory activities. These results showed the biotechnological potential of *B. holophylla* to produce bioactive compounds of economic and medicinal interest and suggest more detailed studies aimed at the isolation and identification of phenolic compounds.

Keywords: quercetin derivatives, Brazilian Cerrado, medicinal plant, phenolic compounds

INTRODUCTION

For many years, medicinal plants have played an important role in the prevention and treatment of human diseases and continue to be a source of innovation for the discovery of new drugs (Dar et al. 2017; Salehi et al. 2019). The Brazilian territory has a great diversity of plants and ranks sixth in the world in terms of biodiversity. With an area of approximately 2 million km^2, the Cerrado biome covers about 22% of the country, being the second largest of the six biomes in Brazil, behind only the Amazon Forest (Ribeiro and Walter 1998). Among the countless plant species of medicinal interest that inhabit the Cerrado, plants of the genus *Bauhinia* belonging to the Fabaceae family (*Leguminosae*) stand out, comprising approximately 300 species popularly known as "pata-de-vaca," "unha-de-vaca," "unha-de-boi" or "bauinia"(Vaz and Tozzi 2003; Vaz et al. 2010).

Several species of *Bauhinia* are used in Brazil for the treatment of infections, with analgesic, hypoglycemic, antidiarrheal, anti-inflammatory, and diuretic properties (Cechinel Filho 2009; Bonilha et al. 2015). Phytochemical studies indicate the presence of triterpenes, lactones, steroids, tannins, quinines, and flavonoids, to which therapeutic properties are attributed (Farag et al. 2015; Pinheiro et al. 2017; Schmidt et al. 2018). The hypoglycemic property of *Bauhinia holophylla* has been widely studied (Silva et al. 2015; Pinheiro et al. 2017; Camaforte et al. 2019), in addition to other activities such as antiulcerogenic (Rozza et al. 2015), antiviral (dos Santos et al. 2019) and cytotoxic, apoptotic and mutagenic/antimutagenic activity (Ribeiro et al. 2018).

Flavonoids are considered one of the most prevalent classes of metabolites in *Bauhinia* species, with flavonols being the main subclass, followed by flavones, flavans and flavanones. Bioguided studies of extracts and fractions of the leaves of *B. holophylla* have led to the identification of several glycosylated flavonoids and aglycone (Rozza et al. 2015; Camaforte et al. 2019).

Flavonoids are almost exclusively present as glycosides in plants. The substitution of sugars in the aglycone generally occurs through phenolic hydroxyl groups forming *O*-glycosides or linked through C-C bonds forming *C*-glycosides (Plazonic et al. 2009). There are complex glycosylated flavonoids with up to five sugar residues and some studies have shown that different glycosylated flavonoids perform different biological activities (Rochfort et al. 2006). Therefore, we highlight the importance of obtaining accurate structural information, including glycosylation positions, interglycoside bonds and the nature of aglycones and sugar rings (Qin et al. 2017; Lei et al. 2018).

Flavonoids have characteristic absorption spectra in the ultraviolet (UV) spectrum determined by the common core of benzopyran, with two absorption maximums, one occurring between 240–285 nm (band II) and the other at 300–400 nm (band I). In general, band II can be considered as derived from ring A and band I as derived from ring B. The more hydroxyl groups in the nucleus, the greater the bathochromic effect; and, consequently, the spectra shift towards the longest lengths of the wave. Regarding the methylation or esterification of the hydroxyl groups, these groups do not alter, in general, the spectra, except in the characteristic hydroxyl of the flavonoids (in C-3) or C-4', when a hypsochromic effect is observed in the band of higher wavelengths (Zuanazzi and Montanha 2010).

In addition to the characterization of flavonoids by UV spectroscopy, nuclear magnetic resonance (NMR) and mass spectroscopy (MS) techniques are generally the most used for structural elucidation of flavonoids and their glycosylated derivatives. However, in phytochemical analysis, it is difficult to isolate enough flavonoids for complete characterization by NMR. This is mainly due to the small amount of sample available and in this case, MS-based methods are required since they are quite sensitive (Vukics and Guttman 2008). Thus, several methodologies based on liquid chromatography-tandem mass spectroscopy (LC-MS/MS) have been developed to identify this class of secondary metabolites (Cuyckens and Claeys 2004; Vukics and Guttman 2008; Lei et al. 2018).

Among the most used techniques for obtaining mass spectra and structural investigation of flavonoids, electrospray ionization (ESI) stands out as it allows the cleavage of the C ring by the Retro-Diels-Alder (RDA) fragmentation mechanism, generating the ions $^{ij}A^{-/+}$ and $^{ij}B^{-/+}$ and providing the diagnostic fragments for identification of the flavonoids, as well as the number and type of constituents of the A and B ring (Ma et al. 1997; Cuyckens and Claeys 2004).

Therefore, this study aimed to evaluate and identify the flavonoids present in the leaves of *B. holophylla* using an HPLC-DAD-ESI-MS method and to relate them to the medicinal properties attributed to this species.

Material and Methods

Plant Material

Leaves of *Bauhinia holophylla* (Bong.) Steud. (Fabaceae: Cercideae) were collected in the Brazilian Cerrado, located in the municipality of Ijaci, south of the State of Minas Gerais, Brazil (21º09'97"S and 44º55'65" W GRW, at 835 m altitude) in March 2014 (SISBIO n° 24542-3). The samples were identified by Dr. Andreia Fonseca Silva and the vouchers were deposited at the PAMG Herbarium (PAMG 57021) of the Agricultural Research Company of Minas Gerais (EPAMIG). This study has permission to access the components of the plant genetic heritage (n° 010500/2014-6/CNPq/CGEN/MMA) and is registered on the SisGen Platform (Register A12A940), according to the Brazilian Biodiversity Law (13.123/2015). The plant material was dried in a ventilated oven at 40°C for 24h and pulverized in a knife mill.

Obtaining the Extract and Fractions from *B. holophylla* Leaves

The dried and powdered leaves (200 g) of *B. holophylla* were extracted by exhaustive percolation using 70% ethanol as the extraction solvent for 7 days. The solvent was removed on a rotary evaporator at 50°C under reduced pressure, obtaining the crude hydroethanolic extract (CE, 23.30 g).

The CE was solubilized in 70% ethanol (350 mL) and subsequently 200 mL of this extract was subjected to liquid-liquid partition in a separating funnel with *n*-hexane, chloroform, ethyl acetate, and *n*-butanol. The solvents were removed using a rotary evaporator at 50°C, under reduced pressure resulting in hexane (Hex, 1.48 g), chloroform ($CHCl_3$, 0.14 g), ethyl acetate (EtOAc, 1.21 g), butanolic (ButOH, 0.55 g) and hydroethanolic fractions (EtOH, 2.64 g).

Purification of the EtOAc Fraction from *B. holophylla*

The EtOAc fraction (0.60 g) was purified by Classical Column Liquid Chromatography (CLC, silica gel 60, 70-230 Mesh) using *n*-hexane, ethyl acetate, methanol, and distilled water in a polarity gradient as eluents. A total of 62 fractions of approximately 10 to 20 mL were obtained, which were analyzed by Thin Layer Chromatography (TLC, silica gel 60 F254). Subfractions 26, 38, 46, 49, 50, and 57 were selected for analysis by High Performance Liquid Chromatography (HPLC) coupled to detectors with diode array (DAD) and Mass Spectrometry (MS).

HPLC-DAD-EM Analyses of *B. holophylla* Subfractions

A Shimadzu Prominence Liquid Chromatography (LC-20AD) coupled to a diode array detector (DAD) and a MicrOTOF-Q III mass spectrometer (Bruker Daltonics), monitoring between 200 and 800 nm and operating in positive/negative ionization mode (*m/z* 120–1300), were used to analyze *B. holophylla* subfractions. The mass spectra were obtained by electrospray ionization source and quadrupole and Time-Of-Flight (TOF) analyzers. The chromatographic column was a Kinetex C18 (4.6 x 2.6 mm, 100 Å, Phenomenex). The injection volume was 1 μL of the sample (prepared at a concentration of 1 mg mL^{-1}), the flow rate was 0.3 mL min^{-1}. The mobile phase was composed of acetonitrile (solvent B) and deionized water (solvent A) with the addition of 0.1% formic acid (v/v). The applied elution profile was: 0–2 min – 3% B, 2–25 min – 3 to 25% B, 25–40 min – 25 to 80% B, 40–43 min – 80% B, 43–44 min – 3% B and 44–48 min – 3% B. For mass spectrometry analysis, nitrogen gas was used as a nebulizer (4 Bar), dry (9 L min^{-1}) and collision gas.

Results and Discussion

Identification of the Compounds in the Subfractions of *B. holophylla* Leaves

The subfractions 26, 38, and 46 of *B. holophylla* leaves were analyzed by HPLC-DAD-ESI-MS, and their constituents were identified based on ultraviolet (UV) spectra, high resolution mass spectrometry data, including fragmentation patterns data by ESI (-)-MS/MS, and by comparing results reported in the literature (Ablajan et al. 2006; Farag et al. 2015; Rozza et al. 2015; Camaforte et al. 2019; dos Santos et al. 2019). The representative base peak chromatograms (BPC) are shown in Figure 1 and the main flavonoids identified from subfractions in Table 1.

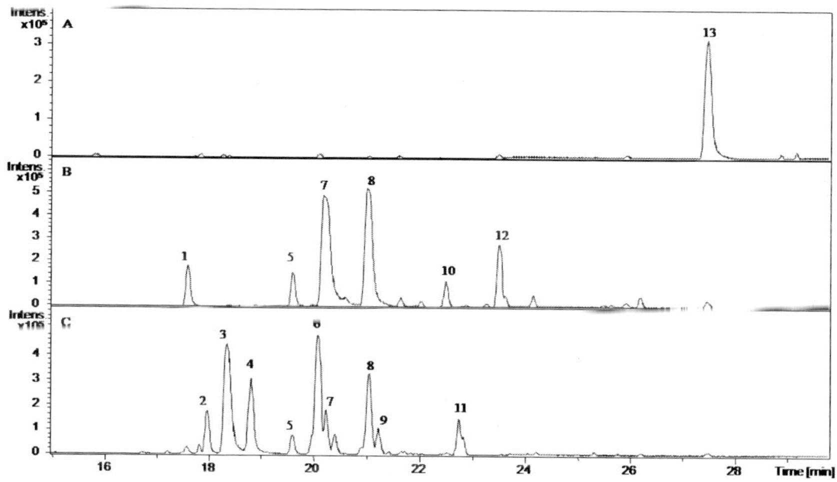

Figure 1. Base peak chromatogram (BPC) of the subfractions 26 (A), 38 (B), and 46 (C) in negative ion mode.

The patterns of fragmentation of markers reported in the literature compared to those obtained in the present study allowed us to infer that the flavonoids identified in the analyzed samples were flavonol-3-*O*-glycosides such as quercetin, myrcetin, and kaempferol. For these types of flavonoids, simple regular cleavages of consecutive glycosidic bonds in the ion $[M - H]^-$ under negative ionization mode are observed (Chen et al. 2015). Thus, flavonol-3-*O*-glycosides were deducted by the consecutive losses of aglycone fragment ions *m/z* 300, *m/z* 271, *m/z* 255, *m/z* 243 to isorhamnetin; and *m/z* 300, *m/z* 271, *m/z* 255, *m/z* 243, *m/z* 179 for quercetin derivatives; *m/z* 316, *m/z* 287, *m/z* 271, *m/z* 179 for myricetin derivatives and *m/z* 285, *m/z* 255, *m/z* 227, *m/z* 179 for kaempferol derivatives from [M – H] glycosylated with different types of sugars.

In addition to the fragmentation patterns observed for the samples analyzed, the UV-visible absorption profiles of the identified substances showed UV λ_{max} absorption bands between 250 and 270 and 340 and 360 for bands II and band I, respectively, which also characterizes 3-*O*-substituted flavonols (Markham 1982).

Using HPLC-PAD-ESI-MS, Rozza et al. (2015) and Camaforte et al. (2019) identified the presence of flavonol-*O*-glycosides derived from quercetin, myricetin, and kaempferol as the main chemical constituents of the hydroethanolic extract (EtOH:H_2O; 7:3 v/v) of *B. holophylla* leaves.

In the analysis of the mass spectrum of subfraction 26 it was possible to observe an intense ion corresponding to the deprotonated molecule of *m/z* 315 assigned to isorhamnetin (Figure 1A), as well as product ions of *m/z* 271 $[M - 44 - H]^-$ and *m/z* 243 $[M - 72 - H]^-$ that indicated the loss of the fragments $[M - H - CO]^-$ and $[M - CO]^-$, respectively, after the loss of the methoxy group (-CH_3) of the C-ring (Chen et al. 2015; Qin et al. 2017; Camaforte et al. 2019).

Justesen (2001) studied the fragmentation pattern of ten deprotonated methoxylated flavonoids in ESI-MS/MS in the negative ionization mode, and found that although these compounds are isomers they showed different behaviors in the fragmentation pattern; they concluded that the exact position of the methoxy group cannot be defined without comparison with Nuclear Magnetic Resonance (NMR) studies of the isolated substances.

Table 1. Metabolites identified in the subfractions of *B. holophylla* leaves using HPLC-DAD-ESI-MS in negative ionization mode

Peak no.	Rt (min.)	UV_{max} (nm)	$[M-H]^-$ (*m/z*)	MS/MS^n	Compound	Subfractions 26	38	46
1	17.6	263/352	449	**316**, 287, 271, 242, 214, 179	Myricetin-*O*-pentoside	-	+	-
2	18.0	264/350	463	**316**, 287, 271, 179	Myricetin-*O*-deoxyhexoside	-	-	+
3	18.4	255/353	463	**300**, 271, 255, 243, 179	Quercetin-*O*-hexoside	-	-	+
4	18.8	255/353	463	**300**, 271, 255, 243, 179	Quercetin-*O*-hexoside	-	-	+
5	19.6	256/353	433	**300**, 271, 255, 243, 227, 179	Quercetin-*O*-xilopyranoside	-	+	+
6	20.1	265/347	447	284, **255**, 227	Kaempferol-3-*O*-glucoside	-	-	+
7	20.2	255/351	433	**300**, 271, 255, 243, 227, 179	Quercetin-*O*-pentoside	-	+	+
8	21.0	255/348	447	**300**, 271, 255, 243, 179	Quercetin-*O*-deoxyhexoside	-	+	+
9	21.2	268/350	477	**314,**299, 285, 271, 257, 243, 227	Isorhamnetin-3-*O*-hexoside	-	-	+
10	22.5	265/356	417	**284**, 255, 227, 211	Kaempferol-*O*-pentoside	-	+	-
11	22.7	268/354	477	**300**, 271, 256, 243	Unidentified	-	-	+
12	23.5	263/343	431	300, **284**, 255, 227, 211	Luteolin-deoxyhexose	-	-	-
13	27.5	254/354	315	**300**, 271, 255, 243, 227	Isorhamnetin	+	-	-

+: present, -: absence and bold numbers: relevant fragments for identification.

Subfraction 38 presented as a mixture with two major compounds and in the analysis of the chromatogram obtained in HPLC-DAD, both substances presented UV-visible spectrum characteristic of 3-*O*-substituted flavonoids ($\lambda_{max} \approx$ 255 and 350 nm). Several peaks were detected in its mass spectrum in negative ionization mode: *m/z* 417, 431, 433, 447, and 449, with ions *m/z* 433 (20.2 min.) and 447 (21.0 min.) predominating (Figure 1B). The MS/MS fragmentation of ions *m/z* 417 and 431 produced mainly the product ion of *m/z* 285, indicating the loss of the pentoside and deoxyhexose units, respectively (Vukics and Guttman 2008). This fragment was also observed by Camaforte et al. (2019) in the analysis of the leaves of *B. holophylla*, being identified as kaempferol-*O*-pentoside $[M - H]^-$ in *m/z* 417 and luteolin-deoxyhexose for $[M - H]^-$ in *m/z* 431. The ions $[M - H]^-$ in *m/z* 433 at the retention times of 19.6 and 20.2 minutes showed the same fragments in *m/z* 300 $[M - 133 - H]^-$, suggesting the loss of a pentoside unit, *O*-substituted in the flavonoid aglycone; in addition to the fragments in *m/z* 271 $[M - 162 - H]^-$; *m/z* 255 $[M - 178 - H]^-$ e *m/z* 179 $[M - 254 - H]^-$, this last product ion $^{1,2}A^-$ typical of RDA fragmentation in flavonols (3-OH) with dihydroxylated A-ring (Ma et al. 1997; Cuyckens and Claeys 2004), indicating the presence of quercetin-*O*-pentoside derivatives. The MS/MS fragmentation of *m/z* 447 was presented the fragment ion of *m/z* 300 $[M - 147 - H]^-$, after the loss of a deoxyhexoside unit moiety and the product ion *m/z* 271 $[Y_o - H - CO]^-$, characteristic of flavonol-3-*O*-monoglycosylated (Ablajan et al. 2006), besides the fragments in *m/z* 255 $[Y_o - 2H - COOH]^-$ and *m/z* 179 $[M - 268 - H]^-$, product ion $^{1,2}A^-$ (Ma et al. 1997; Cuyckens and Claeys 2004). This fragmentation pattern confirms the presence of quercetin-*O*-deoxyhexoside. The deprotonate molecule of *m/z* 449 $[M - H]^-$ produced a fragment of *m/z* 316 $[M - 133 - H]^-$ after the loss of pentoses, as well as the fragments of the product ions *m/z* 179 $[M - H - CO - H_2O]^-$ and *m/z* 271, confirming the presence of myricetin-*O*-pentoside (Sobeh et al. 2017; dos Santos et al. 2018).

In the mass spectrum of subfraction 46 it was possible to identify about nine peaks of deprotonated molecules $[M - H]^-$ in *m/z* 433 (19.6 and 20.2 min.), 447 (20.1 and 21.0 min.), 463 (18.0, 18.4, and 18.8 min.), and 477 (21.2 and 22.7 min.), with the most intense peaks of *m/z* 463 (18.4 and 18.8 min.) and 447 (20.1 and 21.0 min.) (Figure 1C). The chemical profile of subfraction 46 differs from subfraction 38, mainly due to the presence of flavonoids with deprotonated molecules $[M - H]^-$ in*m/z* 463 and 477, in addition to the ion *m/z* 447 in 20.1 minutes that it presents a different fragmentation pattern than that found as predominant at 21.0 minutes in subfraction 38.

The MS/MS fragmentation of *m/z* 463 (18.0 min.) produced a deprotonated aglycone, the ion of *m/z* 316 $[M - 147 - H]^-$, indicating that such ion corresponds to a myricetin-*O*-deoxyhexoside (Camaforte et al. 2019). For the other precursor ions $[M - H]^-$ in *m/z* 463 at 18.4 and 18.8 minutes fragments were observed in *m/z* 300 $[M - 163 - H]^-$, suggesting the loss of a hexoside unit, *m/z* 271 $[M - 192 - H]^-$, corresponding to the presence of the product ion of the fragment $[Y_o - H - CO - H]^-$ indicating the position of the sugar bond in the aglycone, characteristic of 3-*O*-monoglycoside, *m/z* 255 corresponds to the fragment ion $[Y_o - 2H - COOH]^-$ (Ablajan et al. 2006); finally, *m/z* 179 $[M - 284 - H]^-$, product ion $^{1,2}A^-$ (Ma et al. 1997; Cuyckens and Claeys 2004). These data suggest the presence of quercetin-*O*-hexoside derivates. The MS/MS fragmentation of *m/z* 447 at 20.1 minutes produced mainly the product ion of *m/z* 255 $[Y_o - H - CO - H]^-$, confirming the presence of kaempferol-3-*O*-glucoside. The predominant presence of this fragment ion (*m/z* 255) allows differentiating the glycosylated 3-*O* and 7-*O* isomers of kaempferol since the 7-*O* glycosylated derivative presents the fragment ion abundant in *m/z* 257 $[Y_o - CO]^-$ (Ablajan et al. 2006).

For the precursor ion $[M - H]^-$ in *m/z* 477 at 21.2 minutes, ion fragments were observed in *m/z* 314 $[M - 163 - H]^-$, suggesting the loss of a hexoside unit; *m/z* 299 $[M - 178 - H]^-$, after the loss of the methoxy group ($-CH_3$) of the aglycone; and other similar ions (*m/z* 271, 257, 243, and 227) to the fragmentation of isorhamnetin (Farag et al. 2015). The precursor ion *m/z* 477 at 21.7 minutes could not be identified.

Biological Potential of *Bauhinia holophylla*

Some species of the genus *Bauhinia* are being studied regarding their chemical profiles and biological activity and thus several compounds have been isolated and identified, such as fatty acids, steroids, terpenoids, and flavonoids (Bonilha et al. 2015; Farag et al. 2015). In general, studies of species in this genus confirm their therapeutic properties, such as their high antidiabetic potential, and these effects were associated mainly with the presence of high contents of flavonoids, such as quercetin-3-*O*-*α*-(2″-galloyl)-rhamnoside and kaempferol-3-*O*-*α*-(2″-galloyl)-rhamnoside (Gonzalez-Mujica et al. 2003; Estrada et al. 2005; Cechinel Filho 2009).

Among the species most studied for antidiabetic activity is *B. forficata*; aqueous extracts of its leaves have been widely used as an adjuvant in the treatment of diabetes mellitus and the main bioactive compound is the kaempferol 3,7-di-*O*-*α*-L-rhamnoside, also known as kaempferitrin, the chemical and pharmacological marker responsible for hypoglycemic activity (Souza et al. 2018).

The molecular mechanisms of the antidiabetic effect of flavonoid derivatives are not yet fully described, but some studies have shown that flavonoids can reduce the intestinal absorption of dietary carbohydrates, modulating the enzymes involved in glucose metabolism (inhibiting *α*-glycosidase), improving *β*-cell function, and insulin secretion and action, besides their properties as antioxidant and anti-inflammatory agents (Dinneen et al. 1992; Iwai et al. 2006).

In a study ofthe *n*-butanolic extract of the leaves of *B. holophylla*, Pieroni (2013) isolated and identified the flavonoids: litospermoside (gryphonin), myricetin-*O*-hexoside, myricetin-*O*-pentoside, myricetin-3-*O*-*α*-L-rhamnopyranoside, quercetin-3-*O*-*β*-D-galactopyranoside, quercetin-3-*O*-*β*-D-xylopyranoside, quercetin-*O*-pentoside, quercetin-3-*O*-*α*-L-arabinofuranoside, quercetin-3-*O*-*α*-L-rhamnopyranoside, kaempferol-*O*-pentoside, luteolin-*O*-deoxyhexose, quercetin, luteolin and isorhamnetin. These results are consistent with the identification of these substances in the present study.

Subsequently, extracts from the leaves of *B. holophylla* were evaluated for antiulcerogenic (Rozza et al. 2015), hypoglycemic and hypolipidemic (Camaforte et al. 2019), antidiabetic (Silva et al. 2015; Pinheiro et al. 2017), antiviral (dos Santos et al. 2019), cytotoxic, apoptotic and mutagenic/antimutagenic activities (Ribeiro et al. 2018).

Conclusion

The qualitative analysis of the chemical profile of an extract of *B. holophylla* leaves revealed the presence of flavonoid-*O*-glycoside derivatives, such as myricetin-*O*-pentoside, myricetin-*O*-deoxyhexoside, quercetin-*O*-hexoside, quercetin-*O*-xilopyranoside, kaempferol-3-*O*-glucoside, quercetin-*O*-pentoside, quercetin-*O*-deoxyhexoside, isorhamnetin-3-*O*-hexoside, kaempferol-*O*-pentoside, luteolin-deoxyhexose, and isorhamnetin. The results show the biotechnological potential of *B. holophylla* to produce bioactive compounds of economic and medicinal interest, in addition to suggesting further investigations aimed at the isolation and biological evaluation of isolated compounds.

Acknowledgments

We thank Dr. Andreia Fonseca Silva for the botanical identification of this plant. The authors are grateful to the Universidade Federal de São João del-Rei (UFSJ), Conselho Nacional de Desenvolvimento Científico e Tecnológico (CNPq), Coordenação de Aperfeiçoamento de Pessoal de Nível Superior (CAPES), and Fundação de Amparo à Pesquisa do Estado de Minas Gerais (FAPEMIG - APQ-00855-19) for financial support. This study was financed in part by the Coordenação de Aperfeiçoamento de Pessoal de Nível Superior – Brasil (CAPES) - Finance Code 001.

REFERENCES

Ablajan, Keyume, Zeper Abliz, Xiao-Ya Shang, Jiu-Ming He, Riu-Ping Zhang, and Jian-Gong Shi. 2006. "Structural characterization of flavonol 3,7-di-*O*-glycosides and determination of the glycosylation position by using negative ion electrospray ionization tandem mass spectrometry." *Journal of Mass Spectrometry* 41: 352–60. https://doi.org/10.1002/jms.995.

Bonilha, Iuri de F., Camila H. F. Cuelho, Juliana C. Brondani, Jocelene F. Cossetin, Lucas D. Marangoni, and Melânia P. Manfron. 2015. "Biological potential of plants from the genus *Bauhinia*." *Revista Cubana de Farmacia* 49: 583–94. http://scielo.sld.cu/pdf/far/v49n3/far17315.pdf.

Camaforte, Nathalia A. P., Luiz L. Saldanha, Priscilla M. P. Vareda, João M. Rezende-Neto, Mario R. Senger, Aislan Q. Delgado, Henrque J. N. Morgan, et al. 2019. "Hypoglycaemic activity of *Bauhinia holophylla* through GSK3-β inhibition and glycogenesis activation." *Pharmaceutical Biology* 57: 269–79. https://doi.org/10.1080/13880209.2019.1599962.

Cechinel Filho, Valdir. 2009. "Chemical composition and biological potential of plants from the genus *Bauhinia*." *Phytotherapia Research* 23: 1347–54. https://doi.org/10.1002/ptr.2756.

Chen, Yeqing, Hongli Yu, Hao Wu, Yaozong Pan, Luilong Wang, Yangping Jin, and Chengchao Zhang. 2015. "Characterization and Quantification by LC-MS/MS of the Chemical Components of the Heating Products of the Flavonoids Extract in Pollen Typhae for Transformation Rule Exploration." *Molecules* 20:18352–66. https://doi.org/10.3390/molecules201018352.

Cuyckens, Filip and Magda Claeys. 2004. "Mass spectrometry in the structural analysis of flavonoids." *Journal of Mass Spectrometry* 39: 1–15. https://doi.org/10.1002/jms.585.

Dar, Refaz A., Mohd Shahnawaz, and Parvaiz H. Qazi. 2017. "General overview of medicinal plants: A review." *The Journal of*

Phytopharmacology 6: 349–51. http://www.phytopharmajournal.com/Vol6_Issue6_08.pdf.

Dinneen, Sean, John Gerich, and Robert Rizza. 1992. "Carbohydrate metabolism in non insulin-dependent diabetes mellitus." *New England Journal of Medicine* 327: 707–13. https://doi.org/10.1056/NEJM199209033271007.

Dos Santos, Catarina, Renan S. Galaverna, Celio F. F. Angolini, Vania V. A. Nunes, Luiz F. R. Almeida, Ana L. T. G. Ruiz, João E. de Carvalho, Regina M. T. Duarte, Marta, C. T. Duarte, and Marcos N. Eberlin. 2018. "Antioxidative, Antiproliferative and Antimicrobial Activities of Phenolic Compounds from Three *Myrcia* Species." *Molecules* 23: 986–98. https://doi.org/10.3390/molecules23050986.

Dos Santos, Michelli, Thaiz R. Teixeira, Felipe R. da S. Santos, William G. Lima, Ariane C. Ferraz, Nathália L. Silva, Flávio J. Leite, et al. 2019. "*Bauhinia holophylla* (Bong.) Steud. leaves-derived extracts as potent anti-dengue serotype 2." *Natural Product Research* 33: 1–6. https://doi.org/10.1080/14786419.2019.1669030.

Estrada, Omar, Masahisa Hasegawa, Freddy Gonzalez-Mujíca, Norma Motta, Elsa Perdomo, Alejandra Solorzano, Jeannette Méndez, Bernardo Méndez, and Gabriela Zea. 2005. "Evaluation of flavonoids from *Bauhinia megalandra* leaves as inhibitors of glucose-6-phosphatase system." *Phytotherapia Research* 19: 859–63. https://doi.org/10.1002/ptr.1703.

Farag, Mohamed A., Sarah T. Sakna, Nabaweya M. El-Fiky, Marawan M. Shabana, and Ludger A. Wessjohann. 2015. "Phytochemical, antioxidant and antidiabetic evaluation of eight *Bauhinia* L. species from Egypt using UHPLC–PDA–qTOF-MS and chemometrics." *Phytochemistry* 119: 41–50. https://doi.org/10.1016/j.phytochem.2015.09.004.

Gonzalez-Mujica. Freddy, Norma Motta, Ana H. Márquez, and Joana Capote-Zulueta. 2003. "Effects of *Bauhinia megalandra* aqueous leaf extract on intestinal glucose absorption and uptake by enterocyte brush border membrane vesicles." *Fitoterapia* 74: 84–90. https://doi.org/10.1016/S0367-326X(02)00307-6.

Iwai, Kunihisa, Mi-Yeon Kim, Akio Onodera, and Hajime Matsue. 2006. "Alpha-glucosidase inhibitory and antihyperglycemic effects of polyphenols in the fruit of *Viburnum dilatatum* Thunb." *Journal of Agricultural and Food Chemistry* 54: 4588–92. https://doi.org/10.1021/jf0606353.

Justesen, Ulla. 2001. "Collision-induced fragmentation of deprotonated methoxylated flavonoids, obtained by electrospray ionization mass spectrometry." *Journal of Mass Spectrometry* 36: 169–78. https://doi.org/10.1002/jms.118.

Lei, Zhentian, Barbara W. Sumner, Anil Bhatia, Saurav J. Sarna, and Lloyd W. Sumner. 2018. "UHPLC-MS Analyses of Plant Flavonoids." *Current Protocols in Plant Biology* 4: e20085. https://doi.org/10.1002/cppb.20085.

Ma, Yu-Liang, Qimin M. Li, Hilfr Van den Heuvel, and Magda Claeys. 1997. "Characterization of flavone and flavonol aglycones by collision-induced dissociation tandem mass spectrometry." *Rapid Communications in Mass Spectrometry* 11: 1357–64. https://doi.org/10.1002/(SICI)1097-0231(199708)11:12<1357::AID-RCM983>3.0.CO;2-9.

Markham, Ken R. 1982. *Techniques of flavonoid identification*. London: Academic Press.

Pieroni, Laís Goyos. 2013. *Constituintes químicos e atividade antioxidante e alelopática de Bauhinia holophylla (Bong.) Steud.* MSc thesis, Universidade Estadual Paulista "Júlio de Mesquita Filho." [*Chemical constituents and antioxidant and allelopathic activity of Bauhinia holophylla (Bong.) Steud.*]

Pinheiro, Marcelo S., Luhara S. Rodrigues, Leila S. Neto, Rafaianne Q. Moraes-Souza, Thaigra S. Soares, Maldileine F. Américo, Kleber E. Campos, Débora C. Damasceno, and Gustavo T. Volpato. 2017. "Effect of *Bauhinia holophylla* treatment in Streptozotocin-induced diabetic rats." *Annals of the Brazilian Academy of Sciences* 89: 263–72. https://doi.org/10.1590/0001-3765201720160050.

Plazonic, Ana, Franz Bucar, Zeljan Males. Ana Mornar, Biljana Nigovic, and Nikola Kujundzic. 2009. "Identification and Quantification of

Flavonoids and Phenolic Acids in Burr Parsley (*Caucalis platycarpos* L.), Using High-Performance Liquid Chromatography with Diode Array Detection and Electrospray Ionization Mass Spectrometry." *Molecules* 14: 2466–90. https://doi.org/10.3390/molecules14072466.

Qin, Yan, Boyan Gao, Hai,ing Shi, Jie Cao, Chengle Yin, Weiying Lu, Liangli Yu, and Zhihong Cheng. 2017. "Characterization of flavonol mono-, di-, tri- and tetra-*O*-glycosides by ultra-performance liquid chromatography-electrospray ionization-quadrupole time-of-flight mass spectrometry and its application for identification of flavonol glycosides in *Viola tianschanica*." *Journal of Pharmaceutical and Biomedical Analysis* 142: 113–24. https://doi.org/10.1016/j.jpba.2017.05.007.

Ribeiro, Diego L., Heloísa L. Cilião, Ana F. L. Specian, Juliana M. Serpeloni, Marcelo T. de Oliveira, Eliana A. Varanda, Wagner Vilegas, et al. 2018. "Phytochemical study and evaluation of cytotoxicity, mutagenicity, cell cycle kinetics and gene expression of *Bauhinia holophylla* (Bong.) Steud. in HepG2 cells in vitro." *Cytothechonogy* 70: 713–28. https://doi.org/10.1007/s10616-017-0173-5.

Ribeiro, José F., and Bruno M. T. Walter. 1998. "Fitofisionomias do bioma do Cerrado: os biomas do Brasil." In *Cerrado: ambiente e flora*, edited by Sueli M. Sano, and Semíramis P. de Almeida, 89–116. Planaltina: EMBRAPA-CPAC. ["Phytophysiognomies of the Cerrado biome: the biomes of Brazil." In *Cerrado: environment and flora*]

Rochfort, Simone J., Miclaed Imsic, Rod Jones, Victor C. Trenerry, and Bruce Tomkins. 2006. "Characterization of flavonol conjugates in immature leaves of pak choi [*Brassica rapa* L. ssp. *chinensis* L. (Hanelt.)] by HPLC–DAD and LC–MS/MS." *Journal of Agricultural and Food Chemistry* 54: 4855–60. https://doi.org/10.1021/jf060154j.

Rozza, Ariane L., Daniela A. S. Cesar, Laís G. Pieroni, Luiz L. Saldanha, Anne L. Dokkedal, Felipe M. de Faria, Alba R. Souza-Brito, Wagner Vilegas, Regina K. Takahira, and Cláudia H. Pellizzon. 2015. "Antiulcerogenic activity and toxicity of *Bauhinia holophylla* hydroalcoholic extract." *Hindawi Publishing Corporation* 2015: 1–9. https://doi.org/10.1155/2015/439506.

Salehi, Bahare, Athar Ata, Nanjangud V. A. Kumar, Farukh Sharopov, Karina Ramírez-Alarcón, Ana Ruiz-Ortega, Seyed A. Ayatollahi, et al. 2019. "Antidiabetic Potential of Medicinal Plants and Their Active Components." *Biomolecules* 9: 551–672. https://doi.org/10.3390/biom9100551.

Schmidt, Marcella E. P., Fernanda B. Pires, Lucas P. Bressan, Fábio V. da Silva Jr., Osmar Lameira, and Marcelo B. Rosa. 2018. "Some triterpenic compounds in extracts of *Cecropia* and *Bauhinia* species for different sampling years." *Revista Brasileira de Farmacognosia* 28: 21–6. https://doi.org/10.1016/j.bjp.2017.11.005. [*Brazilian Journal of Pharmacognosy*]

Silva, Thais L., Kamirri Savazzi, Jeferson J. S. Sousa, Rafaianne Q. de M. Souza, Thaigra de S. Soares, Marcelo S. Pinheiro, Kleber E. de Campos, and Gustavo T. Volpato. 2015. "Biochemical parameters in diabetic pregnant rats treated with *Bauhinia holophylla* aqueous extract." *Diabetology & Metabolic Syndrome* 7: A78. https://doi.org/10.1186/1758-5996-7-S1-A78.

Sobeh, Mansour, Soha A. Hassan, Mohamed A. El Raey, Wael A. Khalil, Mahmoud A. E. Hassan, and Michael Wink. 2017. "Polyphenolics from Albizia harveyi Exhibit Antioxidant Activities and Counteract Oxidative Damage and Ultra-Structural Changes of Cryopreserved Bull Semen." *Molecules* 22: 1993–2007. https://doi.org/10.3390/molecules22111993.

Souza, Bárbara V. C., Regilda S. dos R. Araújo, Oskar A. Silva, Lucas C. Faustino, Maria F. B. Gonçalves, Mirian L. dos Santos, Grasielly R. Souza, Lindalva M. Rocha, Mônica L. S. Cardoso, and Lívio C. C. Nunes. 2018. "*Bauhinia forficata* in the treatment of diabetes mellitus: a patent review." *Expert Opinion on Therapeutic Patents* 28: 129–38. https://doi.org/10.3390/10.1080/13543776.2018.1409208.

Vaz, Angela M. S. da F., Roseli L. da C. Bortoluzzi, and Luis A. E. da Silva. 2010. "Checklist of *Bauhinia sensu stricto* (Caesalpiniaceae) in

Brazil." *Plant Ecology and Evolution* 143: 212–21. https://doi.org/10.5091/plecevo.2010.391.

Vaz, Angela M. S. da F. and Ana M. G de A. Tozzi. 2003. "*Bauhinia* ser. *Canseria* (Leguminosae: Caesalpinioideae no Brasil)." *Rodriguesia* 54: 55–143. https://doi.org/10.1590/2175-78602003548305.

Vukics, Viktoria and András Guttman. 2008. "Structural characterization of flavonoid glycosides by multi-stage mass spectrometry." *Mass Spectrometry Reviews* 29: 1–16. https://doi.org/10.1002/mas.20212.

Zuanazzi, José A. S., and Montanha, Jarbas A. 2010. "Flavonoides." In *Farmacognosia da planta ao medicamento*, edited by Cláudia M. O. Simões, Eloir P. Schenkel, Grace Gosmann, João C. P. de Mello, Lilian A. Mentz, and Pedro R. Petrovick, 577–614. Porto Alegre: Editora UFRGS UFSC. [*Pharmacognosy of the plant to the medicine*]

In: The Book of Flavonoids
Editors: Chao-Hui Feng et al.
ISBN: 978-1-68507-376-3

Chapter 9

PRODUCTION AND CHARACTERIZATION OF FLAVONOIDS AND PHENOLIC ACID DERIVATIVES IN *BAUHINIA VARIEGATA* CALLI

***Viviana Rodrigues Cardoso*[1], *Mairon César Coimbra*[1], *Thaiz Rodrigues Teixeira*[2] and *Ana Hortência Fonseca Castro*[1,*]**
[1]Universidade Federal de São João del-Rei, Divinópolis, Brazil
[2]Universidade de São Paulo, Faculdade de Ciências Farmacêuticas de Ribeirão Preto, Ribeirão Preto, Brazil

ABSTRACT

Bauhinia variegata L. (Fabaceae), popularly known as "pata-de-vaca," is a woody species that contains flavonoids, triterpenoids, and steroids as active compounds, which are distributed in different organs,

[*] Corresponding Author's E-mail: acastro@ufsj.edu.br.

In folk medicine, the bark is astringent and used for skin diseases and ulcers. The stem has antimicrobial and anti-tumor activity. The roots are used as an anti-inflammatory agent, and the leaves have anti-diabetic and diuretic properties and reduce cholesterol. Plant tissue culture and, more specifically, callus culture techniques are an effective biotechnological tool that makes it possible to obtain bioactive compounds *in vitro* through the optimization of culture conditions, using different types and concentrations of plant growth regulators and light conditions. The aim of this work was to establish calli cultures of *B. variegata* and to produce phenolic compounds, especially flavonoids, in *in vitro* culture. For calli induction, leaf explants were placed on Murashige and Skoog (MS) medium supplemented with 30 g L^{-1} of sucrose, solidified with 7 g L^{-1} agar, added to 2,4-dichlorophenoxyacetic acid (2,4-D) (4.52; 9.05; 18.10 μM) and 6-benzylaminopurine (BAP) (0; 4.44; 8.88; 17.75 μM) and their combination, and grown in the presence and absence of light. The calli growth curve was established on MS medium with 30 g L^{-1} sucrose with 4.52 μM 2.4-D and 4.44 μM BAP in the presence of light. For the phytochemical study, the crude hydroethanolic extract of the dried callus was obtained by maceration. The quantity of phenols and flavonoids were evaluated by standard techniques. The identification of flavonoids and phenolic acid derivatives was performed by high performance liquid chromatography coupled to a diode array detector (HPLC-DAD). Calli induced on MS medium with 4.52 μM 2.4-D and 4.44 μM BAP in the presence of light presented high growth and a high total phenol and flavonoid content. HPLC analysis showed the presence of different flavonoids (flavones, catechins, flavanone, flavonol), phenolic acid derivatives, and anthraquinone. In conclusion, it is possible to obtain different types of flavonoids and phenolic acids derivatives in callus cultures of *B. variegata*.

Keywords: medicinal plant, *in vitro* culture, phenolic compounds, secondary metabolites

INTRODUCTION

The *Bauhinia* L. (Fabaceae) genus contains about 300 species represented by medium-sized trees with white or pink flowers, pod-like fruits, and uncinated leaves and are popularly known as "pata-de-vaca" or "unha-de-vaca." About 57 species occur in Brazil, of which 37 are endemic (Vaz 2015). Species of this genus are widely distributed in most tropical countries and have been used frequently in folk medicine. In Africa, Asia, and America, *Bauhinia* spp. are used as diuretic, hypoglycemic, tonic, and depurative agents (Duarte-Almeida et al. 2015). Phytochemical and pharmacological studies demonstrate the presence of several compounds of medicinal interest, such as triterpenes, lactones, steroids, tannins, quinones, and flavonoids, that have been isolated and identified from some *Bauhinia* spp. (Pinheiro et al. 2017; Schmidt et al. 2018; Souza et al. 2018).

Bauhinia variegata is an ornamental species, originally from India, that grows well in tropical regions all over the world. Chemical studies have demonstrated the presence of flavonoids (narigenin-5,7-dimethoxy-4-ramnoglucoside, kaempferol-3-galactoside, kaempferol-3-ramno-glucoside), triterpenoids (lupeol), and steroids (systeserol) (Reddy et al. 2003) distributed in different parts of the plant. From an ethnopharmacological point of view, the bark is astringent and used for skin diseases and ulcers. The stem has antimicrobial and anti-tumor activity. The root is used as an anti-inflammatory agent (Fang et al. 2010), and its leaves have antidiabetic, diuretic, and hypocholesterolemic action (Lorenzi and Matos 2008).

Among the techniques of plant tissue culture, callus culture generates undifferentiated cells with relatively uniform development, which are not affected by microorganisms, in addition to having a reduced vegetative cycle that enables the production of metabolites of interest in a relatively short period of time (Ochoa-Villarreal et al. 2016; Hidalgo et al. 2018). Calli are partially differentiated structures represented by masses of irregular cells that multiply widely in response to chemical or physical injuries and could differentiate into tissues and organs (Ogita 2015). The growth and development of callus can be influenced by the plant material itself, the nutrient medium, and external factors, such as light and temperature, which act by accelerating, delaying, or even inhibiting cell proliferation (Smith 2012). According to Ikeuchi et al. (2013), for callus induction, it is generally necessary to have an exogenous supply of plant growth regulators. Induction is regulated by the interaction and balance between the regulators provided and the hormones produced internally by the explant (Ji et al. 2015). Auxins have been widely used for the induction of callus, and in general, the use of high concentrations of auxins induce the formation of callus. The interaction between auxins and cytokinins is also widely used and often determinant for induction (Chandler and Werr 2015). Among the growth regulators most used in the induction of calli are the auxins 2,4-D (2,4-dichlorophenoxyacetic acid) and ANA (naphthalene acetic acid) and cytokinins BAP (6-benzylaminopurine) and TDZ (tidiazuron) (George, Hall and De Klerk 2008).

Different plant compounds with broad biological activity have been obtained from callus culture, such as flavonoids (Bernabé-Antônio et al., 2017; Coimbra et al. 2019), phenolic acids (Coimbra et al. 2017), cardenolides (Sahin et al. 2013), isoflavonoids (Gueven and Knorr 2011), lectins (Castro et al. 2018), anthocyanins (Wang et al. 2019), coumarins (Piovan et al. 2014), and rotenoids (Santos et al. 2007).

The high chemical and medicinal potential of *B. variegata* stimulates the use of techniques that allow its sustainable use and the search for alternatives to produce metabolites of pharmaceutical interest, on a large scale, from other sources. The aim of this work was to produce and characterize flavonoid and phenolic acid derivatives in *B. variegata* calli cultures.

MATERIAL AND METHODS

Chemicals

Chlorogenic acid, *trans*-cinnamic acid, caffeic acid, benzoic acid, gallic acid, rutin, (+)-catechin hydrate, quercetin, 2,4-dichlorophenoxyacetic acid, 6-benzilaminopurine, ethanol, formic acid, Folin-Ciocalteu reagent, and agar were purchased from Sigma-Aldrich

Co. (St. Louis, MO, USA). Water was treated in a Purite water purification system (Purite, Oxon, UK).

Plant Material

Leaf explants were obtained from *Bauhinia variegata* L. seedlings cultured *in vitro*. Seeds were collected in Divinópolis, Midwest Minas Gerais State, Brazil (20°10'45.9"S and 44°55'07.2"W GRW) (SISBIO no. 24542-3). Fertile samples were collected, and the vouchers were identified by Andréia Fonseca Silva of the PAMG Herbarium (PAMG 56307) at the Agricultural Research Company of Minas Gerais (EPAMIG). This work was registered in the SisGen Platform (Register A56CD8E), according to Brazilian Biodiversity Law 13.123/2015.

Callus Induction and Growth

Leaf segments were placed on basal medium MS (Murashige and Skoog 1962) with 30 g L^{-1} sucrose and 2,4-D (4.52, 9.05, or 18.10 μM) or BAP (4.44, 8.88, or 17.75 μM) or their combination and solidified with 7 g L^{-1} agar. The pH was adjusted to 5.8 ± 0.1 after adding plant growth regulators, and the medium was sterilized at 120°C (1.37×10^5 Pa) for 20 min. The explants were transferred to a growth chamber and kept at 27 ± 2°C under the presence of light (16:8 h light/dark regime, with a light intensity of 40 μmol m^{-2} s^{-1}) or the absence of light.

After 45 days of inoculation, callus induction (%), color, consistency, fresh and dry weight of calli, and total phenol and flavonoid content were evaluated. The chromatographic profile of phenolic compounds was determined by high performance liquid chromatography with a diode array detector (HPLC-DAD). A completely randomized design in a factorial arrangement of 15 × 2 (15 concentrations of 2,4-D and/or BAP and two conditions of light, presence or absence) was performed. The controls were represented by the absence of growth regulators in the media. Each treatment was composed of 20 replicates.

The growth pattern of callus cultures was followed using a growth curve with an interval of 7 days, for 105 days, corresponding to the decline phase, according to Smith (2012). The growth curve was established in complete MS medium supplemented with 30 g L^{-1} sucrose, solidified with 7 g L^{-1} agar, and included 4.52 μM of 2,4-D + 4.44 μM of BAP. Calli samples were taken to evaluate growth and the total phenol and flavonoid content. For assessing callus growth by determining the fresh and dry matter, the experimental design was completely randomized consisting of 25 repetitions.

Hydroethanolic Extract Preparation

The hydroethanolic extracts were prepared with 1g dry callus mass, extracted with 10 mL of 70% ethanol (70% EtOH), in an ultrasound bath at 35°C, for 30 min. This procedure was repeated 3 times. The extracts were dried at 45°C until the solvent was completely evaporated. The dry crude extract was then resuspended in 10 mL 70% EtOH and used for the following determinations (Brazil, 2019).

Total Phenol Content

Phenols were quantified with 100 μL of hydroethanolic extract according to Pastrana-Bonilla et al. (2003). The total phenol level was calculated using a calibration curve with 100 μg mL^{-1} gallic acid solution as the standard. Determinations were performed in triplicate, and the resultswere given in microgram equivalents of gallic acid per milligram of dry extract (μg GAEq mg^{-1} DE).

Total Flavonoid Content

The total flavonoid assay was performed according to Woisky and Salatino (1998), and flavonoid content was calculated using a calibration curve with 100 μg mL^{-1} quercetin in a ethanol solution of 2% aluminum chloride as a standard. Determinations were performed in triplicate, and the results were given in microgram equivalents of quercetin per milligram of dry extract (μg QEq mg^{-1} DE).

HPLC Analysis

Chromatographic profiles were obtained in modular system liquid chromatography Shimadzu Prominence HPLC (Shimadzu Corp., Kyoto, Japan). The separation of compounds was performed with reverse-phase column Gemini C18 (4.6 X 250 mm, 5 μm, Phenomenex®, Torrance, CA, USA) conditioned at 35°C. The chromatographic conditions were obtained using mobile phases comprising A) water:formic acid (99.9:0.1) and B) methanol:formic acid (99.9:0.1) at the proportion of 10–20% B (0–4 min.); 20–50% B (4–6 min); 50–90% B (6–10 min); 90–10% B (10–11 min); 10% B (11–14 min). A 20 μL injection volume and a flow rate of 1.0 mL min^{-1} were employed. Separations were monitored at three wavelengths to detect phenolic compounds, 254 and 328 nm for phenolic acids and flavan-3-ols, respectively, and 350 nm for flavones, flavonols, and chalcones (Sakakibara et al. 2003). The determination of phenolic compounds was performed by comparing retention times and UV spectra of standards that were previously injected and data from the literature (Matsubara and Rodriguez-Maya 2006; Sakakibara et al. 2003). The reference substances used were benzoic acid, caffeic acid, cinnamic acid, chlorogenic acid, gallic acid, quercetin, (+)-catechin hydrate, and rutin.

Data Analysis

Data were subjected to analysis of variance by an F test at ap ≤ 0.05 significance level using the Variance Analysis System of Balanced Data SISVAR 5.1 Software (Ferreira 2011). Mean rates were further separated by Tukey's Test when differences were significant ($P < 0.05$).

RESULTS AND DISCUSSION

Callus Induction

Callus induction occurred in response to 2,4-D and BAP at different concentrations and combinations in the presence and absence of light, with no induction in the absence of the plant growth regulators employed (Tables 1 and 2). When the balance between the growth regulators added to the medium and the endogenous concentrations of plant hormones is reached, there is a promotion or inhibition of the physiological and morphological processes in the plants (Chandler and Werr 2015).

All treatments provided 100% callus induction (Tables 1 and 2). Calli showed a color that varied between yellow, green, and brown in the presence of light (Table 1) and green and brown in the absence of light (Table 2). In general, calli induced in the presence of light proved to be predominantly friable, and in the absence of light, calli showed compact consistency. Friable calli were observed in all treatments, where different concentrations of 2,4-D were used alone or associated with a higher concentration of BAP in the presence or absence of light and in treatments where BAP was used in the presence of light. According to Beigmohamadi et al. (2019), the color and consistency of the calli depend on the concentration, type and combination of growth regulators, and the constitution of the culture medium.

The results showed that callus induction of *B. variegata* was influenced mainly by the exogenous plant growth regulators employed, as well as by the presence or absence of light during culture. Higher growth of cultures was observed in the media supplemented with 4.52 μM 2.4-D + 4.44 μM BAP (T7) in the presence of light (Table 1) and 18.10 μM 2.4-D (T3) and 4.52 μM 2,4-D + 4.44 μM BAP (T7) in the absence of light (Table 2). The positive effect of plant growth regulators on callus induction has been previously reported for several species. Auxins play an important role in the induction of calli, and cytokinins facilitate its effect (Mirzaee et al. 2016). The synergistic effects of plant growth regulators are critical in cell induction and differentiation (Shoja et al. 2010). Auxins and cytokinins are necessary for cell division, acting on the cell cycle, in the transition of the G1-S and G2-M intervals in plant cell cultures and in plants (Stals 2001). Auxins stimulate cell wall acidification, resulting in increased extensibility, and induce the transcription of specific mRNAs, which code for proteins associated with growth. In comparison, cytokinins act directly in the cell cycle by regulating the synthesis of proteins involved in the formation and operation of the mitotic spindle (Silveira et al. 2004).

Total Phenol and Flavonoid Content

The total phenol and flavonoid content varied according to the type, combination and concentration of plant growth regulators and light conditions ($P < 0.05$). In general, the total phenol content was higher in calli than those found in the initial explant, regardless of the type and concentration of the regulators used both in the presence and absence of light ($P < 0.05$) (Tables 1 and 2). Calli induced in treatments 8 and 9, in the absence of light, showed the highest levels of total phenols (121.01 ± 0.34 μg EqGA mg^{-1} DE and 123.10 ± 3.08 μg EqGA mg^{-1} DE, respectively), with these values being about 8 fold higher than those found in the initial explant (15.94 ± 1.32 μg EqGA mg^{-1} DE) (Table 2). However, the calli induced by these treatments showed lower values of fresh and dry matter when compared to those obtained in treatment 7 in the presence and absence of light, with 71.95 ± 0.59 μg EqGA mg^{-1} DE and 102.52 ± 0.31 μg EqGA mg^{-1} DE, respectively.

In general, the total flavonoid content was lower in calli induced in the absence of light (Table 2). Calli induced in the presence of light predominantly showed higher values than those found in the initial explant, regardless of the type and concentration of the regulators employed ($P < 0.05$) (Table 1). Higher levels of total flavonoids were observed in calli induced in treatment 5 (24.62 ± 0.53 μg EqQ mg^{-1} DE), followed by treatments 6 (18.39 ± 0.41 μg EqQ mg^{-1} DE) and 7 (17.44 ± 0.98 μg EqQ mg^{-1} DE) in the presence of light, with these being 4.25-fold (treatment 5) and 3-fold (treatments 6 and 7) larger than those found in the initial explant (5.80 ± 0.99μg EqQ mg^{-1} DE). However, the calli induced in treatments 5 and 6 showed low values of fresh and dry matter in comparison with those obtained in treatment 7. Therefore, in the presence of light, treatment 7 was chosen for the study of calli growth.

The production of phenolic compounds by *in vitro* culture is well described in the literature, but there are major differences in relation to the production of individual compounds (Dias et al. 2016). The *in vitro* production of secondary metabolites offers several benefits when compared to the usual production in plants grown *ex vitro* (Karakas 2020). As an example can consider the production of flavonoids in callus cultures of *Phaseolus coccineus* and *Glycinemax* (Fabaceae) (Dymarska et al. 2020), as well as the production of steroids, phenols, flavonoids and derivatives of gallic, benzoic acids, ferulic and p-coumaric in callus of *Pyrostegia venusta* (Bignoniaceae) (Coimbra et al. 2017; Coimbra et al. 2019). The addition of different types, combinations, and concentrations of plant growth regulators on the culture medium has been reported to alter the production and accumulation of phenolic compounds (Karakas 2020). The optimization of the culture medium by adding growth regulators, precursors, or elicitation are tools that can promote changes in the biosynthetic routes and cause an increase in the production of phenolic compounds (Gupta et al. 2016).

HPLC Analysis

Samples of the initial explant and calli from treatments 5, 6, 7, 8, and 9 in the presence and absence of light and 14 in the absence of light were subjected to analysis by HPLC-DAD. The results presented in Table 3 suggest the presence of different flavonoids and phenolic acids in the samples by comparison with the retention times and UV spectra of the reference substances and data found in the literature.

The results observed for the initial explant suggest the presence of two phenolic acid derivatives (gallic and chlorogenic acids), three types of flavonoids (flavone, theaflavin, and catechin), and anthraquinone (Table 3).

Two phenolic acid derivatives were detected in calli induced in treatment 5 in the absence of light, which were cinnamic and benzoic acids. In addition to these compounds, flavonol was found in the calli induced in treatment 8 in the absence of light, with all three substances belonging to the group of phenolic compounds. A chromatographic pattern like that of the initial explant was verified for the hydroethanolic extracts of calli induced in treatments 5 (in the presence of light), 6, and 7 (in the absence of light). Gallic and clorogenic acids derivatives and anthraquinone were detected in all samples, both in the presence and absence of light, except in treatment 5 in absence of light. Flavonol had a more restricted occurrence and was detected ony in the hydroethanolic extracts of calli from treatment 8 in absence of light. However, despite theaflavin occurring quite frequently in *B. variegata* calli, this substance was not detected in calli grown in media containing 8.88 μM BAP (treatment 5) in the absence of light and 4.52 μM 2,4-D + 4.44 μM BAP (treatment 7) in the presence of light, as well as catechin, which was not detected in the hydroethanolic extracts of calli from media supplemented with 8.88 μM BAP (treatment 5) in the absence of light and 4.52 μM 2,4-D + 8.88 μM BAP (treatment 8) in the presence or absence of light (Table 3).

Table 1. Color, consistency, callus induction (%), fresh matter (FM), dry matter (DM), total phenol (TP) and total flavonoid (TF) of the *B. variegata* calli induced from leaf explants, using different concentrations and combinations of 2,4-D and BAP, in the presence of light, after 45 days of culture

Treatment	Concentration	Color	Consistency	Induction (%)	FM (g)	DM (g)	TP (µg GAEq mg^{-1} DE)	TF (µg QEq mg^{-1} DE)
T0	No regulator	-	-	-	-	-	102.54 ± 0.04 a	12.70 ± 0.82 c
T1	4.52 µM 2,4-D	Green	Friable	100	0.52 d	0.04 d	18.87 ± 0.30 f	10.47 ± 0.28 c
T2	9.05 µM 2,4-D	Green	Friable	100	0.82 d	0.06 c	19.23 ± 0.24 f	7.33 ± 0.72 d
T3	18.10 µM,2,4-D	Yellow	Friable	100	2.33 b	0.13 b	16.91 ± 0.14 g	10.95 ± 1.09 c
T4	4.44 µM BAP	Yellow	Friable	100	1.01 c	0.06 c	20.32 ± 0.27 f	12.45 ± 0.24 c
T5	8.88 µM BAP	Green	Friable	100	0.32 d	0.02 d	37.14 ± 0.23 e	24.62 ± 0.53 a
T6	17.75 µM BAP	Green	Friable	100	0.79 d	0.05 d	43.91 ± 0.35 e	18.39 ± 0.41 b
T7	4.52 µM 2,4-D + 4.44 µM BAP	Green	Compact	100	3.21 a	0.15 a	71.95 ± 0.59 c	17.44 ± 0.98 b
T8	4.52 µM 2,4-D + 8.88 µM BAP	Green	Compact	100	2.25 b	0.12 b	93.56 ± 1.24 b	7.89 ± 0.63 d
T9	4.52 µM 2,4-D + 17.75 µM BAP	Brown	Compact	100	2.51 b	0.13 b	107.04 ± 1.65 a	6.78 ± 0.34 d
T10	9.05 µM 2,4-D + 4.44 µM BAP	Brown	Compact	100	1.46 c	0.10 c	58.25 ± 0.12 d	10.77 ± 0.32 c
T11	9.05 µM 2,4-D + 8.88 µM BAP	Brown	Compact	100	2.15 b	0.12 b	68.38 ± 0.31 c	9.24 ± 0.30 c
T12	9.05 µM 2,4-D + 17.75 µM BAP	Green	Compact	100	1.58 c	0.06 c	65.04 ± 0.31 c	12.37 ± 0.66 c
T13	18.10 µM 2,4-D + 4.44 µM BAP	Yellow	Friable	100	1.21 c	0.10 c	48.79 ± 0.97 e	9.18 ± 0.99 c
T14	18.10 µM 2,4-D + 8.88 µM BAP	Brown	Friable	100	1.54 c	0.08 c	41.64 ± 0.33 e	10.47 ± 0.38 c
T15	18.10 µM 2,4-D + 17.75 µM BAP	Brown	Friable	100	2.10 b	0.11 b	28.09 ± 1.48 f	9.80 ± 0.22 c
Initial explant	-	-	-	-	-	-	15.94 ± 1.32 g	5.80 ± 0.99 d

Means in the column followed by the same letter are not significantly different at $P < 0.05$ by the Tukey test. *Results given by mean ± standard deviation.

Table 2. Color, consistency, callus induction (%), fresh matter (FM), dry matter (DM), total phenol (TP) and total flavonoid (TF) of the *B. variegata* calli induced from leaf explants, using different concentrations and combinations of 2,4-D and BAP, in the absence of light, after 45 days of culture

Treatment	Concentration	Color	Consistency	Induction (%)	FM (g)	DM (g)	TP* (µg GAEq mg-1 DE)	TF* (µg QEq mg-1 DE)
T0	No regulator	-	-	-	-	-	52.12 ± 1.89 d	8.59 ± 0.94 b
T1	4.52 µM 2,4-D	Brown	Friable	100	0.44 d	0.03 c	16.25 ± 0.23 f	6.20 ± 0.19 b
T2	9.05 µM 2,4-D	Brown	Friable	100	0.49 d	0.04 c	18.03 ± 0.19 f	4.89 ± 0.37 c
T3	18.10 µM,2,4-D	Brown	Friable	100	1.46 a	0.12 a	50.69 ± 0.50 d	3.41 ± 0.12 d
T4	4.44 µM BAP	Green	Compact	100	0.13 e	0.02 d	25.69 ± 0.58e	7.61 ± 0.31 b
T5	8.88 µM BAP	Green	Compact	100	0.12 e	0.01 d	36.85 ± 0.38 e	9.32 ± 0.38 a
T6	17.75 µM BAP	Brown	Compact	100	0.16 e	0.01 d	32.29 ± 0.49 e	4.36 ± 0.13 c
T7	4.52 µM 2,4-D + 4.44 µM BAP	Brown	Compact	100	1.39 a	0.10 a	102.52 ± 0.31 b	9.41 ± 0,43 a
T8	4.52 µM 2,4-D + 8.88 µM BAP	Brown	Compact	100	1.01 c	0.06 c	121.01 ± 0.34 a	6.76 ± 0.44 b
T9	4.52 µM 2,4-D + 17.75 µM BAP	Brown	Compact	100	0.71 c	0.06 c	123.10 ± 3.08 a	7.35 ± 0.32 b
T10	9.05 µM 2,4-D + 4.44 µM BAP	Brown	Compact	100	1.26 b	0.14 a	79.62 ± 1.42 c	3.82 ± 0.21 d
T11	9.05 µM 2,4-D + 8.88 µM BAP	Brown	Compact	100	0.38 d	0.04 c	65.17 ± 0.60 c	6.97 ± 0.25 b
T12	9.05 µM 2,4-D + 17.75 µM BAP	Brown	Compact	100	0.71 c	0.05 c	77.48 ± 1.28 c	5.14 ± 0.24 c
T13	18.10 µM 2,4-D + 4.44 µM BAP	Brown	Friable	100	1.27 b	0.08 b	108.37 ± 0.57 b	4.42 ± 0.43 c
T14	18.10 µM 2,4-D + 8.88 µM BAP	Brown	Friable	100	1.22 b	0.08 b	46.96 ± 1.61 d	10.04 ± 0.20 a
T15	18.10 µM 2,4-D + 17.75 µM BAP	Green	Friable	100	0.64 c	0.04 c	51.55 ± 1.78 d	5.19 ± 0.19 c
Initial explant	-	-	-	-	-	-	15.94 ± 1.32 f	5.80 ± 0.99 c

Means in the column followed by the same letter are not significantly different at $P < 0.05$ by the Tukey test. *Results given by mean ± standard deviation.

Table 3. Compounds present in the hydroethanolic extracts of the initial explants (IE) and *Bauhinia variegata* calli

Peak	Compound	RT (min.)	λ Max (nm)	Treatment												Reference
				IE	5*	5**	6*	6**	7*	7**	8*	8**	9*	9**	14**	
1	Unidentified	1.333	277; 398	+	+	-	+	+	+	+	+	+	+	+	+	-
2	Gallic acid derivative	1.525	288	+	+	-	+	+	+	+	+	+	+	+	+	Sakakibara et al. 2003
3	Clorogenic acid derivative	1.803	273; 388	+	+	-	+	+	+	+	+	+	+	+	+	Reference Substance
4	Cinnamic acid derivative	9.653	239; 288; 348	-	-	+	-	-	-	-	-	-	-	-	-	Sakakibara et al. 2003
5	Benzoic acid derivative	9.882	286; 337	-	-	+	-	-	-	-	-	-	-	-	-	
6	Flavonol	10.239	292; 368	-	-	-	-	-	-	-	-	+	-	-	-	Sakakibara et al. 2003
7	Flavone	10.711	269; 327	+	+	-	-	+	-	+	+	+	-	-	-	
8	Theaflavin	11.127	309; 457; 485	+	+	-	+	+	-	+	+	+	+	+	+	
9	Catechin	11.744	276	+	+	-	+	+	+	+	-	-	+	+	+	
10	Anthraquinone	12.118	221; 280	+	+	-	+	+	+	+	+	+	+	+	+	

Treatments 5 (8.88 μM BAP), 6 (17.75 μM BAP), 7 (4.52 μM 2.4-D + 4.44 μM BAP), 8 (4.52 μM 2,4-D + 8.88 μM BAP) and 9 (4.52 μM 2.4-D + 17.75 μM BAP), in the presence (*) and absence (**) of light and 14 (18.10 μM 2, 4-D + 8.88 μM BAP), in the absence of light by analysis by HPLC-DAD. RT= retention time.

Bauhinia spp. have a chemical constitution containing triterpenes, lactones, steroids, tannins, quinones, and flavonoids (Silva and Cechinel-Filho 2002; Pinheiro et al. 2017; Schmidt et al. 2018). Flavonoids are considered the major compounds responsible for the main therapeutic activities attributed to different *Bauhinia* spp. (Cunha et al. 2010). According to Farag (2015), flavonols represent the most abundant subclass, followed by flavones, flavans, and flavanones. In folk medicine, different *Bauhinia* spp. are used to treat pain, inflammation, infections, anddiabetes (Silva and Cechinel-Filho 2002). Recently, Santos et al. (2019) showed that leaf-derived extracts of *B. holophylla*, a native medicinal plant from the Brazilian Cerrado has potent activity against the Dengue virus. Different *Bauhinia* spp. have demonstrated antidiabetic properties, especially *B. forficata*, whose flavonoid kaempferol, obtained from the leaves, has hypoglycemic and antioxidant effects (Toloza-Zambrano et al. 2015).

Several glycosylated and aglycone flavonoids, such as quercetin, rutin, apigenin, apigenin 7-O-glucoside, and triterpene saponins, have been reported from different plant parts of *B. variegata* (Shahana et al. 2017), which are primarily responsible for antinociceptive, anti-inflammatory (Mohamed et al. 2009), cytotoxic (Sharma et al. 2019), and antidiabetic (Shahana et al. 2017) activities attributed to this species.

Callus Growth

Callus growth followed a sigmoidal pattern with a tendency to gain biomass due to the increase in culture time. The growth curve showed five distinct phases: lag (0–14th day), exponential (15th to 56th day), linear (57th to 70th day), stationary (71th to 91th day), and decline (from 92th day), represented by the regression equation $y= -3E\text{-}07x^3 + 2E\text{-}05x^2 + 0{,}0032x + 0.0006$ ($R^2= 0{,}9509$) (Figure 1). During the establishment and growth of cultures, compact calli were obtained with three distinct colors: yellow, green, and brown.

The *lag* phase occurred until the 14th day of culture with a growth rate of 13%. At this stage, the initial calli were yellow at seven and fourteen days of culture. According to Smith (2012), the *lag* phase is a phase of greater energy production, where cells are prepared for division. The mobilization of metabolites and synthesis of proteins and specific compounds begins. This phase results in low growth of the callus. The exponential phase, the period represented by the highest rates of cell division, extended from the 15th to the 56th day after inoculation with a growth rate of 61%. Calli showed compact consistency and exhibited a green color at 21 days, returning to yellow until the 56th day. At 42 and 49 days of cultivation, the calli had a friable consistency. The exponential phase is biosynthetic with greater growth, due to the highest rate of cell division, characteristic of this period; the number of cells increased (Smith 2012). From the 57th to the 70th day, the calli showed a period of slow growth, represented by the linear phase, with a decrease in cell division and metabolic activity with a growth of only 5% but with an increase in cell volume. At 63 and 70 days of the culture, the calli maintained compact consistency and were yellow with darkened portions of brown, indicating the beginning of the oxidation process of the cells and, consequently, the beginning of calli senescence. According to Plazek and Dubert (2010), the darkening occurs due to oxidation caused by the release of phenolic compounds *in vitro* by the injured tissue, which modifies the composition of the culture medium and the absorption of metabolites.

The deceleration phase, represented by the stationary phase was observed between the 71st and 91st days of culture presenting a growth of 21%. In this phase, the cultures must be transferred to another media due to the reduction of nutrients, production of toxic products, drying of the agar, and reduction of oxygen inside the cells. In the stationary phase, there is greater accumulation of bioactive compounds, especially secondary metabolites (Smith 2012)

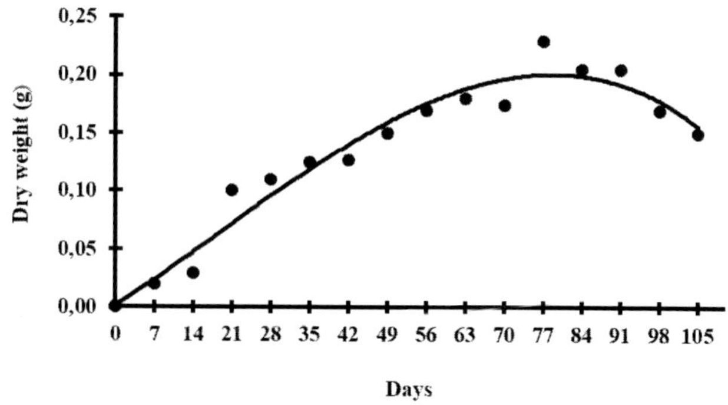

Figure 1. Callus growth curve from *B. variegata* leaf explants on MS medium supplemented with 4.52 μM 2,4-D + 4.44 μM BAP, during 105 days of culture in presence of light.

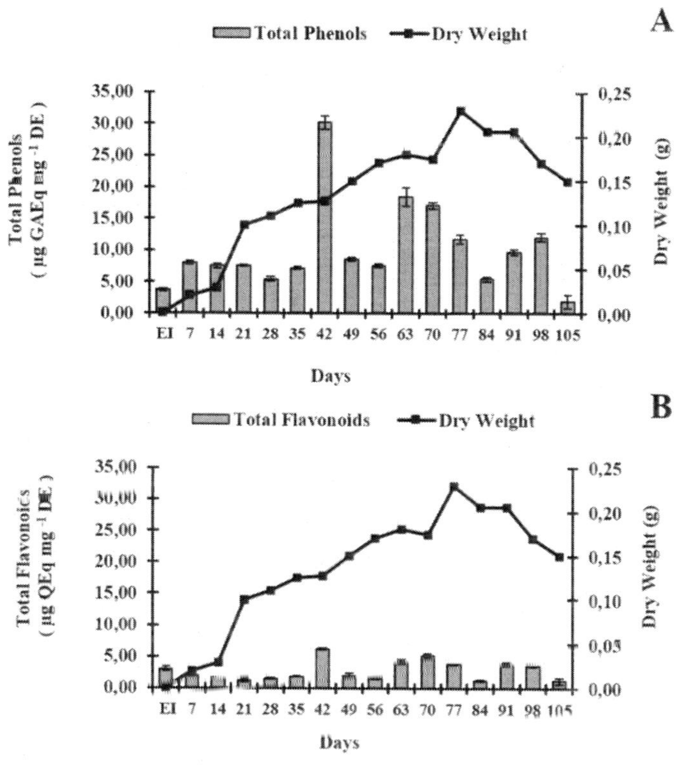

Figure 2. Total phenols (A) and flavonoids (B) of the *B. variegata* calli from leaf explants on MS medium supplemented with 4.52 μM 2,4-D + 4.44 μM of BAP during 105 days of culture in the presence of light.

The calli showed an intensification in brown coloring and compact consistency. The darkening of the tissue, in addition to other characteristics, such as necrosis and drying of the medium, are normal 'signs' present in callus culture that indicate senescence and the reduction of cell growth (Jaskani et al. 2008). After the 92nd day, the calli reached the decline phase, with a 35% reduction in biomass and an increase in the darkening of the callus and the culture medium.

Total Phenol and Flavonoid Content

The total phenol and flavonoid production along the calli growth curve is shown in Figure 2. From a quantitative point of view, a significant difference was observed between the total phenol (Figure 2A) and flavonoid content (Figure 2B), being that the concentrations of these metabolites varied significantly over the 105 days of culture in relation to the initial explant and callus growth phases ($P < 0.05$).

The highest total phenols and flavonoids content were observed in the exponential phase, 42 days after the establishment of the growth curve (31.05 ± 2.02 µg EqGA mg^{-1} DE and 7.01 ± 0.87 µg EqQ mg^{-1} DE, respectively) (Figure 2). According to Santos et al. (2007), it is in the stationary phase that the callus reaches the maximum accumulation of secondary metabolites, since without cell division or weight gain, these compounds are a priority in cell metabolism. In the period when there is a reduction in cell division and the culture remains constant, the primary metabolites produced in the growth phases and the sources of nutrients that remain in the medium are directed towards the production of secondary compounds.

This pattern was not observed for *B. variegata* calli. In the stationary phase (71–91 days) (Figure 1), which coincides with greater callus growth, the total phenol and flavonoid content were low, especially at 84 days of the culture, suggesting an inverse relationship between the phenolic compound levels and the dry weight of the callus, as there is a marked investment of cells in the production of biomass from the callus and the availability of lesser amounts of precursors for phenolic compound production. Thus, the decrease in phenolic compound production during callus growth may be associated with the consumption of sucrose present in the culture medium over the time of establishment of the curve, requiring periodic subcultures for the renewal of carbon sources for the construction of new molecules, especially those originating from secondary metabolism (Naik et al. 2010). Ong et al. (2011) reported that the level and shape of the carbon supply in the culture medium can have a strong effect on the secondary metabolism of cells in culture. The effect of sugars on secondary biosynthesis may result in a specific carbon input or may reflect changes in metabolism due to changes in the carbon/nitrogen balance, depending on the type of sugar involved. In addition, somoclonal variations generated *in vitro*, which occur relatively frequently in callus culture, also determine the production of secondary metabolites that are often variable, thus the need for subculturing (Murthy et al. 2014). In several cases, the accumulation of these compounds, especially phenolic compounds, can be influenced by genetic and biochemical changes in cells under *in vitro* culture conditions (Costa et al. 2015).

Conclusion

Auxin (2,4-D) and cytokinin (BAP), as well as the presence or absence of light, strongly influence the induction, growth, consistency, color, and production of phenolic compounds in *B. variegata* calli. The results encourage additional agronomic and biochemical studies to understand the mechanisms involved in the regulation of bioactive compounds production in *B. variegata* calli, with the perspective of developing new protocols to obtain *in vitro* cell cultures with high concentrations of bioactive flavonoids from a biotechnological approach.

ACKNOWLEDGMENTS

We thank Dr. Andreia Fonseca Silva for the botanical identification of this plant. The authors are grateful to the Universidade Federal de São João del-Rei (UFSJ), Conselho Nacional de Desenvolvimento Científico e Tecnológico (CNPq), and Fundação de Amparo à Pesquisa do Estado de Minas Gerais (FAPEMIG–APQ-00855-19) for financial support This study was financed in part by the Coordenação de Aperfeiçoamento de Pessoal de Nível Superior – Brasil (CAPES) - Finance Code 001.

REFERENCES

Beigmohamadi, Mina, Movafeghi, Ali, Sharafi, Ali, Jafari, Samineh, and Hossein Danafar. 2019. "Cell Suspension Culture of *Plumba goeuropaea* L. towards Production of Plumbagin." *Iran Journal of Biotechnology* 17: e2169. Accessed October 31, 2020. doi: 10.21859/ijb.2169.

Bernabé-Antonio, A., Maldonado-Magaña, A., Ramírez-López, C. B., Salcedo-Pérez, E., Meza-Contreras, J. C., González-García, Y., López-Dellamary Toral, F. A., and F. Cruz-Sosa. 2017. "Establishment of Callus and Cell Suspension Cultures of *Eysenhardtiapo lystachya* (Ortega) and Fungistatic Activity of their Extracts." *South African Journal of Botany* 112 :40-47. Accessed July 28, 2019. doi: https://doi.org/10.1016/j.sajb.2017.05.023.

Brazil. (2019). *Brazilian Pharmacopoeia*. Brasília:Anvisa.

Castro, Ana Hortência Fonsêca, Tavares, Harley da Silva, Pereira, Sara Ribeiro Furtado, Granjeiro, Paulo Afonso, Silva, José Antônio da, and Alexsandro Sobreira Galdino. 2018. "Production and Characterization of Lectin from *Bauhinia holophylla* (Fabaceae: Cercideae) Calli." *Plant Cell Tissue and Organ Culture* 134:423-432. Accessed July 28, 2019. doi: https://doi.org/10.1007/s11240-018-1432-7.

Chandler, John William, and Wolfgang Werr. 2015. "Cytokinin–Auxin Cross Talk in Cell Type Specification." *Trends in Plant Science* 20:291-300. Accessed February 26, 2021. doi: 10.1016/j.tplants.2015.02.003.

Coimbra, Mairon César, Chagas, Rafael César Russo, Vilela, Marlúcia Souza Pádua, and Ana Hortência Fonsêca Castro. 2019. "Growth, Morphology and Bioactive Phenolic Compounds Production in *Pyrostegia venusta* Calli." *Biocatalysis and Agricultural Biotechnology* 18:101036. Accessed February 26, 2021. doi: https://doi.org/10.1016/j.bcab.2019.101036.

Coimbra, Mairon César, Chagas, Rafael César Russo, Duarte-Almeida, Joaquim Maurício, and Ana Hortência Fonsêca Castro. 2017. "Influence of Plant Growth Regulators and Light on Callus Induction and Bioactive Phenolic Compounds Production in *Pyrostegia venusta* (Bignoniaceae)." *Indian Journal of Experimental Biology* 55:584-560. Accessed February 26, 2021. doi: http://nopr.niscair.res.in/handle/123456789/42551.

Costa, Jefferson da Luz, Silva, André Luís Lopes da, Bier, Mário César Jucoski, Brondani, Gilvano Ebling, Gollo, André Luiz, Letti, Luiz Alberto Júnior, Erasmo, Eduardo Andrea Lemus, and Carlos Ricardo Soccol. 2015. "Callus Growth Kinetics of Physic Nut (*Jatropha curcas* L.) and Level of Fatty Acids from Crude Oil Obtained *in vitro*." *Applied Biochemistry and Biotechnology* 176:892-902. Accessed June 17, 2020. doi: 10.1007/s12010-015-1618-y.

Cunha, A. M., Menon, S., Menon, R., Couto, A. G., Bürger, C., and M. W. Biavatti. 2010. "Hypoglycemic Activity of Dried Extracts of *Bauhinia forficata* Link." *Phytomedicine* 17:37-41. Accessed June 17, 2020. doi: 10.1016/j.phymed.2009.06.007.

Dias, Maria Inês, Sousa, Maria João, Alves, Rita C., and Isabel C. R. F. Ferreira. 2016. "Exploring Plant Tissue Culture to Improve the Production of Phenolic Compounds: a Review." *Industrial Crop and Products* 82:9-22. Accessed February 28, 2019. doi: https://doi.org/10.1016/j.indcrop.2015.12.016.

Duarte-Almeida, Joaquim Maurício, Clemente, Milene S., Arruda, Rosani C. O., Vaz, Ângela M. S. F. and Antônio Salatino. 2015. "Glands on the Foliar Surfaces of Tribe Cercideae (Caesapiniodeae, Leguminosae): Distribution and Taxonomic Significance." *Anais da Academia Brasileira de Ciências* 85:787-796. Accessed February 28, 2019. doi: https://doi.org/10.1590/0001-3765201520140151. [*Annals of the Brazilian Academy of Sciences*]

Dymarska, Monika., Janeczko, Tomasz, and Edyta Kostrzewa-Susłow. 2020. "The Callus of *Phaseolus coccineus* and *Glycinemax B*iotransform Flavanones into the Corresponding Flavones." *Molecules* 25:5767. Accessed February 26, 2021. doi: https://doi.org/10.3390/molecules25235767.

Farag, Mohamed A., Sakna, Sarah T., El-Fiky, Nabaweya M., Shabana, Marawan M., and Ludger A. Wessjohann. 2015. "Phytochemical, Antioxidant and Antidiabetic Evaluation of Eight *Bauhinia* L. Species from Egypt Using UHPLC–PDA–qTOF-MS and Chemometrics." *Phytochemistry* 119:41–50. Accessed September 14, 2019. doi: 10.1016/j.phytochem.2015.09.004.

Fang, Evandro Fei, Wong, Jack Ho, Bah, Clara Shui Fern, Lin, Peng, Tsao, Sai Wah, and Tzi Bun Ng. 2010. "*Bauhinia variegata* var. *variegata* Tryps Ininhibitor: from Isolation to Potential Medicinal Applications." *Biochemical and Biophysical Research Communications* 396:806–811. Accessed September 14, 2019. doi: 10.1016/j.bbrc.2010.04.140.

Ferreira, Daniel Furtado. 2011. "Sisvar: a Computer Statistical Analysis System." *Ciência e Agrotecnologia* 35:1039-1042. Accessed October 31, 2020. doi: https://doi.org/10.1590/S1413-70542011000600001. [*Science and Agrotechnology*]

George, Edwin F., Hall, Michael A., and Geert-Jan De Klerk. 2008. *Plant propagation by tissue culture*. Dordrecht: Springer.

Gueven, Alper, and Dietrich Knorr. 2011. "Isoflavonoid Production by Soy Plant Callus Suspension Culture." *Journal of Food Engineering* 103:237–243. Accessed September 14, 2019. doi: https://doi.org/10.1016/j.jfoodeng.2010.10.019.

Gupta, Pratibha, Sharma, Satyawati, and Sanjay Saxena. 2016. "Effect of Abiotic Stress on Growth Parameters and Steviol Glycoside Level in *Stevia rebaudiana* (Bertoni) Raised *in vitro*." *Journal of Applied Research on Medicinal and Aromatic Plants* 3:160-167. Accessed October 31, 2020. doi: 10.1016/j.jarmap.2016.03.004.

Hidalgo, Diego, Sanchez, Raul, Lalaleo, Liliana, Bonfill, Mercedes, Corchete, Purificacion, and Javier Palazon. 2018. "Biotechnological Production of Pharmaceuticals and Biopharmaceuticals in Plant Cell and Organ Cultures." *Current Medicinal Chemistry Journal* 25:3577-3596. Accessed September 14, 2019. doi: 10.2174/09298673 25666180309124317.

Ikeuchi, Momoko, Sugimoto, Keiko, and Akira Iwase. 2013. "Plant Callus: Mechanisms of Iinduction and Repression." *The Plant Cell* 25:3159-3173. Accessed October 31, 2020. doi: https://doi.org/ 10.1105/tpc.113.116053.

Jaskani, Muhammad J., Abbas, Haider, Sultana, R., Khan, M. M., Qasim, M., and Iqrar Khan. 2008. "Effect of Growth Hormones on Micropropagation of *Vitis vinifera* L. cv. Perlette." *Pakistan Journal of Botany* 40:105-109.

Ji, Xiao-Hao, H., Wang, Yan-Ting, Zhang, Rui, Wu, Shu-Jing, An, Meng-Meng, Li, Min, Wang, Chuan-Zeng, Chen, Xiao-Liu, Zhang, Yan-Min, and Xue-Sen Chen. 2015. "Effect of Auxin, Cytokinin and Nitrogen on Anthocyanin Biosynthesis in Callus Cultures of Red-Fleshed Apple (*Malus sieversii* f. niedzwetzkyana)." *Plant Cell Tissue and Organ Culture* 120: 325-337. Accessed October 31, 2020. doi: https://doi.org/10.1007/s11240-014-0609-y.

Karakas, Fatma Pehlivan. 2020. "Efficient Plant Regeneration and Callus Induction from Nodal and Hypocotyls Explants of Gojiberry (*Lycium barbarum* L.) and Comparison of Phenolic Profiles in Calli Formed Under Different Combinations of Plant Growth Regulators." *Plant Physiology and Biochemistry* 146:384-391. Accessed January 23, 2021. doi: https://doi.org/10.1016/j.plaphy.2019.11.009.

Lorenzi, Harri and Francisco José de Abreu Matos. 2008. *Medicinal plants in Brazil*. Nova Odessa:Instituto Plantarum.

Matsubara, Simara, and Delia B. Rodriguez-Maya. 2006. "Teores de Catequinas e Teaflavinas em Chás Comercializados no Brasil." *Ciência e Tecnologia de Alimentos* 26:401-407. Accessed September 14, 2019. doi: https://doi.org/10.1590/S0101-20612006000200024. ["Contents of Catechins and Theaflavins in Teas Marketed in Brazil." *Food Science and Technology*]

Mirzaee, H., Sharafi, A., and H. Hashemi Sohi. 2016. "*In vitro* Regeneration and Transient Expression of Recombinant Sesquiterpene Cyclase (SQC) in *Artemisia annua* L." *South African Journal of Botany* 104:225-231. Accessed October 31, 2020. doi: https://doi.org/10.1016/j.sajb.2015.10.005.

Mohamed, Mona A., Mammoud, Madeha R., and Heiko Hayen. 2009. "Evaluation of Antinociceptive and Anti-Inflammatory Activities of a New Triterpene Saponin *from Bauhinia variegata* Leaves. *Zeitschriftfur Natur for schung C Journal of Biosciences* 64:798-808. Accessed September 14, 2019. doi: 10.1515/znc-2009-11-1208.

Murashige, Toshio, and Folke Skoog. 1962 "A Revised Medium for Rapid Growth and Bioassay with to Tobacco Tissue Cultures." *Physiologia Plantarum* 15:473-497. Accessed October 30, 2019. doi: https://doi.org/10.1111/j.1399-3054.1962.tb08052.x.

Murthy, Hosakatte Niranjana, Lee, Eun-Jung, and Kee-Yoeup Paek. 2014. "Production of Secondary Metabolites from Cell and Organ Cultures: Strategies and Approaches for Biomass Improvement and Metabolite Accumulation." *Plant Cell, Tissue and Organ Culture* 118:1-16. Accessed September 14, 2019. doi: hhhps://doi.org/10.1007/s11240-014-0467-7.

Naik, Poornananda Madhava, Manohar, Shirugumbi Hanamanthagouda, Praveen, Nagella, and Hosakatte Niranjana Murty. 2010. "Effects of Sucrose and pH Levels on *In Vitro* Shoot Regeneration from Leaf Explants of *Bacopamon nieri* and Accumulation of Bacoside A in Regenerated Shoots." *Plant Cell, Tissue and Organ Culture* 100:235-239. Accessed September 14, 2019. doi: https://doi.org/10.1007/s11240-009-9639-2.

Ochoa-Villarreal, Marisol, Howat, Susan., Hong, SunMi, Jang, Mi Ok, Jin, Young Woo, Lee, Eun Kyong, and Garry J. Loake. 2016. "Plant Cell Culture Strategies for the Production of Natural Products." *BMB Reports* 49:49-58. Accessed October 31, 2020. doi: 10.5483/bmbrep.2016.49.3.264.

Ogita, Shinjiro. 2015. "Plant Cell, Tissue and Organ Culture: the Most Flexible Foundations for Plant Metabolic Engineering Applications." *Natural Product Communications* 10:815-820. Accessed October 31, 2020. doi: https://doi.org/10.1177/1934578X1501000527.

Ong, S. L., Ling, A. P. K., Poospooragi, R., and S. Moosa. 2011. "Production of Flavonoid Compounds in Cell Cultures of *Ficus deltoidea* as Influenced by Medium Composition." *International Journal of Medicinal and Aromatic Plants* 1:62-74.

Pastrana-Bonilla, Eduardo, Akoh, Casimir C., Sellappan, Subramani, and Gerard Krewer. 2003. "Phenolic Contet and Antioxidant Capacity of Muscadine Grapes." *Journal of Agricultural and Food Chemistry* 51:5497-5503. Accessed September 14, 2019. doi: 10.1021/jf030113c.

Pinheiro, Marcelo S., Rodrigues, Luahara S., Neto, Leila S., Moraes-Souza, Rafaianne Q., Soares, Thaigra S., Américo, Madileine F., Campos, Kleber E., Damasceno, Débora C., and Gustavo T. Volpato. 2017. "Effect of *Bauhinia holophylla* Treatment in Streptozotocin-Induced Diabetic Rats." *Anais da Academia Brasileira de Ciências* 89:263–272. Accessed October 31, 2020. doi: https://doi.org/10.1590/0001-3765201720160050. [*Annals of the Brazilian Academy of Sciences*]

Piovan, Anna, Filippini, Raffaella, and Gabbriella Innocenti. 2014. "Coumarin Compounds in *Coronilla scorpioides* Callus Cultures." *Natural Product Communications* 9:489-492. Accessed Accessed June 17, 2020. doi: https://doi.org/10.1177/1934578X1400900415.

Plazek, A., and Franciszek Dubert. 2010. "Improvement of Medium for *Miscanthus x Giganteus* Callus Induction and Plant Regeneration." *Acta Biologica Cracoviensia Series Botanica* 52:105-110. Accessed Accessed June 17, 2020. doi: 10.2478/v10182-010-0013-9.

Reddy, Mopuru V. B., Reddy, Muntha K., Gunasekar, Duvvuru, Caux, Cristelle, and Bernard Bodo. 2003. "Flavanone and a Dihydrodibenzoxepin from *Bauhinia variegata*." *Phytochemistry* 64:879-882. Accessed Accessed June 17, 2020. doi: 10.1016/s0031-9422(03)00416-3.

Sahin, G., Verma, S. K., and E. Gurel. 2013. "Calcium and Magnesium Elimination Enhances Accumulation of Cardenolides in Callus Cultures of Endemic *Digitalis* Species of Turkey." *Plant Physiology and Biochemistry* 73:139-143. Accessed June 17, 2020. doi: 10.1016/j.plaphy.2013.09.007.

Sakakibara, Hiroyuki, Honda, Yoshinori, Nakagawa, Satoshi, Ashida, Hitoshi, and Kazuki Kanazawa. 2003. "Simultaneous Determination of All Polyphenols in Vegetables, Fruits, and Teas. *Journal of Agricultural and Food Chemistry* 51:571-581. Accessed July 18, 2019. doi: https://doi.org/10.1021/jf020926l.

Santos, Alberdan Silva, Araújo, Silvana Fernandes, Goulart, Henrique Fonseca, Caetano, Luiz Carlos, Arruda, Mara Sílvia Pinheiro, Santos, Lourivaldo da Silva, and Antônio Euzébio Goulart Santa'ana. 2007. "Dehydrorotenoid Produced by Callus Tissue Culture and Wild Plant Roots of *Boerhaavia coccinea*." *Revista Brasileira de Farmacognosia* 17:538-541. Accessed October 31, 2020. doi: https://doi.org/10.1590/S0102-695X2007000400011. [*Brazilian Journal of Pharmacognosy*]

Santos, Michelli dos., Teixeira, Thaiz Rodrigues, Santos, Felipe Rocha as Silva, Lima, William Gustavo, Ferraz, Ariane Coelho, Silva, Nathália Lucca, Leite, Flávio José, Siqueira, João Máximo, Luyten, Walter, Castro, Ana Hortência Fonsêca, Magalhães, José Carlos de, and Jaqueline Maria Siqueira Ferreira. 2019. "*Bauhinia holophylla* (Bong.) Steud. Leaves-Derived Extracts as Potent Anti-Dengue Serotype 2." *Natural Product Research* 33:1-6. Accessed October 31, 2020. doi: 10.1080/14786419.2019.1669030.

Schmidt, Marcella Emília Pietra, Pires, Fernanda Brum, Bressan, Lucas Paines, Silva Júnior, Fábio Vieira da, Lameira, Osmar, and Marcelo Barcellos da Rosa. 2018. "Some Triterpenic Compounds in Extracts of *Cecropia* and *Bauhinia* Species for Different Ampling Years." *Revista Brasileira de Farmacognosia* 28:21–26. Accessed July 18, 2019. doi: https://doi.org/10.1016/j.bjp.2017.11.005.

Shahana, Syeda, Nikalje, Anna Pratima G., and G. Nikalje. 2017. "A Brief Review on *Bauhinia variegata*: Phytochemistry, Antidiabetic and Antioxidant Potential." *American Journal of PharmTech Research* 7:186–197.

Sharma, Neha, Sharma, Anket, Bhatia, Gaurav, Landi, Marco, Brestic, Marian, Singh, Bikram, Singh, Jatinder, Kaur, Satwinderjeet, and Renu Bhardwaj. 2019. "Isolation of Phytochemicals from *Bauhinia variegata* L. Bark and their *In Vitro* Antioxidant and Cytotoxic Potential. *Antioxidants (Basel)* 8:492. Accessed July 18, 2020. doi: 10.3390/antiox8100492.

Shoja, A. M., Hassanpouraghdam, M. B., Khosrowshahli, M., and A. Movafeghi. 2010. "Callogenesis capability and callis on a clonal variation of costmary (*Tanacetum balsumlta* L.). *Romanian Biotechnology Letters* 15:5120-5124.

Silva, Karine Luize, and Valdir Cechinel Filho. 2002. "Plantas do Gênero *Bauhinia*: Composição Química e Potencial Farmacológico." *Química Nova* 25:449-454. Accessed July 18, 2019. doi: https://doi.org/10.1590/S0100-40422002000300018. [Plants of the Bauhinia Genus: Chemical Composition and Pharmacological Potential. *New Chemistry*]

Silveira, Vanildo, Floh, Eny Iochevet Segal, Handro, Walter, and Miguel Pedro Guerra. 2004. "Effect of Plant Growth Regulators on the Cellular Growth and Levels of Intracelular Protein, Starch and Polyamines in Embryogenic Suspension Cultures of *Pinnus taeda*." *Plant Cell, Tissue and Organ Culture* 76:53-60. Accessed October 31, 2020. doi: https://doi.org/10.1023/A:1025847515435.

Smith, Roberta H. 2012. *Plant tissue culture: techniques and experiments*. San Diego. Academic Press.

Souza, Bárbara Verônica Cardoso de, Araújo, Regilda Saraiva dos Reis Moreira, Silva, Oskar Almeida, Faustino, Lucas Costa, Gonçalves, Maria Fabrícia Beserra, Santos, Mirian Lima dos, Souza, Grasielly

Rocha, Rocha, Lindalva Moura, Cardoso, Mônica Larissa Sousa, and Lívio César Cunha Nunes. 2018. "*Bauhinia forficata* in the Treatment of Diabetes Mellitus: a Patent Review." *Expert Opinion on Therapeutic Patents* 28:129-138. Accessed July 28, 2020. doi: 10.1080/13543776.2018.1409208.

Stals, Hilde, and Dirk Inzé. 2001. "When Plant Cells Decide to Divide." *Trends in Plant Science* 6:359-364. Accessed July 28, 2020. doi: https://doi.org/10.1016/S1360-1385(01)02016-7.

Toloza-Zambrano, Pamela, Avello, Márcia, and Pola Fernandez. 2015. "Determination of rutin and trigonelline in extracts of *Bauhinia forficate* subsp. *pruinosa* and *hypoglycemic* effect on diabetic and prediabetic patients humans." *Boletín Latinoamericano y del Caribe de Plantas Medicinales y Aromaticas* 14:21–32.

Vaz, Ângela Maria S. F, and Ana Maria G. A. Tozzi. 2005. "Sinopse de *Bauhinia* sect. *Pauletia* (Cav.) DC. (Leguminosae: Caesalpinioideae: Cercidae) no Brasil." *Revista Brasileira de Botânica* 28:477-491. Accessed July 28, 2019. doi: https://doi.org/10.1590/S0100-84042005000300006. ["Synopsis of Bauhinia sect. Pauletia (Cav.) DC. (Leguminosae: Caesalpinioideae: Cercidae) in Brazil." *Brazilian Journal of Botany*]

Wang, Yicheng, Sun, Jingjing, Wang, Nan, Xu, Haifeng, Qu, Changzhi, Jiang, Shenghui, Fang, Hongcheng, Su, Mengyu, Zhang, Zongying, and Xuesen Chen. 2019. "MdMYBL2 Helps Regulate Cytokinin-Induced Anthocyanin Biosynthesis in Red-Fleshedapple (*Malussieversii* f. niedzwetzkyana) Callus." *Functional Plant Biology* 46:187-196. Accessed February 28, 2021. doi: 10.1071/FP17216.

Woisky, Ricardo G., and Antônio Salatino. 1998. "Analysis of Propolis: Some Parameters and Procedures for Chemical Quality Control." *Journal of Apicultural Research* 37:99-105. Accessed October 31, 2020. doi: https://doi.org/10.1080/00218839.1998.11100961

In: The Book of Flavonoids
Editors: Chao-Hui Feng et al.
ISBN: 978-1-68507-376-3

Chapter 10

THE CHEMICAL DIVERSITY AND BIOLOGICAL ACTIVITIES OF FLAVONOIDS OF TRIBE BIGNONIEAE (BIGNONIACEAE, LAMIALES)

Juliana C. S. Carvalho, Miguel Peña-Hidalgo, Lúcia G. Lohmann and Marcelo J. P. Ferreira*
Universidade de São Paulo, Instituto de Biociências,
Departamento de Botânica, São Paulo, SP, Brasil

ABSTRACT

Flavonoids constitute a large group of polyphenolic compounds with a benzo-γ-pyrone structure that are ubiquitously found in plants. The tribe Bignonieae is the most diverse and abundant clade of lianas in the Neotropics comprising around 393 species and 20 genera and constituting the largest tribe in the Bignoniaceae. Phytochemical studies of a variety of Bignonieae species have documented a large chemical diversity of flavonoids with a plethora of biological activities. This

* Corresponding Author's E-mail: marcelopena@ib.usp.br.

chapter summarizes the occurrence of flavonoids in Bignonieae and describes the biological activities investigated for these structures in the literature. We further investigate the distribution of flavonoids through major clades of tribe Bignonieae based on the most recent phylogeny of the group. Almost one hundred flavonoids characterized by polymethoxylated structures have been documented in Bignonieae. Dimeric flavonoids are restricted to two genera, *Fridericia* and *Mansoa*. Antiprotozoal, antiradical, and antitumoral properties are the main activities described for Bignonieae flavonoids.

Keywords: flavones, dimeric flavonoids, lianas, bioactivities

INTRODUCTION

The Bignoniaceae comprises 82 genera and approximately 840 species (Lohmann and Ulloa, 2006 onwards) that are distributed through the Tropical regions of the globe, especially in dry and rainforests, with 72% of species found in the Neotropics (Gentry, 1980). Species of this family are commonly used as ornamentals and wood sources, as well as are used in folk medicine and indigenous rituals (Gentry, 1992). Due to the medicinal and economic importance of some species of Bignoniaceae, phytochemical studies have been carried out and several classes of natural products have been described (Cipriani et al., 2012).

The family includes eight main clades: the tribes Bignonieae, Catalpeae, Jacarandeae, Oroxyleae, Tecomeae, Tourrettieae and the informally named "Paleotropical clade" and "Tabebuia alliance" (Olmstead et al., 2009). Bignonieae is the largest tribe, with around 393 species and 20 genera (Lohmann & Taylor, 2014; Lohmann & Fonseca, 2019). The monophyly of the tribe is supported by molecular data and morphological synapomorphies (Lohmann & Taylor, 2014). Species of Bignonieae generally have 3-foliolate leaves or 2-foliolate leaves with the terminal leaflet modified into a tendril (Souza-Baena et al., 2014). The showy Bignonieae flowers have a cupular calyx and 5-lobed corollas, typically infundibuliform or occasionally tubular, associated with several pollinators such as bats, bees, butterflies, hawkmoths, hummingbirds, and wasps (Lohmann, 2004). The wood anatomy composed of 4-32-discontinuous phloem wedges in cross sections is a synapomorphy of Bignonieae (Pace et al., 2009). Fruits are septicidal capsules and seeds are generally winged and wind dispersed or in some cases corky and water dispersed (Lohmann & Taylor, 2014).

Some species of Bignonieae are used in folk medicine to treat illness. For example, *Fridericia chica* (Bonpl.) L.G.Lohmann known to treat inflammation and skin infections in the Brazilian Amazon (Gemelli et al., 2015). *Mansoa alliacea* (Lam.) A.H.Gentry, on the other hand, is traditionally used for the treatment of rheumatism, fever, and influenza (Faccin et al., 2017). *Martinella iquitoensis* A. Samp. is used to treat inflammation and eye irritation by Amazon tribes in Peru and Brazil (Gentry & Cook, 1984).

Phytochemical and pharmacological studies of Bignonieae have shown a large diversity of natural products and a plethora of bioactivities. Among these compounds, flavonoids are common secondary metabolites identified in several species of the tribe. Flavonoids are the largest group of natural polyphenols, with more than 9.000 compounds described (Yonekura-Sakakubara et al., 2019). Flavonoids are biosynthesized through two distinct routes (shikimate and acetate pathways), resulting in the formation of basic structure of two benzene rings (rings A and B) linked by a three-carbon chain (usually found as ring C), simplified as C6-C3-C6 (Figure 1).

Flavonoids are formed from a cinnamoyl-CoA starter unit, with chain extension using three molecules of malonyl-CoA. These reactions are catalyzed by a particular PKS-III enzyme yielding a polyketide structure that can be cyclized in two different ways furnishing stilbenes or chalcones. Chalcones act as precursors for a vast range of flavonoid derivatives. Most flavonoids contain a six-membered heterocyclic ring, formed by Michael-type nucleophilic attack of a phenol group onto the α,β-unsaturated ketone to give flavanones, such as hesperetin (36). Subsequently, several enzymes can modify flavanone structures, resulting in diversified flavonoid classes like flavones, flavonols, anthocyanidins, and catechins (Dewick, 2009; Figure 2). Eight groups of skeletons (anthocyanidin, 3-desoxyanthocyanidin, catechin, dihydroflavonol, flavanone, flavone, flavonol, and isoflavone) generate flavonoid structures based on oxidation, unsaturation, and cyclization. These basic structures are usually modified by enzymes, resulting in additional hydroxylation, glycosylation, methylation, prenylation, and sulfation (Tsimogiannis et al., 2007; Yonekura-Sakakubara et al., 2019).

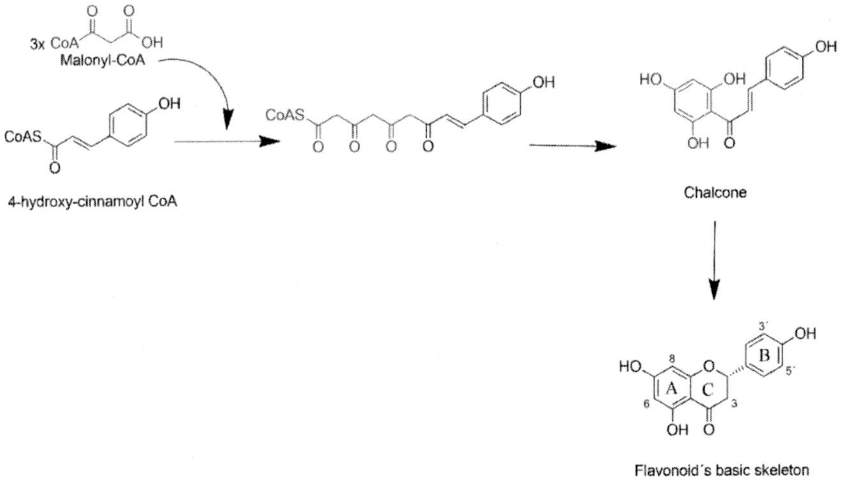

Figure 1. Scheme of flavonoid biosynthetic routes.
Adapted from Santos et al., 2020.

In plants, the production of flavonoids is related to multiple biological functions, such as UV protection, phytopathogenic response, signalization of nodulation, fertilization, auxin transportation, and pollinator attraction (Koes et al., 2005; Treutter, 2006). Thereby, flavonoids are widely distributed in the plant kingdom, including microalgae lineages (i.e., Cyanobacteria, Chlorophyta, Rhodophyta, Haptophyta, and Ochrophyta) (Goiris et al., 2014). The distribution of these compounds among several clades suggests a diversification in biosynthetic genes, which has occurred in a series of steps during the evolutionary history of plants (Yonekura et al., 2019), turning flavonoids into useful chemophenetic characters to understand relationship among taxa.

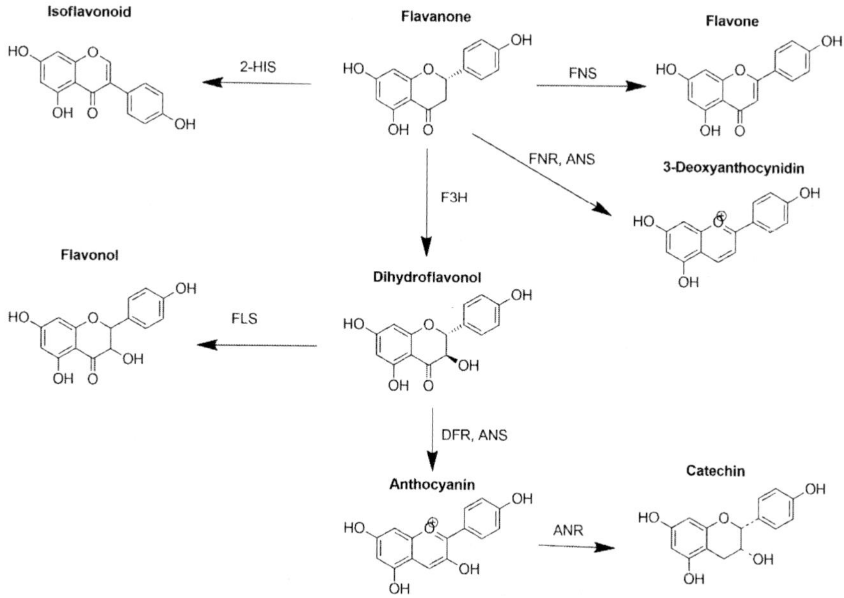

Figure 2. Scheme of flavonoid classes formation. ANS: Anthocyanidin synthase, ANR: Anthocyanidin reductase, DFR: dihydroflavonol-4-reductase, F3H= flavanone 3β-hydroxylase, FLS: flavonol synthase, FNR: Flavanone 4-reductase, FNS=flavone synthase, 2HIS-2-hydroxyisoflavone synthase, IFD=isoflavone dehydratase. Adapted from Santos et al, 2020.

Due to the importance of the Bignonieae for the Neotropical flora, its broad diversity of flavonoids, and importance in folk medicine, this review aims to synthesize the current knowledge and biological activity of Bignonieae flavonoids documented to date. The meaning of these metabolites for taxonomy is evaluated in the light of the most recent phylogeny of tribe Bignonieae.

METHODS

A systematic search of flavonoids and their biological activities was carried out in July 2021 in the SciFinder database, using each of the 20 genera of Bignonieae as keywords. All articles published in Portuguese, English, and Spanish were compiled. Following this search, all flavonoids recovered in members of Bignonieae were classified into six groups: (i) anthocyanidin and anthocyanins, (ii) chalcones, (iii) dimeric flavonoids, (iv) flavones, (v) flavanones, and (vi) flavonols. Compound names were synthesized and organized in alphabetical order (Table 1). The chemical structure of the flavonoids identified were represented using ChemDraw Ultra 12.0. The distribution of compounds per species is summarized in Table 2.

Table 1. Flavonoids identified in Bignonieae species

Structures	Compounds	Species	Plant parts	References
Anthocyanins and anthocyanidins				
1	3'-Hydroxy-carajurone	*Fridericia chica*	Leaves	(Devia et al., 2002)
2	3'-Hydroxy-carajurin	*Fridericia chica*	Leaves	(Devia et al., 2002)
3	Carajurin	*Fridericia chica*	Leaves and Flowers	(Devia et al., 2002) (Scogin, 1980)
4	Carajurone	*Fridericia chica*	Leaves	(Devia et al., 2002)
5	Cyanidin-3-O-glucoside	*Amphilophium buccinatorium*	Flowers	(Scogin, 1980)
		Bignonia capreolata	Flowers	(Scogin, 1980)
		Dolichandra dentata	Flowers	(Scogin, 1980)
		Dolichandra unguis-cati	Flowers	(Scogin, 1980)
		Fridericia chica	Leaves and Flowers	(Devia et al., 2002), (Scogin, 1980)
		Fridericia platyphylla	Flowers	(Scogin, 1980)
		Mansoa alliacea	Flowers	(Scogin, 1980)
6	Cyanidin-3-O-rutinoside	*Amphilophium buccinatorium*	Flowers	(Scogin, 1980)
		Fridericia chica	Flowers	(Scogin, 1980)
		Fridericia platyphylla	Flowers	(Scogin, 1980)
7	Pelargonidin-3-O-glucoside	*Amphilophium buccinatorium*	Flowers	(Scogin, 1980)
8	Pelargonidin-3-O-rutinoside	*Amphilophium buccinatorium*	Flowers	(Scogin, 1980)

Structures	Compounds	Species	Plant parts	References
Chalcones				
9	3',4'-Dihydroxy-3,4,5'-trimethoxy-chalcone	*Fridericia platyphylla*	Flowers	(Rezende-Júnior et al., 2020)
10	3,4-Dimethoxy-chalcone	*Fridericia platyphylla*	Flowers	(Rezende-Júnior et al., 2020)
11	3'-Hydroxy-3-acetate-4-methoxy-chalcone	*Fridericia platyphylla*	Flowers	(Rezende-Júnior et al., 2020)
12	4'-Hydroxy-3,4-dimethoxy-chalcone	*Fridericia platyphylla*	Flowers	(Rezende-Júnior et al., 2020)
Dimeric flavonoids				
13	Brachydin A	*Fridericia platyphylla*	Roots	(Rocha et al., 2014)
14	Brachydin B	*Fridericia platyphylla*	Roots	(Rocha et al., 2014)
15	Brachydin C	*Fridericia platyphylla*	Roots	(Rocha et al., 2014)
16	Brachydin D	*Fridericia platyphylla*	Roots	(Rocha et al., 2017)
17	Brachydin E	*Fridericia platyphylla*	Roots	(Rocha et al., 2017)
18	Brachydin F	*Fridericia platyphylla*	Roots	(Rocha et al., 2017)
19	Brachydin G	*Fridericia platyphylla*	Roots	(Rocha et al., 2017)
20	Brachydin H	*Fridericia platyphylla*	Roots	(Rocha et al., 2017)
21	Brachydin I	*Fridericia platyphylla*	Roots	(Rocha et al., 2017)
22	Brachydin J	*Fridericia platyphylla*	Roots	(Rocha et al., 2017)
23	Catuabin A	*Anemopaegma arvense*	Stem bark	(Tabanca et al., 2007)
24	Cinchonain Ia	*Anemopaegma arvense*	Stem bark	(Tabanca et al., 2007)
25	Cinchonain IIa	*Anemopaegma arvense*	Stem bark	(Tabanca et al., 2007)
26	Kandelin AI	*Anemopaegma arvense*	Stem bark	(Tabanca et al., 2007)
27	Mansoin A	*Mansoa hirsuta*	Fruits	(Campana et al., 2014)
28	Mansoin B	*Mansoa hirsuta*	Fruits	(Campana et al., 2014)

Table 1. (Continued)

Structures	Compounds	Species	Plant parts	References
Dimeric flavonoids				
29	Mansoin C	*Mansoa hirsuta*	Fruits	(Campana et al., 2016)
30	Mansoin D	*Mansoa hirsuta*	Fruits	(Campana et al., 2016)
31	Mansoin E	*Mansoa hirsuta*	Fruits	(Campana et al., 2016)
32	Mansoin F	*Mansoa hirsuta*	Fruits	(Campana et al., 2016)
Flavanones				
33	5,5'-Dihydroxy-3',4',7-trimethoxyflavanone	*Dolichandra quadrivalvis*	Stem bark	(Lima et al., 2005)
34	Alpinetin	*Fridericia speciosa*	Leaves	(Milani et al., 2020)
		Fridericia triplinervia	Leaves	(Leite et al., 2006)
35	Hesperidin	*Pyrostegia venusta*	Roots	(Ferreira et al., 2000)
36	Hesperetin	*Amphilophium crucigerum*	Seeds	(Prá et al., 2017)
Flavones				
37	7-O-Glucuronide 4′-methoxy-scutellarein	*Adenocalymma imperatoris-maximilianii*	Leaves, stems, and roots	(Oliveira et al., 2017)
38	3',4'-Dihydroxy-5,6,7-trimethoxyflavone	*Fridericia platyphylla*	Leaves	(Alcerito et al., 2002)
39	4'-Hydroxywogonin	*Bignonia callistegioides*	Leaves	(Castillo et al., 2013)
40	4'-O-Methyl scutellarin 6-O apiosyl galactoside	*Dolichandra unguis-cati*	Unflowering aerial part	(Aboutabl et al., 2008)
41	5-O-Methyl-scutellarein	*Fridericia chica*	Leaves	(Lima et al., 2020)
42	6-Hydroxyapigenin-7-O-glucoside	*Mansoa difficilis*	Leaves	(Guilhon et al., 2012)
43	6-Hydroxyluteolin	*Fridericia chica*	Leaves	(Siraichi et al., 2013)
44	6-Hydroxyluteolin-glucoside	*Cuspidaria pulchra*	Aerial parts	(Alvarenga et al., 2015)

Structures	Compounds	Species	Plant parts	References
Flavones				
45	6-Methoxy-apigenin-7-O-glucoside	*Dolichandra unguis-cati*	Unflowering aerial part	(Aboutabl et al., 2008)
46	6-Methoxy-acacetin-7-O-glucoside	*Dolichandra unguis-cati*	Unflowering aerial part	(Aboutabl et al., 2008)
47	7-O-Methyl-scutellarein	*Mansoa alliacea*	Flowers	(Rao & Rao, 1982)
48	8-Methoxy-acacetin-7-O-glucoside	*Dolichandra unguis-cati*	Unflowering aerial part	(Aboutabl et al., 2008)
49	Acacetin	*Bignonia callistegioides*	Leaves	(Castillo et al., 2013)
50	Acacetin-7-O-glucoside	*Pyrostegia venusta*	Flowers	(Veloso et al., 2010)
51	Acacetin-8C-rutinoside	*Amphilophium paniculatum*	Leaves	(Samy et al., 2015)
52	Apigenin	*Amphilophium paniculatum*	Leaves	(Nassar et al., 2013)
		Fridericia chica	Leaves	(Siraichi et al., 2013)
		Fridericia platyphylla	Leaves	(Blatt et al., 1998)
		Fridericia speciosa	Leaves	(Milani et al., 2020)
		Mansoa alliacea	Leaves	(Faccin et al., 2017)
53	Apigenin 7-O-glucuronide	*Mansoa alliacea*	Flowers	(Rao & Rao, 1980)
54	Apigenin 7-O-β-D-glucuronopyranoside-methyl ester	*Mansoa alliacea*	Flowers	(Rao & Rao, 1980)
55	Carajuflavone	*Fridericia chica*	Leaves	(Takemura et al., 1995)
56	Chrysin	*Fridericia formosa*	Leaves, stems, and fruits	(Brandão et al., 2017)
		Fridericia samydoides	Leaves and stems	(Pauletti et al., 2003)
57	Chrysoeriol	*Fridericia chica*	Leaves	(Silva-Silva et al., 2021)
58	Cirsiliol	*Cuspidaria convoluta*	Leaves	(Torres et al., 2018)
		Fridericia platyphylla	Leaves	(Alcerito et al., 2002)

Table 1. (Continued)

Structures	Compounds	Species	Plant parts	References
Flavones				
59	Cirsimarin	*Dolichandra unguis-cati*	Aerial parts	(Liu et al., 2015)
		Fridericia chica	Leaves	(Vasconcelos et al., 2019)
60	Cirsimarin B	*Dolichandra unguis-cati*	Aerial parts	(Liu et al., 2015)
61	Cirsimaritin	*Dolichandra unguis-cati*	Aerial parts	(Chen et al., 2017)
		Fridericia platyphylla	Leaves	(Alcerito et al., 2002)
62	Corymboside	*Dolichandra unguis-cati*	Leaves and Stems	(Duarte et al., 2010)
63	Cosmosiin (Apigenin-7-O-glucoside)	*Amphilophium paniculatum*	Leaves	(Samy et al., 2015)
		Mansoa alliacea	Flowers	(Rao & Rao, 1980)
		Mansoa difficilis	Leaves	(Guilhon et al., 2012)
64	Diosmetin	*Amphilophium crucigerum*	Seeds	(Prá et al., 2017)
		Fridericia speciosa	Leaves	(Milani et al., 2020)
65	Galangustin	*Bignonia callistegioides*	Leaves	(Castillo et al., 2013)
66	Hispidulin	*Fridericia chica*	Leaves	(Vasconcelos et al., 2019)
		Fridericia platyphylla	Leaves	(Alcerito et al., 2002)
67	Isoscutellarein	*Fridericia chica*	Leaves	(Siraichi et al., 2013)
68	Isovitexin	*Martinella obovata*	Leaves	(Arevalo et al., 2011)
69	Katchimoside	*Tynanthus panurensis*	Bark	(Plaza et al., 2005)
70	Linarin (Acacetin-7-O-rutinoside)	*Amphilophium elongatum*	Leaves	(Simões et al., 2011)
		Amphilophium paniculatum	Leaves	(Nassar et al., 2013)

Structures	Compounds	Species	Plant parts	References
Flavones				
71	Luteolin	*Cuspidaria convoluta*	Leaves	(Torres et al., 2018)
		Fridericia chica	Leaves	(Takemura et al., 1995)
		Fridericia platyphylla	Leaves	(Blatt et al., 1998)
		Fridericia speciosa	Leaves	(Milani et al., 2020)
		Mansoa alliacea	Flowers	(Rao & Rao, 1980)
72	Luteolin-7-O-glucoside	*Amphilophium paniculatum*	Leaves	(Samy et al., 2015)
73	Orientin (Luteolin-8C-glucoside)	*Adenocalymma imperatoris-maximilianii*	Leaves, stems, and roots	(De Oliveira et al., 2017)
		Martinella obovata	Leaves	(Arevalo et al., 2011)
74	Pectolinarin	*Amphilophium elongatum*	Leaves	(Simões et al., 2013)
75	Phegopolin (Apigenin-7-methyl ether-4′-glucoside)	*Amphilophium paniculatum*	Leaves	(Nassar et al., 2013)
76	Scutellarein	*Fridericia chica*	Leaves	(Siraichi et al., 2013)
77	Scutellarin (Scutellarein-7-O-glucuronide)	*Mansoa alliacea*	Flowers	(Rao & Rao, 1980)
		Fridericia chica	Leaves	(Siqueira et al., 2019)
78	Sorbarin	*Fridericia elegans*	Leaves	(Krebs, 1987)
79	Vicenin-2	*Adenocalymma imperatoris-maximilianii*	Leaves, stems, and roots	(Oliveira et al., 2017)
		Amphilophium paniculatum	Leaves	(Nassar et al., 2013)
		Bignonia binata	Leaves	(Samy et al., 2021)
		Dolichandra unguis-cati	Unflowering aerial part	(Aboutabl et al., 2008)
		Fridericia chica	Leaves	(Barbosa et al., 2008)
80	Vitexin	*Martinella obovata*	Leaves	(Arevalo et al., 2011)

Table 1. (Continued)

Structures	Compounds	Species	Plant parts	References
Flavonol				
81	4'-Hydroxy-3,7-dimethoxyflavone	*Fridericia chica*	Leaves	(Barbosa et al., 2008)1
82	4'-Methyl-6-methoxy kaempferol-7O,8C-diglucoside	*Dolichandra unguis-cati*	Unflowering aerial part	(Aboutabl et al., 2008)
83	Acacetin-7-O-glucosyl-8-C rhamnosyl 3-O-α-arabinofuranoside	*Dolichandra unguis-cati*	Unflowering aerial part	(Aboutabl et al., 2008)
84	Cacticin (Isorhamnetin-3-O-galactoside)	*Adenocalymma peregrinum*	Leaves and roots	(Grassi et al., 2005)
85	Hyperin (Quercetin-3-O-galactoside)	*Adenocalymma peregrinum*	Leaves and roots	(Grassi et al., 2005)
86	Isoquercetin (Quercetin-3-O-glucoside)	*Fridericia platyphylla*	Leaves	(Blatt et al., 1998)
87	Kaempferol	*Fridericia chica*	Leaves	(Barbosa et al., 2008)1
88	Quercetin	*Bignonia aequinoctialis*	Leaves	(Harborne, 1967)
		Dolichandra unguis-cati	Aerial parts	(Attia, 1999)
		Fridericia platyphylla	Aerial parts	(Bertanha et al., 2020)
89	Quercetin 3-O-robinobioside	*Anemopaegma arvense*	Leaves	(Costanzo et al., 2013)
		Pyrostegia venusta	Flower	(Pereira et al., 2014)
90	Quercetin-3-O-methyl ether	*Dolichandra unguis-cati*	Aerial parts	(Attia, 1999)
91	Quercitrin	*Dolichandra unguis-cati*	Leaves and Steams	(Duarte et al., 2010)
		Mansoa alliacea	Leaves	(Faccin et al., 2017)

Structures	Compounds	Species	Plant parts	References
Flavonol				
92	Rutin	*Anemopaegma arvense*	Leaves	(Costanzo et al., 2013)
		Fridericia chica	Aerial parts	(Gemelli et al., 2015)
		Fridericia platyphylla	Leaves	(Blatt et al., 1998)
		Mansoa alliacea	Leaves	(Faccin et al., 2017)
		Pyrostegia venusta	Leaves	(Blatt et al., 1998)
93	Spireoside	*Martinella obovata*	Leaves	(Arevalo et al., 2011)

Table 2. Distribution of compounds identified in Bignonieae species

Genera	Species	Compounds
Adenocalymma	*A. imperatoris-maximilianii* (Wawra) L.G.Lohmann	37, 73, 79
	A. peregrinum (Miers) L.G.Lohmann	84, 85
Amphilophium	*A. buccinatorium* (DC.) L.G. Lohmann	5-8
	A. crucigerum (L.) L.G. Lohmann	36, 64
	A. elongatum (Vahl) L.G. Lohmann	70, 74
	A. paniculatum (L.) Kunth	51, 52, 63, 70, 72, 75,79
Anemopaegma	*A. arvense* (Vell.) Stellfeld ex J.F. Souza	23-26, 89, 92
Bignonia	*B. aequinoctialis* L.	88
	B. binata Thunb.	79
	B. callistegioides Cham.	39, 49, 65
	B. capreolata L.	5
Cuspidaria	*C. convoluta* (Vell.) A.H. Gentry	58, 71
	C. pulchra (Cham.) L.G. Lohmann	44
Dolichandra	*D. dentata* (K. Schum.) L.G. Lohmann	5
	D. quadrivalvis (Jacq.) L.G. Lohmann	33
	D. unguis-cati (L.) L.G. Lohmann	5, 40, 45, 46, 48, 59-62,79, 82, 83, 88, 90, 91
Fridericia	*F. chica* (Bonpl.) L.G. Lohmann	1-6, 41, 43, 52, 55, 57, 59, 66, 67, 71, 76,77, 79, 81, 87, 92
	F. elegans (Vell.) L.G. Lohmann	78
	F. formosa (Bureau) L.G. Lohmann	56
	F. platyphylla (Cham.) L.G. Lohmann	5, 6, 9-22, 38, 52, 58, 61, 66, 71, 86, 88, 92
	F. samydoides (Cham.) L.G. Lohmann	56

Table 2. (Continued)

Fridericia	*F. speciosa* Mart.	34, 52, 64, 71
	F. triplinervia (Mart. ex DC.) L.G. Lohmann	34
Mansoa	*M. alliacea* (Lam.) A.H. Gentry	5, 47, 52-54, 63, 71, 77, 91, 92
	M. difficilis (Cham.) Bureau & K. Schum.	42, 63
	M. hirsuta DC.	27-32
Martinella	*M. obovata* (Kunth) Bureau & K. Schum.	68, 73, 80, 93
Pyrostegia	*P. venusta* (Ker Gawl.) Miers	35, 50, 89, 92
Tynanthus	*T. panurensis* (Bureau) Sandwith	69

Results and Discussion

Flavonoids of Bignonieae

Flavonoids were reported in the literature in 55% of all Bignonieae genera recognized (11 out of 20 genera), i.e., *Adenocalymma* Mart. ex Meisn. emend L.G. Lohmann, *Amphilophium* Kunth. emend L.G. Lohmann, *Anemopaegma* Mart. ex Meisn., *Bignonia* L., *Cuspidaria* DC., *Dolichandra* Cham. emend L.G.Lohmann, *Fridericia* Mart. emend L.G. Lohmann, *Mansoa* DC., *Martinella* Baill., *Pyrostegia* C. Presl, and *Tynanthus* Miers. Overall, 93 flavonoids were documented in Bignonieae, representing 133 occurrences of multiple types of flavonoids including anthocyanins and anthocyanidins, chalcones, dimeric flavonoids, flavanones, flavones, and flavonols (Table 1). Overall, flavonoids occur either as aglycones, C-glycosyl, and O-glycosyl derivatives. Their structures contain hydroxyl, methoxyl, and glycosyl groups mainly located at C-3, C-5, C-7, C-3’’, and C-4’’. Several compounds with oxidation at the C-6 position were identified, this feature is characteristic of members of the Bignoniaceae (Alcerito et al., 2002)

Anthocyanins and Anthocyanidins

Anthocyanins and anthocyanidins (1–8; Figure 3) are reported in species of *Amphilophium, Bignonia, Dolichandra, Fridericia*, and *Mansoa*. In *Fridericia chica* (Bonpl.) L.G. Lohmann, an unusual 3-deoxyanthocyanidins (1–4) that are hydroxylated and methoxylated at C6, C2', C3', and C4' were documented (Devia et al., 2002; Zorn et al., 2001). While these compounds have only been described for *F. chica*, three varieties of *F. chica* showed qualitative differences in anthocyanidin profiles (Moraga-Tellis et al., 2020; Schiozer et al., 2012). It is worth noting that different extraction methods can enhance the content of such compounds. For example, extraction with supercritical carbon dioxide (scCO_2) leads to highly selective extraction of carajurin (3), although resulting in a low yield extraction (Paula et al., 2013, 2014). On the other hand, enzymatic treatment with xylanase on leaves yielded higher contents of anthocyanidins (Taffarello et al., 2013). It is interesting that the anthocyanidins documented in *F. chica* were described from leaves of *F. chica*, an uncommon feature since this class of flavonoids is generally described from flowers. The accumulation of foliar anthocyanins occurs under specific conditions such as in young and expanding foliage, in autumnal foliage of deciduous species, in response to nutrient deficiency or ultraviolet (UV) radiation exposure, and under damage or defense against browsing herbivores or pathogenic fungal infection (Close & Beadle, 2003). Macerated leaves of *Fridericia chica* produce a strong red dye that are commonly used as body paint or to dye basketry fibers (Gentry, 1992). The red coloring might be explained by the anthocyanins and anthocyanidins found in its leaves.

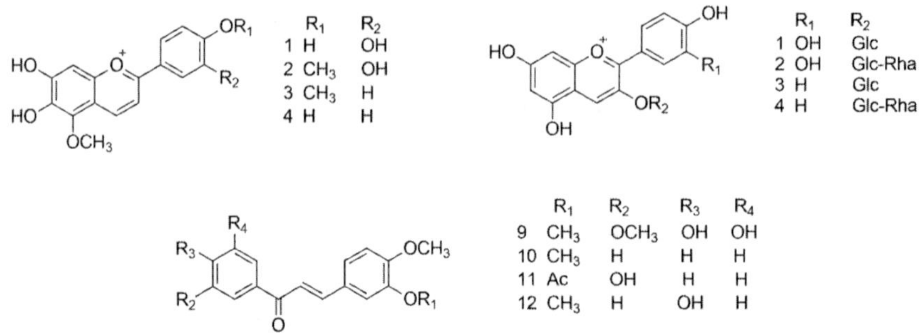

Figure 3. Anthocyanins (1-4), anthocyanidins (5-8) and chalcones (9-12) identified from Bignonieae species.

Additional compounds (5–8) were found in flowers of *Amphilophium, Bignonia, Dolichandra, Fridericia*, and *Mansoa*. These floral pigments were responsible for the corolla color observed in the studied species (Scogin, 1980).

Chalcones

Four chalcones (9–12; Figure 3) were reported exclusively in *Fridericia platyphylla* (Cham.) L.G. Lohmann. These compounds were isolated from the dichloromethane phase of ethanol extract of flowers, and their structures are methoxylated and hydroxylated in the B ring. Chalcones play an important role in plant defense, especially against phytopathogens (Dao et al., 2011).

Dimeric Flavonoids

Bignonieae is a rich source of dimers of flavonoids, with 20 different compounds (13–32; Figure 4) described in *Anemopaegma, Fridericia*, and *Mansoa.* Biflavonoids include two units of flavonoids linked through C-C or C-O bonds from symmetric or asymmetric forms, providing a great diversity of these compounds (Gonjito et al., 2017). In Bignonieae, each one of the three genera produced distinguished skeletons and structures.

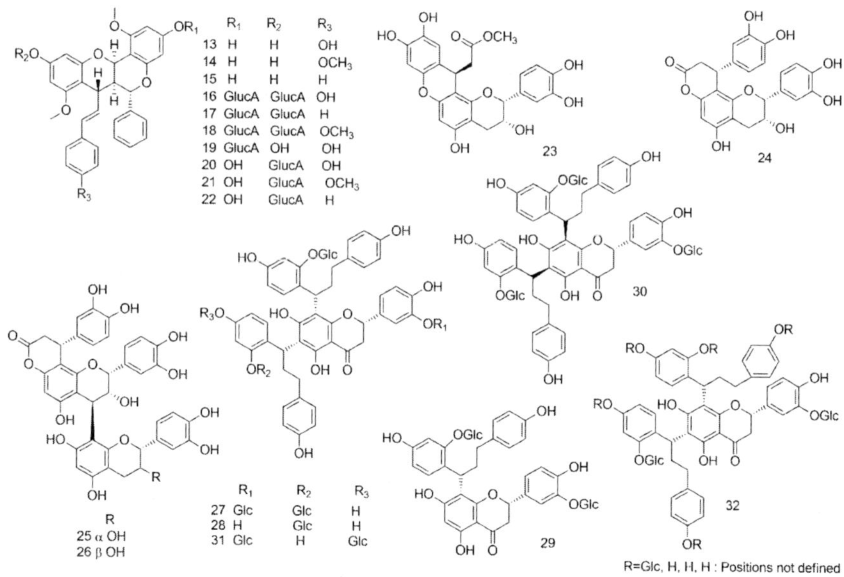

Figure 4. Dimeric flavonoids (13-32) identified from Bignonieae species.

Four compounds (23–26) identified in *Anemopaegma arvense* (Vell.) Stellfeld ex J.F. Souza are composed by a linkage between C8 and C3 positions of two catechins. Also, eleven biflavonoids (13–22) described from *F. platyphylla* were structurally distinguished by a rare configuration [benzopyran[4,3-b]benzopyran] that was only known from *Uvaria dependens* Engl. & Dies (Annonaceae). Dimeric and trimeric flavonoids were also reported in *Mansoa hirsuta* DC., six of which (27–32) show flavanone core linked through C-6 and C-9 positions with a chalcone unit.

Flavanones

Only four flavanones (33–36; Figure 5) were reported in species of *Amphilophium*, *Dolichandra*, *Fridericia*, and *Pyrostegia*. These compounds were identified in roots and aerial parts, such as leaves, stem barks, and seeds. A unique glycosylated compound (35) was identified from *Pyrostegia venusta* (Ker Gawl.) Miers. All flavanones recovered from Bignonieae show at least one methoxy group.

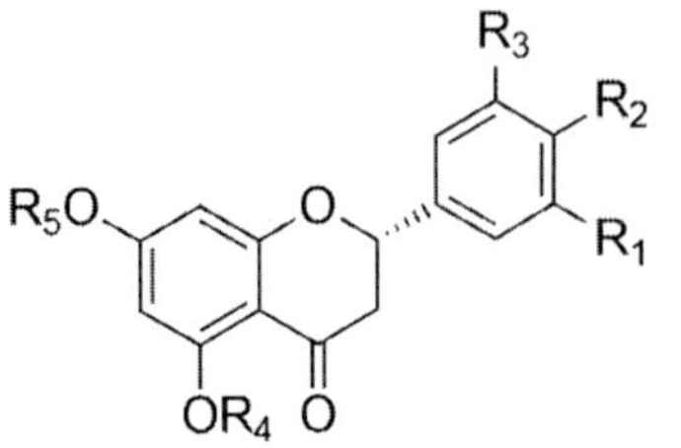

	R_1	R_2	R_3	R_4	R_5
33	OCH_3	OCH_3	OH	H	CH_3
34	H	H	H	CH_3	H
35	OH	OCH_3	H	H	Glc-Rha
36	OH	OCH_3	H	H	H

Figure 4. Flavanones (33-36) identified from Bignonieae species.

Flavones

Flavones (37–80; Figure 6) are ubiquitously distributed in 10 genera: *Adenocalymma*, *Amphilophium*, *Bignonia*, *Cuspidaria*, *Dolichandra*, *Fridericia*, *Mansoa*, *Martinella*, and *Tynanthus*. The same genera in which flavonoids were identified except *Anemopaegma*. However, this genus is poorly known chemically as a single species out of 45 (i.e., *A. arvense*) was studied phytochemically thus far; additional studies are still needed to investigate whether flavones are indeed lacking in *Anemopaegma*. Among the 43 flavones identified in the tribe, glycosylated flavones (25 compounds) occur more frequently than aglycones (18 compounds); glycosylated flavones were found in the 10 genera. Overall, O-glycosylated flavones are more frequently found in Bignonieae than C-glycosylated flavones. On the other hand, sugar moieties are linked to aglycones through a C-C glycosidic bond in compounds like corymboside (62) identified in *Dolichandra unguis-cati*, isovitexin (68) in *Martinella obovata*, orientin (73) in *Adenocalymma imperatoris-maximilianii* and *M. obovata*, vicenin-2 (79) in *A. imperatoris-maximilianii, Amphilophium paniculatum, Bignonia binata, Dolichandra unguis-cati* and *Fridericia chica*, and vitexin (80) in *M. obovata*. The prevalence of flavones over other flavonoids has been previously noted in Bignonieae (Cipriani et al., 2012).

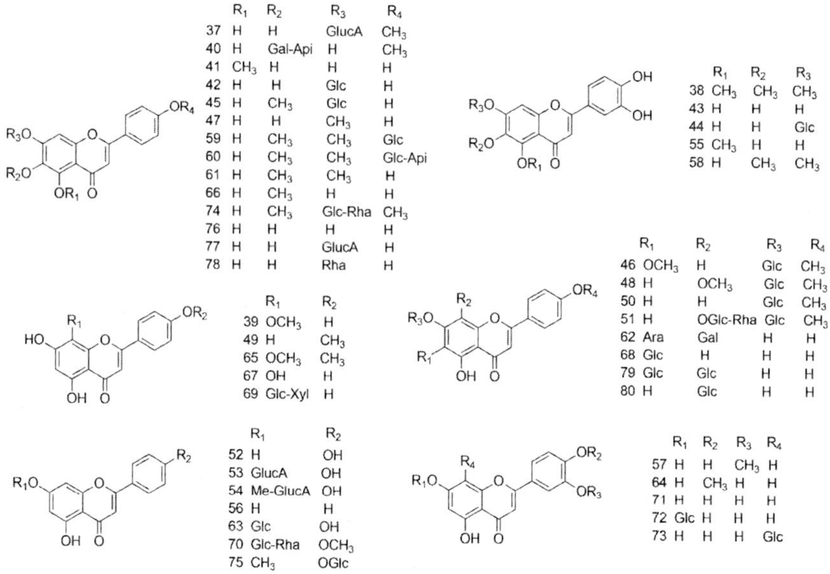

Figure 6. Flavones (37–80) identified from Bignonieae species.

Flavonols

Flavonols (81–93; Figure 7) were found in *Adenocalymma*, *Anemopaegma*, *Bignonia*, *Dolichandra*, *Fridericia*, *Mansoa*, and *Martinella*. Most of these compounds were identified in aerial parts, such as leaves and flowers, although cacticin (84) and hyperin (85) were described in root extracts of *Adenocalymma peregrinum* (Miers) L.G. Lohmann.

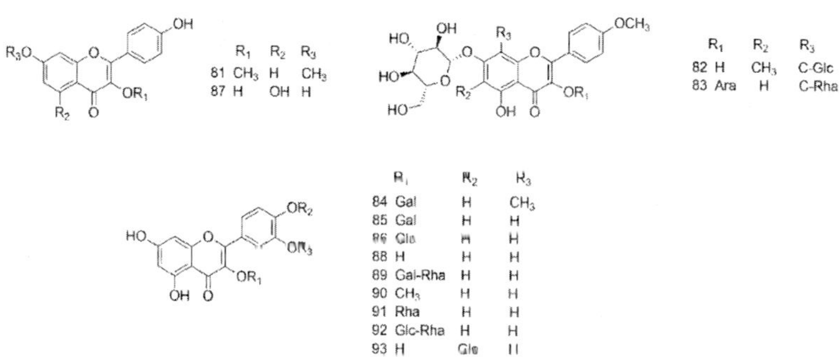

Figure 7. Flavonols (81–93) identified from Bignonieae species.

Biological Activities of Flavonoids from Bignonieae

Flavonoids are known to possess multiple biological activities such as antioxidant, anti-inflammatory, and antitumoral, presenting an important medicinal potential (Tungmunnithum et al., 2018; Ullah et al., 2020). Flavonoids isolated from Bignonieae were considered effective agents in inflammatory diseases, as well as useful for antibiotic, antifungal, antioxidant, and cytotoxic activities. These compounds also have shown antiparasitic effects on *Leishmania amazonensis* Lainson & Shaw, and *Trypanossoma cruzi* Chagas. In addition, insecticidal properties were reported in *Bignonia callistegioides* Cham. These studies highlight the importance of members of Bignonieae as a source of new bioactive flavonoids.

Antibiotic Activity

Natural products represent important sources of new antibiotics (Igarashi, 2019). A high potential of *F. platyphylla* to inhibit NorA and MepA efflux pumps that act in different antimicrobial agents was detecting, highlighting the potential of this species against resistant strains of *Staphylococcus aureus* Rosenbach (Andrade et al., 2020). The chemical composition of *F. platyphylla* is well-documented, which is composed mainly of chalcones (9–12) and dimeric flavonoids (13–22) (Rezende-Junior et al., 2020; Rocha et al., 2014, 2017). Although chalcones had shown low activity against *S. aureus*, these compounds potentialized the effect of Norfloxacin action against SA1199B strain. The chalcone 3',4'-dihydroxy-3,4,5'-trimethoxy-chalcone (9) showed the highest modulate effect for both norfloxacin and ethidium bromide resistance, indicating this compound as an inhibitor of NorA. Molecular docking also confirmed the binding of chalcone with the hydrophobic cavity of NorA and MepA. These results indicated that inhibition of NorA by chalcones can be higher with an increase of methoxylation on such compounds (Rezende-Júnior et al., 2020). Similarly, the presence of methoxylation in the B ring of brachydins also influences NorA resistance, since brachydin B (14) was more efficient to modulate norfloxacin-resistance, reducing four-fold the Minimum inhibitory concentration (MIC) of norfloxacin, when compared to brachydins A and C (Andrade et al., 2020).

The synergic effects of antibiotics such as ampicillin, gentamicin, and oxacillin with flavonoids was tested against resistant strains of *Staphylococcus aureus*. The combination of luteolin (71) and ampicillin showed a strong antibacterial activity against methicillin-resistant *S. aureus* strains (25, 50 and 100 $\mu g.mL^{-1}$ respectively), leading to a reduction between 4 and 8-fold when these strains were co-incubated (0.031, 0.063 and 0.25 $\mu g.mL^{-1}$) (Torres et al. 2019).

Furthermore, scutellarein (76) and a carajurone-enriched fractions of *F. chica* were tested against non-resistance strains of *Staphylococcus aureus* Rosenbach, *Streptococcus pyrogenes* Rosenbach, and *Staphylococcus epidermidis* (Winslow and Winslow) Evans. While compound 76 was inactive against all bacteria tested, the carajurone-enriched fraction showed antibacterial effects against all bacteria evaluated (MIC=100, 50, and 50 $\mu g.mL^{-1}$, respectively) (Violante et al., 2020).

Antifungal Activity

The antifungal capacity of five flavonoids isolated from *A. arvense, F. platyphylla*, and *P. venusta* were investigated. Quercetin-3-O-robinoside (89) was active against wild and mutant strains of *Trichophyton rubrum* (Castell.) Sabour. (MIC=0.5 $mg.mL^{-1}$) and against *Candida albicans* (Robin) Berkhout OF M7-19 strain (MIC= 6 $\mu g.mL^{-1}$) (Costanzo et al., 2013; Pereira et al., 2014). In addition, rutin (92) also showed activity against the same strains of *Trichophyton rubrum* (MIC= 0.25 $mg.mL^{-1}$) (Costanzo et al., 2013) indicating that glycosylated flavonoids can be promising antifungal agents. Similarly, the flavones 3',4'-dihydroxy-5,6,7-trimethoxyflavone (38), cirsiliol (58), and cirsimaritin (61) (1μg of each) inhibited the growth of *Cladosporium sphaerospermum* Penz.. However, a dose ten times higher of hispidulin (66) was less active against the same fungi (Alcerito et al., 2002).

Anti-Inflammatory Activity

Anti-inflammatory and antinociceptive agents have been reported from species of Bignonieae. For example, *F. chica* is traditionally used by Amazon tribes to treat inflammatory diseases. More specifically, hydroethanolic extract and isolated compounds from *F. chica* decreased the modulation of inflammatory processes (Zorn et al., 2001; Lima et al., 2020; Vasconcelos et al., 2019; Takenaka et al., 2020). Furthermore, the flavone 5-O-methylscutellarein (41) reduced total leukocytes and leukocyte migration into the peritoneal cavity, decreasing TNF-α and IL-1β (Lima et al., 2020). Molecular docking of flavonoids reported in the extract showed a good interaction (hydrogen bond and van der Waals interaction) with active site of COX-2 (Vasconcelos et al., 2019). Moreover, bioguided-fraction studies of this species furnished four new anthocyanidins (1–4), including carajurin (3) which inhibited the DNA binding activity of NF-κB (Zorn et al., 2001).

Anti-inflammatory effects were also identified in extracts of *Amphilophium paniculatum* (L.) Kunth. and *Bignonia binata* Thunb. *In vivo* study of ethanol extracts from leaves of *B. binata* significantly inhibited proinflammatory cytokines CCl4-mediators (p<0.001). The ethanol extract of *B. binata* was composed by flavonoid glycosides such as schaftoside and/or isoschaftoside and vicenin (79). On the other hand, acacetin-8C-rutinoside (51) was the main compound in *A. paniculatum* extract suggesting this glycosylated flavone as a promisor anti-inflammatory agent.

Occurrences of dimeric flavonoids like brachydins (14, 15, and 16) and mansoins (27, 28, and 32) were documented in *Fridericia platyphylla* (Cham.) L.G. Lohmann and *Mansoa hirsuta* DC. Despite structural differences between brachydin and mansoin skeletons, both compounds have anti-inflammatory activity. Mansoins A (27) and B (28) inhibited TNF-α release (IC_{50} of 48.1 ± 1.8 μM and 20.0 ± 1.4 μM, respectively). Likewise, mansoin F (32) inhibited TNF-α release by lipopolysaccharide-stimulated THP-1 cells (IC_{50} of 19.3 ± 1.3 μM) and mansoin A (27) reduced the phosphorylation levels of p-65-NF-κB, when assayed at 50 μM (Campana et al. 2016).

Among the three brachydin aglycones (14–16), brachydin A (14) was less active (IC_{50}=33g.mL^{-1}) than brachydin B (15) and brachydin C (16) (IC_{50}=9 and 10 g.mL^{-1}, respectively) in the anti-inflammatory activity assay with human fibroblast like synoviocytes (HFLS). Hydroxylation on the C ring in the structure 14, ascribed a higher polarity than compounds 15 and 16, affecting bioavailability and bioactivity (Salgado et al., 2020). These results provide additional support for using *F. platyphylla* to treat pain and inflammation. Oral treatment in male Swiss mice with dichloromethane fraction from roots (30 mg.Kg^{-1}), mainly composed by brachydin aglycones, further showed an antinociceptive response, apparently related to the opioid system (Rodriguez et al., 2017).

The dichloromethane fraction of *A. crucigerum* seed extract was composed by diosmetin (64) and hesperetin (36) as major compounds (Prá et al., 2017). It is reported that hesperetin may induce antinociceptive effect by antagonizing the transient receptor potential melastatin-3 (TRPM3) channel (Aswar et al., 2014). In turn, the antinociceptive properties of diosmetin seems to occur via the transient receptor potential vanilloid 1 (TRPV1) antagonist in mice (Adamante et al., 2019).

During the inflammation process, lipoxygenases were produced to metabolize arachidonic acid, playing an important role in the pathogenesis of inflammatory diseases (Bertanha et al., 2020). Alvarenga et al. (2015) studied the inhibitory activity of isolated compounds from *Cuspidaria pulchra* (Cham.) L.G. Lohmann against the 15-lipoxygenase (15-LOX), a lipoxygenase that belongs to a class of oxygenases that metabolize polyunsaturated fatty acids. The authors described IC_{50} values of 2.35 mM for the flavone 6-hydroxyluteolin-glucoside (44), close to those reported for the positive control Zileuton®, which had an IC_{50} of 1.54 mM.

Anti-Insect Properties

The flavonoids acacetin (49), 4'-hydroxywogonin (39), and galangusin (65) recovered from methanol extracts obtained from leaves of *Bignonia callistegioides* Cham showed anti-insect properties (Castillo et al., 2013). Biological assays against *Myzus persicae* Sulzer and *Rhopalosiphum padi* Linnaeus (Hemiptera: Aphididae) revealed the significant activity of acacetin against *R. padi* (PI = 0.3 ± 0.1). Further, galangustin (65) exhibited a better deterrence activity on the settling of both aphids (PI = 0.4 ± 0.1 and 0.8 ± 0.1 for *M. persicae* and *R. padi*, respectively), meanwhile, 4'-hydroxywogonin did not exhibit significant activity against neither of aphids (Castillo et al., 2013).

Antioxidant Activity

Antioxidant activity in flavonoids is ascribed by its complex ring system, which provides the capacity to transfer electrons of free radicals (Agati et al., 2020; Tungmunnithum et al., 2018). In Bignonieae, five flavones (52, 58, 69, 71, 76) were evaluated by their antioxidant properties through different assays, as DPPH [Diphenyl-1-picrylhydrazyl radical] and $ABTS^+$ [2,2'-azinobis-(3-ethylbenzothiazoline-6-sulfonate radical] scavenging activities, and TRAP (Total reactive antioxidant potential) assay.

Extracts of *Cuspidaria convoluta* (Vell.) A.H. Gentry, *Fridericia caudigera* (S. Moore) L.G. Lohmann, and *F. chica* containing flavones showed high scavenging activity through the $ABTS^+$ method (IC_{50}= 77.93, 130.18 and 57.84 $\mu g.mL^{-1}$, respectively). Two flavones with antioxidant activities were identified from *C. convoluta*, luteolin (71) and cirsiliol (58) (Torres et al., 2018). The antioxidant activity of these flavones has been primarily attributed to a 3',4'-dihydroxy structure in the B ring (catechol group), recognized as the most active moiety in flavonoid molecules (Ahmadi et al., 2020). Similarly, different reactive antioxidant potentials were found in apigenin (52) and scutellarein (76), both flavones isolated from *F. chica* (2.33 ± 0.09 and 3.54 ± 0.14 μM in the Trolox assay). This difference was attributed to electronic parameters on the A ring of scutellarein, which contains a hydroxyl group at C-6, enabling a higher scavenging on the reactive oxygen species (ROS) (Siraichi et al., 2013). Furthermore, glycosylated compounds, such as luteolin-7-O-glucoside (72) (IC_{50}=4.70 ± 0.90 μM) and katchimoside (69) (0.855 ± 0.033 TEAC, Trolox equivalent antioxidant capacity) were identified as highly reactive compounds in scavenging DPPH and $ABTS^{+}$ radicals, respectively (Plaza et al. 2005; Samy et al. 2015).

Antiparasitic Activity

Fridericia was the only genus of Bignonieae ever tested for antiparasitic activity. Flavonoids with trypanocidal and leishmanicidal properties were recovered from extracts, fractions, and isolated compounds of *F. chica*, *F. platyphylla*, *F. speciosa* Mart., and *F. triplinervia* (Mart. ex DC.) L.G. Lohmann (Leite et al., 2006; Da Rocha et al., 2014; Rocha et al., 2018; Silva-Silva et al. 2021).

Dimeric flavonoids like brachydins A, B, and C were tested as treatments for *Trypanosoma* and *Leishmania* parasites. Brachydin A (14) (IC_{50}<20μM) was inactive against the Y-strain of *T. cruzi* (trypomastigote and amastigote forms), *L. amazonensis*, *L. braziliensis*, and *L. infantum*. However, brachydins B and C (15 and 16) showed considerable activity against *T. cruzi* trypomastigote (IC_{50}= 5.3±1.15 and 6.6±0.39 μM, respectively) and amastigote forms (IC_{50}=6.0±0.33 and 6.8±0.41 μM, respectively), when compared to benznidazole, the reference drug. Furthermore, brachydin B reduced blood parasitemia *in vivo* in 92%, representing a promising alternative treatment for Chagas disease (da Rocha et al., 2014).

Similarly, the intense activities against promastigote forms of *L. amazonensis*, *L. braziliensis*, and *L. infantum* were identified for brachydin B (15) and C (16) at the concentration of 0.25μM. Brachydin B was considered highly active against amastigote forms of *L. amazonensis* (IC_{50}=6.25± 1.28) and ultrastructural analyses indicated that this compound induces cell lesion on parasites. When the three brachydins were compared (14 to 16), the outstanding activity on 15 was attributed to higher lipophilicity, considering the presence of methoxy group in the C ring of this compound (Rocha et al., 2019). This data corroborated the earlier finding that the increasing of lipophilicity improved the membrane penetration of the compounds (Buates et al., 1999; Jaramillo et al., 2011; Rocha et al., 2019). Despite that, *in vivo* studies did not reproduce the elevated activity found in the *in vitro* assay (Rocha et al., 2018).

Alpinetin (34), the only flavanone identified in *Fridericia*, showed high activity against *L. amazonensis*, inhibiting promastigote forms in 40.2%, which is a much higher efficiency than that observed in flavones luteolin (71; 26.8 ± 2.5%), apigenin (51; 24.8 ± 4.1%), and diosmetin (64; 1.0 ± 1.4%) (Milani et al., 2020). However, these compounds did not show any activity when tested against Y and CL strains of *T. cruzi* (Leite et al., 2006).

The four anthocyanidins (1–4) recovered from *F. chica* have not been tested against *T. cruzi*. However, leishmanicidal tests showed high correlation between this bioactivity and the presence of carajurin on extracts of *F. chica* (Moragas-Tellis et al., 2020). Also, two flavones identified in *F. chica*, apigenin (52) and luteolin (71), showed an anti-leishmania activity against promastigote form (IC_{50}= 168.7 µM; 110.08 µM, respectively), being 71 highly active against intracellular amastigote (IC_{50}= 41.15 µM) (Silva-Silva et al. 2021). In addition, the anthocyanidins 3'-hydroxy-carajurone (1), carajurin (3) and carajurone (4) were also considered active against promastigotes (IC_{50}= 22.70, 3.66 and 28.28 µg.mL^{-1}, respectively), however only carajurin was consider a potential biological marker of antileshmanial activity, by its high activity against intracellular amastigote (IC_{50}=7.05 µg.mL^{-1}).

Antiviral Activity

Molecular docking with SARS-CoV-2 were tested with *Amphilophium paniculatum* (L.) Kunth compounds. Among the chemical classes, flavonoids were the most effective with bind affinities with Energy scores (kcal/mol) of -9.54 for luteolin-7-O-glucoside (72), -8.54 linarin (70), and -8.34 luteolin (71) (Sami et al., 2021).

Ethanol extracts from leaves, fruits, and stems of *Amphilophium elongatum* (Vahl) L.G. Lohmann showed antiviral activity against human herpes virus type 1 (HSV-1) and dengue virus 2 (DENV-2). These effects were attributed to pectolinarin flavonoids (74) and acacetin-7-O-rutinoside (70) isolated from fruits. The mixture of both compounds showed a high anti-DENV-2 activity (EC_{50} 11.1 ± 1.6 μg.mL^{-1}) (Simões et al., 2011).

Cytotoxic Activity

Two species of Bignonieae (*F. platyphylla* and *Dolichandra unguis-cati* (L.) L.G. Lohmann) were investigated for their cytotoxic activity over cancer cell lineages. The dimeric flavonoids brachydins A, B, and C (14–16) showed, respectively, IC_{50} of 23.41, 4.28, and 4.44 μM and were shown to represent cytotoxic agents against human prostate cells (PC-3 cell line), due to its potential to induce apoptosis and necrosis (Nunes et al., 2020).

Five compounds were recovered from bioguided cytotoxic assays with aerial portions of *D. unguis-cati*, two of which were flavonoids (59 and 61). The flavonoid cirsimaritin (61) displayed a low cytotoxic activity against Bel7402 (liver cancer cell line) and Chang liver cell lines (normal hepatocyte Chang-liver cells) with IC_{50} of 234.55 ± 23.98 μM and 226.88 ± 35 μM, respectively (Chen et al., 2017).

Mutagenicity, cytotoxicity, and genotoxicity are important parameters to guarantee safety use of plant extracts for medicinal purposes (Nasri and Shirzad, 2013). Hydroethanolic extract of *F. chica* was considered safe as a medicinal treatment. Cell viability tests of CHO-K1 (ovary cell, hamster) showed that carajurone lacks cytotoxicity effects (IC_{50}> 40 μg.mL^{-1}) though (Violante et al., 2020). Furthermore, the compounds alpinetin (34), apigenin (51), diosmetin (64), and luteolin (71) showed low toxicity for human lung fibroblast cells, indicating low toxicity for normal cells (Milani et al., 2020).

Gastroprotective Effects

Gastroprotective (included hepatoprotective) effects were identified in *Fridericia platyphylla* and *Bignonia binata* Thunb. (Samy et al., 2015). The root extract of *F. platyphylla* was shown to represent a powerful gastroprotective in studies treating gastric injury *in vivo* (Rocha et al., 2017). This activity was associated with the higher accumulation of brachydins (Rocha et al., 2017). On the other hand, ethanol extracts from leaves of *B. binata* were administered to rats along with the drug silymarin for six consecutive days showing a significant ($p<0.001$) reduction in the serum levels of alanine aminotransferase (ALT), aspartate aminotransferase (AST), lactate dehydrogenase (LDH), bilirubin, cholesterol, and triglycerides, while increasing albumin levels, which is indicative of improved liver functions. These effects were attributed to the high contents of antioxidants such as flavonoid glycosides vicenin-2 (79) and schaftoside and/or isoschaftoside (Samy et al., 2021).

Distribution of Flavonoids in Bignonieae

Overall, 66 studies describing flavonoids from Bignonieae species were reviewed. The number of occurrences was obtained by counting the number of times a compound had been described for each species. If a unique compound was isolated of two different species, consequently it was counted twice (Ferreira et al., 2004). In these studies, 93 compounds and 133 occurrences were distributed in 29 Bignonieae species belonging to 11 genera. Flavonoids are widely distributed in the plant kingdom and their biosynthetic genes reflect their evolutionary history (Yonekura-Sakakibara et al., 2019). Thus, the distribution of these compounds can provide valuable information about phylogenetic relationships. The flavonoid distribution on Bignonieae is presented in Table 2 and summarized on a Heatmap (Figure 7).

Flavones are the main flavonoid class found in Bignonieae, with 43 compounds, followed by dimeric flavonoids with 19 structures, 12 flavonols, 9 anthocyanins and anthocyanidins, four flavanones, and four chalcones. Within the tribe, the predominant compound is cyanidin-3-O-glucoside (5), which has been described in seven species (Table 1), followed by apigenin (52), rutin (92), vicenin-2 (79), and luteolin (71) which were identified in five species each.

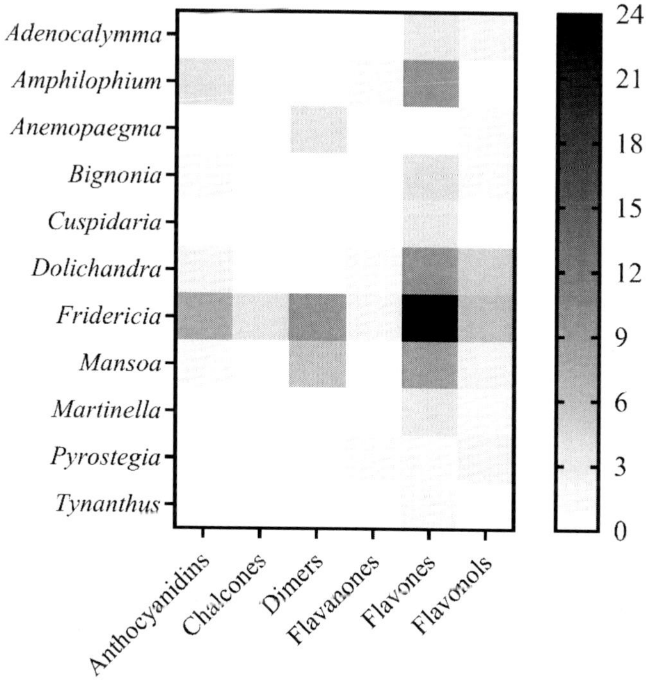

Figure 8. Heatmap of flavonoids identified from different Bignonieae genera per flavonoid class.

Among the genera of Bignonieae, *Fridericia* has the highest number of flavonoids with 51 compounds, including all classes described here. *Fridericia* belongs to the "*Arrabidaea* and allies clade" (Lohmann, 2006) which also comprise *Cuspidaria*, *Tynanthus* Miers, *Lundia* DC., *Tanaecium,* and *Xylophragma*. Flavonoids were reported in *Cuspidaria*, but not in *Tanaecium*, *Tynanthus*, or *Xylophragma*; no studies have been conducted with *Lundia*. The "multiples of four clade" represents another major lineage in Bignonieae (Lohmann, 2006). This clade is composed of six genera (i.e., *Amphilophium, Anemopaegma, Bignonia, Dolichandra*, *Mansoa*, and *Pyrostegia*), four of which showed the production of anthocyanidins and anthocyanins (i.e., *Amphilophium, Bignonia, Dolichandra*, and *Mansoa*). The two genera in which anthocyanidins and anthocyanins were not found (i.e., *Anemopaegma* and *Pyrostegia*) have been under investigated chemically; indeed, the flavonoid composition has only been investigated in *A. arvense* and *P. venusta*.

Dimeric flavonoids were identified in *Anemopaegma, Fridericia*, and *Mansoa*. Although methoxylated flavonoids are commonly found in the plant kingdom (Buckingham, 2009), the highly production of such compounds, specially methoxylated flavones, could be a typical feature in Bignonieae. Furthermore, the oxidation position in the ring A of the flavone structure and the ratio flavonol/flavone are of chemophenetic importance. The presence of oxidation at C-8 position and the high ratio flavonol/flavone are thought to be plesiomorphic (Harborne, 1977). The preponderance of oxidation on the position C-6 instead of the position C-8 in the studied species of Bignonieae, as well as the high proportion of flavones in its flavonoid chemistry might represent a synapomorphy of the tribe. Although some flavonoids identified into Bignonieae are distinctly distributed among the tribe, the scarce of phytochemical studies among members of the tribe prevents detailed analyses of flavonoids in Bignonieae.

CONCLUSION

Here, we summarized the current state of knowledge of flavonoids from Bignonieae. Overall, 133 occurrences from 93 flavonoids were reported to Bignonieae, belonging to six distinct flavonoid classes: (i) anthocyanidins and anthocyanins, (ii) chalcones, (iii) dimers, (iv) flavanones, (v) flavones, and (vi) flavonols. A summary of the biological activity of these compounds highlights their pharmacological activity and the medicinal potential of members of tribe Bignonieae. Nevertheless, new phytochemical studies are still needed for a comprehensive understanding of the chemical diversity and biological activities of flavonoids of tribe Bignonieae.

REFERENCES

Aboutabl, E. A., Hashem, F. A., Sleem, A. A., and Maamoon, A. A. (2008). Flavonoids, anti-inflammatory activity and cytotoxicity of *Macfadyena unguis-cati* L. *African Journal of Traditional, Complementary and Alternative Medicines*, *5*, 18–26. https://doi.org/10.4314/ajtcam.v5i1.31251.

Adamante, G., Almeida, A. S., Rigo, F. K., Silveira, E.S., Coelho, Y. O., Prá, S. D. T., Milioli, A. M., Camponogara, C., Casoti, R., Bellinaso, F., Desideri, A. V., Santos, M. F. C., Ferreira, J., Oliveira, S. M., Trevisan, G. (2019). Diosmetin as a novel transient receptor potential vanilloid 1 antagonist with antinociceptive activity in mice. *Life Sciences*, *216*, 215–226. https://doi.org/10.1016/j.lfs.2018.11.029.

Agati, G., Brunetti, C., Fini, A., Gori, A., Guidi, L., Landi, M., Sebastiani, F., and Tattini, M. (2020). Are flavonoids effective antioxidants in plants? Twenty years of our investigation. *Antioxidants*, *9*, 1–17. https://doi.org/10.3390/antiox9111098.

Ahmadi, S. M., Farhoosh, R., Sharif, A., and Rezaie, M. (2020). Structure-antioxidant activity relationships of luteolin and catechin. *Journal of Food Science*, *85*, 298–305. https://doi.org/10.1111/1750-3841.14994.

Alcerito, T., Barbo, F. E., Negri, G., Santos, D. Y. A. C., Meda, C. I., Young, M. C. M., Chávez, D., and Blatt, C. T. T. (2002). Foliar epicuticular wax of *Arrabidaea brachypoda*: Flavonoids and antifungal activity. *Biochemical Systematics and Ecology*, *30*, 677–683. https://doi.org/10.1016/S0305-1978(01)00149-1.

Alvarenga, T. A., Bertanha, C. S., Oliveira, P. F., Tavares, D. C., Gimenez, V. M. M., Silva, M. L. A., Cunha, W. R., Januário, A. H., and Pauletti, P. M. (2015). Lipoxygenase inhibitory activity of *Cuspidaria pulchra* and isolated compounds. *Natural Product Research*, *29*, 1083–1086. https://doi.org/10.1080/14786419.2014.981182.

Andrade, L. M., Oliveira, A. B. M., Leal, A. L. A. B., Oliveira, F. A. A., Portela, A. L., de Neto, J. S. L., Siqueira-Júnior, J. P., Kaatz, G. W., Rocha, C. Q., and Barreto, H. M. (2020). Antimicrobial activity and inhibition of the NorA efflux pump of *Staphylococcus aureus* by extract and isolated compounds from *Arrabidaea brachypoda*. *Microbial Pathogenesis*, *140* 103935. https://doi.org/10.1016/j.micpath.2019.103935.

Arevalo, C., Ruiz, I., Piccinelli, A. L., Campone, L., and Rastrelli, L. (2011). Phenolic derivatives from the leaves of *Martinella obovata* (Bignoniaceae). *Natural Product Communications*, *6*, 957–960. https://doi.org/10.1177/1934578x1100600708.

Aswar, M., Kute, P., Mahajan, S., Mahajan, U., Nerurkar, G., and Aswar, U. (2014). Protective effect of hesperetin in rat model of partial sciatic nerve ligation induced painful neuropathic pain: an evidence of anti-inflammatory and anti-oxidative activity. *Pharmacology Biochemistry and Behavior*, *124*, 101–107. https://www.sciencedirect.com/science/article/pii/S0091305714001543.

Attia, A. A. (1999). Flavonoidal glycoside and alkaloid beside other constituents from *Bignonia unguis-cati* L. *Bulletin of Pharmaceutical Sciences*, *22*, 139–145. https://doi.org/10.21608/BFSA.1999.66114.

Barbosa, W. L. R., Pinto, L. D. N., Quignard, E., Vieira, J. M. D. S., Silva, J. O. C., and Albuquerque, S. (2008). *Arrabidaea chica* (HBK) Verlot: Phytochemical approach, antifungal and trypanocidal activities. *Revista Brasileira de Farmacognosia*, *18*, 544–548. https://doi.org/10.1590/S0102-695X2008000400008.

Bertanha, C. S., Gimenez, V. M. M., Furtado, R. A., Tavares, D. C., Cunha, W. R., Silva, M. L. A., Januario, A. H., Borges, A., Kawano, D. F., Parreira, R. L. T., and Pauletti, P. M. (2020). Isolation, in vitro and in silico evaluation of phenylethanoid glycoside from *Arrabidaea brachypoda* as lipoxygenase inhibitor. *Journal of the Brazilian Chemical Society*, *31*, 849–855. https://doi.org/10.21577/0103-5053.20190248.

Blatt, C. T. T., Santos, M. D., and Salatino, A. (1998). Flavonoids of Bignoniaceae from the "cerrado" and their possible taxonomic significance. *Plant Systematics and Evolution*, 210, 289–292. https://doi.org/10.1007/BF00985673.

Brandão, G. C., Kroon, E. G., Filho, J. D. S., and Oliveira, A. B. (2017). Antiviral activity of *Fridericia formosa* (Bureau) L.G. Lohmann (Bignoniaceae) extracts and constituents. *Journal of Tropical Medicine*, *2017*, 6106959. https://doi.org/10.1155/2017/6106959.

Buates, S., Matlashewski, G. (1999). Treatment of experimental leishmaniasis with the immunomodulators imiquimod and S-28463: Efficacy and mode of action. *The Journal of Infectious Disease*. 179, 1485–1494. https://doi.org/10.1086/314782.

Campana, P. R. V., Coleman, C. M., Sousa, L. P., Teixeira, M. M., Ferreira, D., and Braga, F. C. (2016). Mansoins C-F, oligomeric flavonoid glucosides isolated from *Mansoa hirsuta* fruits with potential anti-inflammatory activity. *Journal of Natural Products*, *79*, 2279–2286. https://doi.org/10.1021/acs.jnatprod.6b00390.

Campana, P. R. V., Coleman, C. M., Teixeira, M. M., Ferreira, D., and Braga, F. C. (2014). TNF-α inhibition elicited by mansions A and B, heterotrimeric flavonoids isolated from *Mansoa hirsuta*. *Journal of Natural Products*, *77*, 824–830. https://doi.org/10.1021/np400929g.

Castillo, L., Díaz, M., González-Coloma, A., and Rossini, C. (2013). Differential activity against aphid settling of flavones obtained from *Clytostoma callistegioides* (Bignoniaceae). *Industrial Crops and Products*, *44*, 618–621. http://dx.doi.org/10.1016/j.indcrop.2012.09.012.

Chen, L., Chen, D., Zheng, Z., Liu, S., Tong, Q., Xiao, J., Lin, H., and Ming, Y. (2017). Cytotoxic and antioxidant activities of *Macfadyena unguis-cati* L. aerial parts and bioguided isolation of the antitumor active components. *Industrial Crops and Products*, *107*, 531–538. http://dx.doi.org/10.1016/j.indcrop.2017.05.018.

Cipriani, F. A., Figueiredo, M. R., Soares, G. L. G., and Kaplan, M. A. C. (2012). Implicações químicas na sistemática e filogenia de Bignoniaceae [Chemical implications in the systematics and phylogeny of Bignoniaceae]. *Quimica Nova*, *35*, 2125–2131. https://doi.org/10.1590/S0100-40422012001100005.

Close, D. C., and Beadle, C. L. (2003). The ecophysiology of foliar anthocyanin. *Botanical Review*, *69*, 149–161. https://doi.org/10.1663/0006-8101(2003)069[0149:TEOFA]2.0.CO;2.

Costanzo, C. D. G., Fernandes, V. C., Zingaretti, S., Beleboni, R. O., Pereira, A. M. S., Marins, M., Taleb-Contini, S. H., Pereira, P. S., and Fachin, A. L. (2013). Isolation of flavonoids from *Anemopaegma arvense* (Vell) Stellf. Ex de Souza and their antifungal activity against *Trichophyton rubrum*. *Brazilian Journal of Pharmaceutical Sciences*, *49*, 559–565. https://doi.org/10.1590/S1984-82502013000300017.

Dao, T. T. H., Linthorst, H. J. M., and Verpoorte, R. (2011). Chalcone synthase and its functions in plant resistance. *Phytochemistry Reviews*, *10*, 397–412. https://doi.org/10.1007/s11101-011-9211-7.

Devia, B., Llabres, G., Wouters, J., Dupont, L., Escribano-Bailon, M. T., Pascual-Teresa, S., Angenot, L., and Tits, M. (2002). New 3-deoxyanthocyanidins from leaves of *Arrabidaea chica*. *Phytochemical Analysis*, *13*, 114–120. https://doi.org/10.1002/pca.632.

Dewick, P. M. (2009). *Medicinal Natural Products: A Biosynthetic Approach* (3rd Edition). Wiley, 168-170.

Duarte, D. S., Dolabela, M. F., Salas, C. E., Raslan, D. S., Oliveiras, A. B., Nenninger, A., Wiedemann, B., Wagner, H., Lombardi, J., and Lopes, M. T. P. (2010). Chemical characterization and biological activity of *Macfadyena unguis-cati* (Bignoniaceae). *Journal of Pharmacy and Pharmacology*, *52*, 347–352. https://doi.org/10.1211/0022357001773904.

Faccin, H., Loose, R. F., Viana, C., Lameira, O. A., and Carvalho, L. M. (2017). Determination of phenolic compounds in extracts of Amazonian medicinal plants by liquid chromatography-electrospray tandem mass spectrometry. *Analytical Methods*, *9*, 1141–1151. https://doi.org/10.1039/c6ay02937j.

Ferreira, M. J. P., Brant, A. J. C., Rufino, A. R., Alvarenga, S. A. V., Magri, F. M. M., and Emerenciano, V. P. (2004). Prediction of occurrences of diverse chemical classes in the Asteraceae through artificial neural networks. *Phytochemical Analysis*, *15*, 389–396. http://doi.org/10.1002/pca.799.

Ferreira, D. T., Alvares, P. S. M., Houghton, P. J., and Braz-Filho, R. (2000). Constituintes químicos das raízes de *Pyrostegia venusta* e considerações sobre a sua importância medicinal [Chemical constituents of *Pyrostegia venusta* roots and considerations on its medicinal importance]. *Quimica Nova*, *23*, 42–46. https://doi.org/10.1590/s0100-40422000000100010.

Gemelli, T. F., Prado, L. S., Santos, F. S., Souza, A. P., Guecheva, T. N., Henriques, J. A. P., Ferraz, A. B. F., Corrêa, D. S., Dihl, R. R., and Picada, J. N. (2015). Evaluation of safety of *Arrabidaea chica* Verlot (Bignoniaceae), a plant with healing properties. *Journal of Toxicology and Environmental Health – Part A: Current Issues*, *78*, 1170–1180. https://doi.org/10.1080/15287394.2015.1072070.

Gentry, A. H., and Cook, K. (1984). *Martinella* (Bignoniaceae): A widely used eye medicine of South America. *Journal of Ethnopharmacology*, *11*, 337–343. https://doi.org/10.1016/0378-8741(84)90079-5.

Gentry, A. H. (1992). A synopsis of Bignoniaceae ethnobotany and economic botany. *Annals of the Missouri Botanical Garden*, 79, 53–64. https://doi.org/10.2307/2399809.

Goiris, K., Muylaert, K., Voorspoels, S., Noten, B., De Paepe, D., E Baart, G. J., and Cooman, L. (2014). Detection of flavonoids in microalgae from different evolutionary lineages. *Journal of Phycology*, *50*, 483–492. https://doi.org/10.1111/jpy.12180.

Gontijo, V. S., Santos, M. H., and Viegas, C. (2016). Biological and chemical aspects of natural biflavonoids from plants: A brief review. *Mini-Reviews in Medicinal Chemistry*, *17*, 834–862. https://doi.org/10.2174/1389557516666161104130026.

Grassi, R. F., Resende, U. M., Silva, W., Macedo, M. L. R., Butera, A. P., Tulli, E. O., Saffran, F. P., and Siqueira, J. M. (2005). Estudo fitoquímico e avaliação alelopática de *Memora peregrina* – "ciganinha" – Bignoniaceae, uma espécie invasora de pastagens em Mato Grosso do Sul [Phytochemical study and allelopathic evaluation of *Memora peregrina* – "ciganinha" – Bignoniaceae, an invasive species of pasture in Mato Grosso do Sul]. *Quimica Nova*, *28*, 199–203. https://doi.org/10.1590/S0100-40422005000200006.

Guilhon, G. M. S. P., Silva, E. S., Santos, L. S., Zoghbi, M. G. B., Araújo, I. S., and Uetanabaro, A. P. T. (2012). Volatile and non-volatile compounds and antimicrobial activity of *Mansoa difficilis* (Cham.) Bureau & K.Schum. (Bignoniaceae). *Quimica Nova*, *35*, 2249–2253. https://doi.org/10.1590/S0100-40422012001100030.

Harborne, J. B. (1967). Comparative Biochemistry of the Flavonoids VI. *Phytochemistry*, *6*, 1643–1651. https://doi.org/10.1016/S0031-9422(00)82897-6.

Harborne, J. B. (1977). Flavonoids and the evolution of the angiosperms. *Biochemical Systematics and Ecology*. 5, 7-22. https://doi.org/10.1016/0305-1978(77)90013-8.

Igarashi, M. (2019). New natural products to meet the antibiotic crisis: a personal journey. *The Journal of Antibiotics*, *72*, 890–898. https://doi.org/10.1038/s41429-019-0224-6.

Jaramillo, M., Gomez, M. A.; Larsson, O., Shio, M. T., Topisirovic, I., Contreras, I., Luxenburg, R., Rosenfeld, A., Colina, R., and McMaster, R. W. (2011) *Leishmania* repression of host translation through mTOR cleavage is required for parasite survival and infection. *Cell Host Microbe*. 9, 331–341. https://doi.org/10.1016/j.chom.2011.03.008.

Koes, R., Verweij, W., and Quattrocchio, F. (2005). Flavonoids: A colorful model for the regulation and evolution of biochemical pathways. *Trends in Plant Science*, 10, 236–242. https://doi.org/10.1016/j.tplants.2005.03.002.

Krebs, H. C. (1987). Inhaltsstoffe aus *Pseudocalymma elegans* (Vell.) Kuhlm. 1. Isolierung von allantoin, lamiid und sorbarin [Ingredients from *Pseudocalymma elegans* (Vell.) Kuhlm. 1. Isolation of allantoin, lamiide and sorbarin]. *Zeitschrift Fur Naturforschung – Section B Journal of Chemical Sciences*, *42*, 1361–1364. https://doi.org/10.1515/znb-1987-1023.

Leite, J. P. V., Oliveira, A. B., Lombardi, J. A., Filho, J. D. S., and Chiari, E. (2006). Trypanocidal activity of triterpenes from *Arrabidaea triplinervia* and derivatives. *Biological and Pharmaceutical Bulletin*, *29*, 2307–2309. https://doi.org/10.1248/bpb.29.2307.

Lima, J. C. S., Oliveira, R. G., Silva, V. C., Sousa, P. T., Violante, I. M. P., Macho, A., and Martins, D. T. O. (2020). Anti-inflammatory activity of 4′,6,7-trihydroxy-5-methoxyflavone from *Fridericia chica* (Bonpl.) L.G.Lohmann. *Natural Product Research*, *34*, 726–730. https://doi.org/10.1080/14786419.2018.1495636.

Lima, C. S. D. A., Amorim, E. L. C., Nascimento, S. C., Araújo, C. F., Agra, M. F., Barbosa-Filho, J. M., Silva, M. S., Da-Cunha, E. V. L., Vieira, I. J. C., and Braz-Filho, R. (2005). Cytotoxic pyranonaphthoquinones from *Melloa quadrivalvis* (Bignoniaceae). *Natural Product Research*, *19*, 217–222. https://doi.org/10.1080/1478641042000223808.

Liu, S., Zheng, Z., Zheng, G., Chen, L., Zhong, T., and Ming, Y. (2015). A new flavonoside from the invasive plant *Macfadyena unguis-cati*. *Chemistry of Natural Compounds*, *51*, 844–846. https://doi.org/10.1007/s10600-015-1430-5.

Lohmann, L. G., and Taylor, C. M. (2014). A new generic classification of tribe Bignonieae (Bignoniaceae). *Annals of the Missouri Botanical Garden*, *99*, 348–489. https://doi.org/10.3417/2003187.

Lohmann, L.G., Ulloa, C.U., 2006 onwards. *Bignoniaceae in iPlants prototype checklist.* Available from: http://www.iplants.org (accessed on 13 April 2021).

Milani, R. R., Alves, O. J. A., Bertanha, C. S., Gimenez, V. M. M., Squarisi, I. S., Nicolella, H. D., Pagotti, M. C., Tavares, D. C., Silva, M. L. A., Cunha, W. R., Magalhães, L. G., Januario, A. H., and Pauletti, P. M. (2020). Antipromastigote and cytotoxic activities of flavonoids from *Fridericia speciosa* leaves. *Quimica Nova*, *43*, 1404–1407. https://doi.org/10.21577/0100-4042.20170628.

Moragas-Tellis, C. J., Almeida-Souza, F., Chagas, M. S. S., de Souza, P. V. R., Silva-Silva, J. V., Ramos, Y. J., Moreira, D. L., Calabrese, K.

S., and Behrens, M. D. (2020). The influence of anthocyanidin profile on antileishmanial activity of *Arrabidaea chica* morphotypes. *Molecules*, *25*, 1–13. https://doi.org/10.3390/molecules25153547.

Nasri, H., and Hedayatollah, S. (2013). Toxicity and safety of medicinal plants. *Journal of Herb Med Pharmacology*, *2*, 21–22. https://doi.org/10.1016/j.bjp.2014.03.006.

Nassar, M. I., Aboutabl, E. S. A., Eskander, D. M., Grace, M. H., El-Khrisy, E. D. A., and Sleem, A. A. (2013). Flavonoid glycosides and pharmacological activity *of Amphilophium paniculatum*. *Pharmacognosy Research*, *5*, 17–21. https://doi.org/10.4103/0974-8490.105643.

Nunes, H. L., Tuttis, K., Serpeloni, J. M., Nascimento, J. R., Rocha, C. Q., Silva, V. A. O., Lengert, A. V. H., Reis, R. M., and Cólus, I. M. S. (2020). Characterization of the in vitro cytotoxic effects of brachydins isolated from *Fridericia platyphylla* in a prostate cancer cell line. *Journal of Toxicology and Environmental Health, Part A*, *83*, 547–558. https://doi.org/10.1080/15287394.2020.1784339.

Olmstead, R. G., Zjhra, M. L., Lohmann, L. G., Grose, S. O., and Eckert, A. J. (2009). A molecular phylogeny and classification of Bignoniaceae. *American Journal of Botany*, *96*, 1731–1743. https://doi.org/10.3732/ajb.0900004.

Oliveira, G. G., Neto, F. C, Demarque, D. P., Pereira-Junior, J. A. S., Filho, R. C. S. P, Melo, S. J., Almeida, J. R. G. S., Lopes, J. L. C., and Lopes, N. P. (2017). Dereplication of flavonoid glycoconjugates from *Adenocalymma imperatoris-maximilianii* by untargeted tandem mass spectrometry-based molecular networking. *Planta Medica*, *83*, 636–646. https://doi.org/10.1055/s-0042-118712.

Pace, M., Lohmann, L. G., and Angyalossy, V. (2009). The rise and evolution of the cambial variant in Bignonieae (Bignoniaceae). *Evolution and Development*, *11*, 465–479. https://doi.org/10.1111/j.1525-142X.2009.00355.x.

Paula, J. T., Paviani, L. C., Foglio, M. A., Sousa, I. M. O., Duarte, G. H. B., Jorge, M. P., Eberlin, M. N., and Cabral, F. A. (2014). Extraction of anthocyanins and luteolin from *Arrabidaea chica* by sequential extraction in fixed bed using supercritical CO_2, ethanol and water as solvents. *Journal of Supercritical Fluids*, *86*, 100–107. https://doi.org/10.1016/j.supflu.2013.12.008.

Paula, J. T., Paviani, L. C., Foglio, M. A., Sousa, I. M. O., and Cabral, F. A. (2013). Extraction of anthocyanins from *Arrabidaea chica* in fixed bed using CO_2 and CO_2/ethanol/water mixtures as solvents. *Journal of Supercritical Fluids*, *81*, 33–41. https://doi.org/10.1016/j.supflu.2013.04.009.

Pauletti, P. M., Bolzani, V. S., and Young, M. C. M. (2003). Constituintes químicos de *Arrabidaea samydoides* (Bignoniaceae) [Chemical constituents of *Arrabidaea samydoides* (Bignoniaceae)]. *Quimica Nova*, *26*, 641–643. https://doi.org/10.1590/S0100-40422003000500003.

Pereira, A. M. S., Hernandes, C., Pereira, S. I. V., Bertoni, B. W., França, S. C., Pereira, P. S., and Taleb-Contini, S. H. (2014). Evaluation of anticandidal and antioxidant activities of phenolic compounds from *Pyrostegia venusta* (Ker Gawl.) Miers. *Chemico-Biological Interactions*, *224*, 136–141. http://dx.doi.org/10.1016/j.cbi.2014.10.023.

Plaza, A., Montoro, P., Benavides, A., Pizza, C., and Piacente, S. (2005). Phenylpropanoid glycosides from *Tynanthus panurensis*: Characterization and LC-MS quantitative analysis. *Journal of Agricultural and Food Chemistry*, *53*, 2853–2858. https://doi.org/10.1021/jf0479867.

Prá, S. D. T., Ferro, P. R., Milioli, A. M., Rigo, F. K., Chipindo, O. J., Camponogara, C., Casoti, R., Manfron, M. P., Oliveira, S. M., Ferreira, J., and Trevisan, G. (2017). Antinociceptive activity and mechanism of action of hydroalcoholic extract and dichloromethane fraction of *Amphilophium crucigerum* seeds in mice. *Journal of Ethnopharmacology*, *195*, 283–297. http://dx.doi.org/10.1016/j.jep.2016.11.032.

Rao, E. V., and Rao, M. A. (1982). Apigenin-7-O-methylglucoronide from *Adenocalymma alliaceum*. *Current Science*, *51*, 1040.

Rao, M. A., and Rao, E. V. (1980). Flavonoids of the flowers of *Adenocalymma alliaceum*. *Current Science*, *40*, 468–469.

Rezende-Júnior, L. M., Andrade, L. M. de S., Leal, A. L. A. B., Mesquita, A. B. de S., Santos, A. L. P. de A., Neto, J. de S. L., Siqueira-Júnior, J. P., Nogueira, C. E. S., Kaatz, G. W., Coutinho, H. D. M., Martins, N., Rocha, C. Q., and Barreto, H. M. (2020). Chalcones isolated from *Arrabidaea brachypoda* flowers as inhibitors of nora and mepa multidrug efflux pumps of *Staphylococcus aureus*. *Antibiotics*, *9*, 1–12. https://doi.org/10.3390/antibiotics9060351.

Rocha, V. P. C., Rocha, C. Q., Queiroz, E. F., Marcourt, L., Vilegas, W., Grimaldi, G. B., Furrer, P., Allémann, É., Wolfender, J. L., and Soares, M. B. P. (2019). Antileishmanial activity of dimeric flavonoids isolated from *Arrabidaea brachypoda*. *Molecules*, *24*, 1-13. https://doi.org/10.3390/molecules2401000.

Rocha, C. Q., De-Faria, F. M., Marcourt, L., Ebrahimi, S. N., Kitano, B. T., Ghilardi, A. F., Ferreira, A. L., Almeida, A. C. A., Dunder, R. J., Souza-Brito, A. R. M., Hamburger, M., Vilegas, W., Queiroz, E. F., and Wolfender, J. L. (2017). Gastroprotective effects of hydroethanolic root extract of *Arrabidaea brachypoda*: Evidences of cytoprotection and isolation of unusual glycosylated polyphenols. *Phytochemistry*, *135*, 93–105. https://doi.org/10.1016/j.phytochem.2016.12.002.

Rocha, C. Q., Queiroz, E. F., Meira, C. S., Moreira, D. R. M., Soares, M. B. P., Marcourt, L., Vilegas, W., and Wolfender, J. L. (2014). Dimeric flavonoids from *Arrabidaea brachypoda* and assessment of their anti- *Trypanosoma cruzi* activity. *Journal of Natural Products*, *77*, 1345–1350. https://doi.org/10.1021/np401060j.

Rodrigues, V. P., Rocha, C. Q., Périco, L. L., Santos, R. C., Ohara, R., Nishijima, C. M., Queiroz, E. F., Wolfender, J. L., Rocha, L. R. M., Santos, A. R. S., Vilegas, W., and Hiruma-Lima, C. A. (2017). Involvement of opioid system, TRPM8, and ASIC receptors in antinociceptive effect of *Arrabidaea brachypoda* (DC) Bureau.

International Journal of Molecular Sciences, *18*, 1–12. https://doi.org/10.3390/ijms18112304.

Salgado, C., Morin, H., Aquino, N. C., Neff, L., Rocha, C. Q., Vilegas, W., Marcourt, L., Wolfender, J. L., Jordan, O., Ferreira Queiroz, E., and Allémann, E. (2020). In vitro anti-inflammatory activity in arthritic synoviocytes of *A. brachypoda* root extracts and its unusual dimeric flavonoids. *Molecules 25*, 1–13. https://doi.org/10.3390/molecules25215219.

Samy, M. N., Attia, E. Z., Shoman, M. E., Khalil, H. E., Sugimoto, S., Matsunami, K., and Fahim, J. R. (2021). Phytochemical investigation of *Amphilophium paniculatum;* an underexplored Bignoniaceae species as a source of SARS-CoV-2 M^{pro} inhibitory metabolites: Isolation, identification, and molecular docking study. *South African Journal of Botany*, 141, 421–430. https://doi.org/10.1016/j.sajb.2021.05.023.

Samy, M. N., Hamed, A. N. E., Mahmoud, B. K., Attia, E. Z., Abdelmohsen, U. R., Fawzy, M. A., Attya, M. E., and Kamel, M. S. (2021). LC-MS-based identification of bioactive compounds and hepatoprotective and nephroprotective activities of *Bignonia binata* leaves against carbon tetrachloride-induced injury in rats. *Natural Product Research*, *Ahead of Print,* 1–5. https://doi.org/10.1080/14786419.2021.1873982.

Samy, M. N., Khalil, H. E., Sugimoto, S., Matsunami, K., Otsuka, H., and Kamel, M. S. (2015). Amphipaniculosides A-D, triterpenoid glycosides, and amphipaniculoside E, an aliphatic alcohol glycoside from the leaves of *Amphilophium paniculatum*. *Phytochemistry*, *115*, 261–268. http://dx.doi.org/10.1016/j.phytochem.2015.02.020.

Santos, D. Y. A. C. (2020). *Biossíntese, Funções e Aplicações dos Metabólitos Secundários de Plantas* (1st ed.) [*Biosynthesis, Functions and Applications of Plant Secondary Metabolites*]. Appris. Pp. 11-45.

Schiozer, A. L., Cabral, E. C., Godoy, L. A. F., Chaves, F. C. M., Poppi, R. J., Riveros, J. M., Eberlin, M. N., and Barata, L. E. S. (2012). Electrospray ionization mass spectrometry fingerprinting of extracts of the leaves of *Arrabidaea chica*. *Journal of the Brazilian Chemical Society*, *23*, 409–414. https://doi.org/10.1590/s0103-50532012000300006.

Scogin, R. (1980). Anthocyanins of the Bignoniaceae. *Biochemical Systematics and Ecology*, *8*, 273–276. https://doi.org/10.1016/0305-1978(80)90058-7.

Silva-Silva, J. V., Moragas-Tellis, C. J., Chagas, M. S. S., Souza, P. V. R., Souza, C. S. F., Hardoim, D. de J., Taniwaki, N. N., Moreira, D. de L., Dutra Behrens, M., Calabrese, K. S., and Almeida-Souza, F. (2021). Antileishmanial activity of flavones-rich fraction from *Arrabidaea chica* Verlot (Bignoniaceae). *Frontiers in Pharmacology*, 12, 1–14. https://doi.org/10.3389/fphar.2021.703985.

Silva-Silva, J. V., Moragas-Tellis, C. J., Chagas, M. S. S., Souza, P. V. R., Moreira, D. L., Souza, C. S. F., Teixeira, K. F., Cenci, A. R., Oliveira, A. S., Almeida-Souza, F., Behrens, M. D., and Calabrese, K. S. (2021). Carajurin: an anthocyanidin from *Arrabidaea chica* as a potential biological marker of antileishmanial activity. *Biomedicine and Pharmacotherapy*, *141*, 111910 https://doi.org/10.1016/j.biopha.2021.111910.

Simões, L. R., Maciel, G. M., Brandão, G. C., Filho, J. D. S., Oliveira, A. B., and Castilho, R. O. (2013). Chemical constituents of *Distictella elongata* (Vahl) Urb. (Bignoniaceae). *Anais da Academia Brasileira de Ciencias*, *85*, 873–879. https://doi.org/10.1590/S0001-37652013000300003. [*Annals of the Brazilian Academy of Sciences*]

Simões, L. R., Maciel, G. M., Brandão, G. C., Kroon, E. G., Castilho, R. O., and Oliveira, A. B. (2011). Antiviral activity of *Distictella elongata* (Vahl) Urb. (Bignoniaceae), a potentially useful source of anti-dengue drugs from the state of Minas Gerais, Brazil. *Letters in Applied Microbiology*, *53*, 602–607. https://doi.org/10.1111/j.1472-765X.2011.03146.x.

Siqueira, F. C., Leitão, D. S. T. C., Mercadante, A. Z., Chisté, R. C., and Lopes, A. S. (2019). Profile of phenolic compounds and carotenoids of *Arrabidaea chica* leaves and the in vitro singlet oxygen quenching capacity of their hydrophilic extract. *Food Research International, 126*, 108597. https://doi.org/10.1016/j.foodres.2019.108597.

Siraichi, J. T. G., Felipe, D. F., Brambilla, L. Z. S., Gatto, M. J., Terra, V. A., Cecchini, A. L., Cortez, L. E. R., Rodrigues-Filho, E., and Cortez, D. A. G. (2013). Antioxidant capacity of the leaf extract obtained from *Arrabidaea chica* cultivated in Southern Brazil. *PloS ONE, 8*, 1–9. https://doi.org/10.1371/journal.pone.0072733.

Sousa-Baena, M. S., Sinha N. R., and Lohmann., L. G. 2014. Evolution and development of tendrils in Bignonieae (Bignoniaceae, Lamiales). *Annals of the Missouri Botanical Garden, 99*, 323–347. https://doi.org/10.3417/2011018.

Tabanca, N., Pawar, R. S., Ferreira, D., Marais, J. P. J., Khan, S. I., Joshi, V., Wedge, D. E., and Khan, I. A. (2007). Flavan-3-ol-phenylpropanoid conjugates from *Anemopaegma arvense* and their antioxidant activities. *Planta Medica, 73*, 1107–1111. https://doi.org/10.1055/s-2007-981563.

Taffarello, D., Jorge, M. P., Sousa, I. M. O., Duarte, M. C. T., Figueira, G. M., Queiroz, N. D. C. A., Rodrigues, R. A. F., Carvalho, J. E., Goes, A. L. T. R., Foglio, M. A., Riveros, J. M., Eberlin, M. N., and Cabral, E. C. (2013). Atividade de extratos *de Arrabidaea chica* (Humb. & Bonpl.) Verlot obtidos por processos biotecnológicos sobre a proliferação de fibroblastos e células tumorais humanas [Activity of extracts of *Arrabidaea chica* (Humb. & Bonpl.) Verlot obtained by biotechnological processes on the proliferation of fibroblasts and human tumor cells]. *Quimica Nova, 36*, 431–436. https://doi.org/10.1590/S0100-40422013000300014.

Takemura, O. S., Iinuma, M., Tosa, H., Miguel, O. G., Moreira, E. A., and Nozawa, Y. (1995). A flavone from leaves of *Arrabidaea chica* f. *cuprea*. *Phytochemistry, 38*, 1299–1300. https://doi.org/10.1016/0031-9422(94)00786-S.

Takenaka, I. K. T. M., Amorim, J. M., Barros, P. A. V., Brandão, G. C., Contarini, S. M. L., Melo, É. L. S. S., Almeida-Leite, C. M., Martins, F. D. S., Cardoso, V. N., Castilho, R. O., and Fernandes, S. O. A. (2020). Chemical characterization and anti-inflammatory assessment of the hydroethanolic extract of *Fridericia chica*. *Revista Brasileira de Farmacognosia*, *30*, 559–567. https://doi.org/10.1007/s43450-020-00085-7.

Torres, C., Sturla, M., Romero, A., and Judis, M. (2019). Bioguided isolation of antimicrobial polyphenols from *Cuspidaria convoluta* leaves and their synergistic effect with antibiotics. *Asian Pacific Journal of Tropical Biomedicine*, *9*, 434–442. https://doi.org/10.4103/2221-1691.269525.

Torres, C. A., Zamora, C. M. P., Nuñez, M. B., and Gonzalez, A. M. (2018). In vitro antioxidant, antilipoxygenase and antimicrobial activities of extracts from seven climbing plants belonging to the Bignoniaceae. *Journal of Integrative Medicine*, *16*, 255–262. https://doi.org/10.1016/j.joim.2018.04.009.

Treutter, D. (2006). Significance of flavonoids in plant resistance: A review. *Environmental Chemistry Letters*, 4, 147–157. https://doi.org/10.1007/s10311-006-0068-8.

Tungmunnithum, D., Thongboonyou, A., Pholboon, A., and Yangsabai, A. (2018). Flavonoids and other phenolic compounds from medicinal plants for pharmaceutical and medical aspects: An overview. *Medicines*, *5*, 93. https://doi.org/10.3390/medicines5030093.

Ullah, A., Munir, S., Badshah, S. L., Khan, N., Ghani, L., Poulson, B. G., Emwas, A., and Jaremko, M. (2020). Important flavonoids and their role as a therapeutic agent. *Molecules*, *25*, 1–39. https://doi.org/10.3390/molecules25225243.

Vasconcelos, C. C., Lopes, A. J. O., Sousa, E. L. F., Camelo, D. S., Lima, F. C. V. M., Rocha, C. Q., Silva, G. E. B., Garcia, J. B. S., and Cartágenes, M. D. S. D. S. (2019). Effects of extract of *Arrabidaea chica* Verlot on an experimental model of osteoarthritis. *International Journal of Molecular Sciences*, *20*, 4717. https://doi.org/10.3390/ijms20194717.

Violante, I. M. P., Carollo, C. A., Silva, L. I., Oliveira, A. Q. C., Pardinho, F. C., Garcez, W. S., Garcez, F. R., Arunachalam, K., and Martins, D. T. O. (2020). Cytotoxicity and antibacterial activity of scutellarein and carajurone-enriched fraction obtained from the hydroethanolic extract of the leaves of *Fridericia chica* (Bonpl.) L.G. Lohmann. *Natural Product Research*, *Ahead of print* 1–7. https://doi.org/10.1080/14786419.2020.1753050.

Yonekura-Sakakibara, K., Higashi, Y., and Nakabayashi, R. (2019). The origin and evolution of plant flavonoid metabolism. *Frontiers in Plant Science*, *10*, 1–16. https://doi.org/10.3389/fpls.2019.00943.

Zorn, B., García-Pieres, A. J., Castro, V., Murillo, R., Mora, G., & Merfort, I. (2001). 3-Desoxyanthocyanidins from *Arrabidaea chica*. *Phytochemistry*, *56*, 831–835. https://doi.org/10.1016/S0031-9422(01)00038-3.

In: The Book of Flavonoids
Editors: Chao-Hui Feng et al.
ISBN: 978-1-68507-376-3

Chapter 11

Bitter Oranges of Seville: Essential Oils and Bioflavonoids

Francisco Javier Alés-Álvarez [*]
and Paloma Álvarez-Mateos
Department of Chemical Engineering,
University of Seville, Seville, Spain

Abstract

Since ancient times, the different parts of the bitter orange of Seville have been used as food or for healing. One of its current main applications is the obtaining of a wide range of flavonoids which, after years of research and development, have proved to provide numerous benefits for human health including action against COVID-19, which is under research at laboratory level. Among the flavonoids found in the bitter orange of Seville naringin can be highlighted, which is very beneficial for human health. On the other hand, essential oils have been extracted and used since ancient times, from ritual and religious uses to the current uses in cosmetics, perfumery, food and health. This chapter overviews the history of the bitter oranges of Seville, the essential oils

[*] Corresponding Author's E-mail: fales@us.es.

and naringin (a bioflavonoid) that can be extracted from them and their potential applications.

Keywords: flavonoids, bitter oranges, essential oils, naringin

1. Introduction

The smell of orange blossom that surrounds and characterizes the city of Seville (Spain) is indisputable. It is something that defines the Andalusian capital and is part of its identity. Orange tree is one of the trees that we can most appreciate in the Sevillian countryside, giving light and color to its green meadows. All of this is thanks to the bitter oranges of Seville. In this chapter we want to convey the origins of the orange cultivation in Seville, i.e., when and why it started to be planted.

In Greek mythology, we can find the origin of this fruit, the bitter orange of Seville, and its relationship with Andalusia, especially its place in the capital. Hercules, the founder of Seville in Greek mythology, carried out a series of tasks ordered by Euristeo. One of the tasks was to cross into Africa and get the golden apple of immortality, which is linked to the orange. Today we know that the orange was introduced by some sailors from Genoa that came from Asia. In Asia, the orange was associated with a state of happiness, spirituality and physical satisfaction. During the Arab reign in southern Spain and knowledgeable about this belief, Arabs planted numerous orange trees to achieve the desired happiness. At the time, the Arab population wanted the current capital of Andalusia to be a benchmark of perfumery worldwide. In 1970 there were more than five thousand orange trees planted across our lands. That is why in the 12th century the Patio de Los Naranjos of Seville (orange gardens) was built. Today we can visit the orange gardens inside the cathedral. Therefore, the orange tree, from which bitter orange is extracted, is a tree of Asian origin that in the 10th century was introduced in our country and that gradually acquired a key significance in the city. It was also discovered that it had great medicinal properties among others. This use is still given to the fruit of bitter orange nowadays. In addition, like the bitter orange of Seville, there are many other citrus fruits that also characterize the city; these are sweet orange, lemon, grapefruit, lime and so on.

Bioflavonoids are biologically active members of a group of polyphenolic compounds that come from plants. These polyphenolic compounds are found in citrus fruit, mainly in the peel and the pulp. Each fruit contains a type of flavonoid in different proportion and quantity. Among the different flavonoids that can be found in the bitter orange, neohesperidin can be highlighted. This compound is a very potent sweetener in the food industry that stands out if we compare it with other sweeteners such as aspartame, sucralose, etcetera. Neohesperidin, like many other flavonoids, is extracted from the peel of the citrus fruit and can be found in different formats: fine cut, thick cut, quarters, strips, like granulated peel, diced and so on [1].

The following compounds and health benefits can be obtained from the citrus peel:

- Hesperidin, which is a type of flavonoid that has the fundamental characteristic of helping to metabolize fluids in the blood and to reduce fat to make it easier to be eliminated from the body.
- Naringin, which gives bitterness to food products.
- Diosmin, which has heart-healthy benefits.
- Neohesperidin, the main flavonoid present in the bitter orange.
- Pectins, which are natural fibers that preserve stable sugar levels and protect the body from stomach problems.
- A very high vitamins content that helps the immune system of the human being, being a great antiseptic.
- A high content of essential oils, which have important applications in the cosmetic, perfume and agro-food sectors.

Under different formats, there is a wide range of uses of citrus peel in the food sector such as infusions, citrus chips, candies, candied peel and jams.

The pulp is key in the orange industry. This part of the fruit is separated from the peel and seeds through a manual or industrial process. It is made through a combination of puree, pieces of the fruit itself or even the peel. The pulp is obtained from fresh fruits, selected, cleaned and in an optimal stage of maturation. There are numerous benefits that this food product has for the health of the human being, since the organism will receive both the pulp and the peel contributions. One of its best-known and most recognized applications is marmalade, to be specific bitter orange marmalade [1]. Besides, it also has many other uses such as for fruit juices, ice creams, smoothies, salad toppings, sweets, purees, as well as frozen citrus pulp. As it can be seen, the pulp is such a rich product that it can be used for numerous products in the food sector. To be specific, citrus fruit, as the raw material for the production of fruit juice, jams and jellies industry, is responsible for millions of jobs around the world, providing huge benefits [2, 3].

2. Essential Oils

Essential oils have been flavoring human life since the beginning of civilizations. Their goodness and therapeutic power have been more than demonstrated over time. In the Mediterranean countries we can find plants, fruits and trees from which essential oils are extracted, among which we can highlight the bitter and sweet oranges, lavender, rosemary, aniseed, clove, lavender, mint, fennel, laurel and lemon.

Nature in Spain is described by the aroma of its fields, by the Mediterranean Sea and by the sunlight. All this encourages the growth and cultivation of certain fruits, plants and trees, from which the essential oils that characterize the Mediterranean basin are extracted.

The history of essential oils has accompanied the development of humanity. It was in the Middle Ages when these oils with healing properties were used, then they were exported to Egypt, China, Greece and Rome. The Arabs were also very interested in these products and their great properties and, finally, in the Europe of the Modern Age until the present time.

The human being has used these oils in both religious and magical rites, as well as a form of cure and remedy. In ancient Egypt, their antiseptic properties were used in the embalming of mummies. Similarly, they were used in fumigation. The use of aromatherapy for medicine, cosmetics and even religion then began. Each civilization gave its particular use to essential oils that were very important for the population. They have been used as healing elements, also to drive away bad spirits. The Chinese population, later, made use of aromatic plants and herbs, although they were used mainly for medicinal purposes. The Greeks included aromatic plants in their baths, food, hygiene, medicine and religious rites. Of note is the insatiable search that there was at that time for finding the elixir of eternal youth, therapies in which essential oils were key. Later, the Romans bet on hygiene to achieve better health, so they turned to aromatherapy and its benefits. Then the Arabs were who specialized in the extraction of essential oils from plants. They even improved the distillation of essential oils by introducing refrigeration throughout the production process.

Already in the 20th century, the Frenchman Gattefossé introduced aromatherapy to natural medicine [4]. He discovered the great healing properties of lavender essential oil due to a serious incident in his laboratory. It is well known that in many cultures, plants have been used to cure illnesses and to calm wounds. By studying ancient knowledge, much progress has been made in natural medicine and in the therapy of the body and mind. New vital ingredients are still found in nature today, so that traditions that have been practiced for centuries are still being reinforced.

Essential oils are currently being rediscovered and therefore new applications are being found to them. The main use of these is, nowadays, in perfumery. There is a wide range of essential oils, among which the following ones can be highlighted:

- The essential oil of sweet orange.
- The essential oil of bitter orange.
- The essential oil of clove.
- The essential oil of lemon.
- The essential oil of rosemary.
- The essential oil of lavender.
- The essential oil of laurel.
- The essential oil of mint.

The most romantic vision of aromatherapy and essential oils tells how they define the life force of plants and this force is equivalent to the human spirit. Numerous companies research on aromatherapy, since it has become one of the most natural techniques to calm the ailments of the human organism; it serves to cure and to prevent diseases and disorders. Before explaining the benefits and properties of the main essential oils of the Mediterranean basin, it is necessary to define what an essential oil is. This is a product obtained from a vegetable raw material, either by distillation with water or steam or from the outer layer of citrus fruit through a mechanical process [4]. Distillation can also be dry, so the only way to extract this essential oil is by a physical means. Most essential oils are obtained by hydrodistillation. Essential oils are very aromatic, volatile, light and non-greasy chemicals; hence, all skin types and people of any age tolerate them. Essential oils can be obtained from different parts of plants: from the flowers, the leaves, the root, the wood, the resin, the bark of the fruit and even from the whole tree. There are plants from which more than one essential oil can be extracted. Each oil extracted has a different name depending on the area of the plant from which it was obtained. With regard to the bitter orange of Seville, for example, the essential oil of orange blossom is extracted by distillation of its flowers. Furthermore, the essential oil of Petitgrain, the essential oil of the orange tree, as well as the essential oil of Neroli are obtained by distillation of the fruit and from the peel of the orange.

In Seville, the fruit par excellence is the orange from which this pleasant and beneficial essential oil is obtained. This oil relaxes the mind and spirit, improves circulation and nourishes the skin. The essential oil of the orange tree is extracted from the peel of the fruit and some of its most well-known benefits are:

- For the skin, it improves oily skin and acne. It also helps the disappearance of blemishes.
- For coughs and flu, the inhaling of steam from this essential oil improves congestion.
- For tension and nervousness, it helps to calm the nervous system and to remove stress.
- For fluid retention, a gently massage with orange tree essential oil improves the circulation of body fluids.

There are numerous applications for the essential oil of the Seville orange because it is an oil that blends very well with almost all other essential oils, thus increasing its possible uses. It is very common to find it in creams, gels, soaps and as a formulation of aesthetic treatments, since it largely helps to the nutrition of the skin in cosmetics. As perfumes additive it is, likewise, very much in demand.

Rosemary should also be highlighted as an essential oil, as it is key in the Mediterranean basin, as well as lavender. The essential oil of rosemary is extracted from its leaves and flowers. With regard to its properties, it should be noted the benefit that this oil gives to the brain function and the relaxation it provides to the muscles, also improving blood circulation, relieving pain in the joints, raising the mood and helping hair growth by stimulating it, among other benefits.

Finally, a special mention must be dedicated to lavender. Its aroma has as many benefits as its essential oil, which has healing and relaxing properties. It is one of the most demanded scents and oils in the market due to its versatility and polyvalence. Those who are starting to get into the world of aromatherapy start with lavender and its essential oil. The good reputation of lavender essential oil as a natural relaxant goes back to ancient times, when it was already used for aromatic baths, to calm nervousness and for insomnia. The skin also receives great improvements from the use of lavender because this essential oil is both revitalizing and soothing.

3. NARINGIN

Naringin is a yellow powdered flavonoid responsible for the bitterness of citrus fruits such as the bitter orange of Seville [1], grapefruit, lemon... so it is used in soft drinks and in bakery and confectionery products. Like the rest of the flavonoids, it has antioxidant, anti-inflammatory and antimicrobial properties. In addition to protecting the heart and its rhythm, this bioflavonoid promotes thermogenesis. For all these reasons, naringin is incorporated into supplement formulations in certain weight loss diets. In the food and animal welfare industry, this flavonoid is considered essential for enhancing the taste of food products and is found primarily in grapefruit and bitter orange peel, but is also present in lemon and sweet orange peel [5] and in the pulp and juice of these. Naringin gives citrus fruits a bitter taste and the amount of naringin in citrus depends on the fruit in question. It is more abundant, generally, in immature fruits. This flavonoid is used, on the other hand, in perfumery, cosmetics and to give flavor to sweets, apart from the uses already mentioned. Naringin helps to improve the lipid profile, reducing the levels of LDL cholesterol and triacylglycerols, while increasing the levels of HDL cholesterol and thus protecting against atherosclerosis. It has similar properties to insulin, such as lipid reduction. All this indicates that it is a very important product, and very much in demand, in the market.

Naringin improves the metabolic syndrome, which is a disease in which there is an insulin disorder in the metabolism, so that glucose is altered and can cause obesity or hypertension. In the same way, it is related to the development of type two diabetes resulting in cardiovascular disease. Naringine helps to correct these disorders and considerably improves this type of disease. Naringine is also a powerful antioxidant and has anti-inflammatory properties.

Flavonoids, including naringin, are bioactive natural products, which have been used as functional foods for years. The citrus fruit with the highest concentration of naringin is grapefruit. These fruits also contain fiber and pectin, which help to reduce the level of cholesterol in the blood and to improve gut motility. On the other hand, it is a great source of magnesium, vitamin B6, lycopene, potassium, calcium, folic acid and iron, among other nutritional contributions.

Some of the potential uses of the citrus peel in the food industry are: infusions, citrus fruit chips, candies, candied peel, jams and all the formats in which peel can be present, as mentioned previously.

4. Bioflavonoids against COVID-19

Citrus bioflavonoids are a type of flavonoid extracted from citrus fruits. These flavonoids, like all other flavonoids, provide various health benefits to animals and humans. One of the bioflavonoid mixtures available in the marked is the Citrus Bioflavonoid Complex (CBC) [6]. It has been proven that it helps to prevent COVID-19, especially because of its vitamin C content, as well as to fight this disease by strengthening the human immune system.

Citrus Bioflavonoid Complex is tailor-made, so it has become a great attraction for different industries. They are designed and manufactured with care and attention to meet all industry customer requirements and to provide the desired function within health. Citrus bioflavonoids have pharmacological, biochemical applications and are used for therapeutic purposes.

CBC contain as active principle bioflavonoids with numerous properties, the most significant of which are:

- Cardiovascular protection.
- Regulation of peripheral circulatory disorders.
- Improvement of cholesterol.
- Weight control.
- Antioxidant power.
- Sun protection.
- Richness in vitamin C.

Citrus fruits are recognized as a source of significant amounts of vitamin C, but they are also an essential source of natural bioflavonoids. Each citrus fruit contains different types of flavonoids, which are found in different amounts and concentrations. The citrus fruits mentioned above contain high concentrations of the following bioflavonoids:

- Diosmin.
- Hesperidin.
- Neohesperidin.
- Rutin.
- Naringin.
- Quercetin.
- Nobiletin.
- Narirutin.
- Tangerine.
- Nobiletin.

The versatility of these typse of bioflavonoid makes them a perfect complement for different applications such as food, dietary supplements, beverages, personal care products and animal feed formulations. The functions of each CBC will depend on the individual characteristics of each flavonoid, as well as the synergies created by forming the blend.

Another way to refer to citrus bioflavonoids may be vitamin P, citrus flavonoids or citroflavonoids. These nomenclatures should be kept in mind as they can be found in the food and pharmaceutical label if we are looking to purchase this product. This type of bioflavonoid stands out for being a non-nitrogenous plant substance present mainly in citrus fruits. It is present in almost all citrus fruits, however, the content will vary greatly from one fruit to another, as well as, within the fruit, from one part to another (peel, pulp, seeds). Most of the bioflavonoid are in the white part of the inside of the fruit skin. Its ability to help improve the consumer's health, so that it is highly demanded and widely used at present.

The beneficial effects associated with citrus bioflavonoids are very numerous. Among these:

- They strengthen the walls of blood vessels, as well as provide them with flexibility. This makes it possible to fight against cardiovascular diseases and diseases of the lymphatic system. For example, they are beneficial for hemorrhoids, varicose ulcers, nosebleeds, lymphedema following breast cancer surgery, chronic venous insufficiency, appearance of bruises...
- They provide a great antioxidant action, necessary to prevent the negative effects of free radicals. On the other hand, this contribution helps prevent oxidation of blood vessels and promotes the absorption of vitamin C by the human body.

The antioxidant and anti-inflammatory properties of citrus bioflavonoids stand out because they favor the expulsion of free radicals, thus preventing and reducing oxidative stress.

Mention should also be made to their synergistic power. Vitamin C is usually extracted from the same citrus fruits as the bioflavonoids, so both have antioxidant activity separately and this means that this activity is considerably increased when they act together.

In addition, they stimulate the immune system. Numerous studies have already shown that citroflavonoids contribute to maintaining and improving the state and functions of the immune system. Among these, the ability to provide a correct and rapid immune response by the human body to a possible disease or virus is now a reality. This is currently being demonstrated with COVID-19, as the citroflavonoids, with their vitamin C supply, are helping to avoid contracting the disease, apart from fighting it.

Healthy circulation is another great advantage of citrus bioflavonoids that provides general health benefits for the circulatory system, as it prevents and treats various vascular diseases, and improves blood circulation.

The anti-allergic potential of citroflavonoids should also be highlighted. Diseases caused by allergies are usually associated with an excess of oxidative stress in the human body, and citrus bioflavonoids help stabilize reactive oxygen species, eliminating free radicals from the body.

Capillary resistance is another of the contributions that bioflavonoids give to the industry. It has been shown that certain citrus bioflavonoids can increase capillary resistance and strengthen the walls of blood capillaries. This is directly related to their ability to protect collagen.

Finally, the benefits of citroflavonoids in people with high cholesterol and obesity should be taken into account. They have the ability to regulate blood cholesterol levels and, therefore, produce beneficial effects on human obesity.

CONCLUSION

There are compounds in nature, such as the flavonoids and essential oils in bitter oranges of Seville, that have been used throughout history without really knowing why, including healing, relaxing and religious applications. With the development of knowledge through research, the actual wide benefits of these compounds have been known and demonstrated. Therefore, the bitter orange of Seville is not only a fruit that grows in its trees but it should be regarded as a source of different compounds that are nowadays prime products for both health and nutrition.

ACKNOWLEDGMENTS

The authors thank Susana Alés Álvarez for her research on flavonoids and insightful suggestions for the writing of this chapter.

REFERENCES

[1] Feng, C. H., García-Martín, J. F., Broncano Lavado, M., López-Barrera, M. D. C., Álvarez-Mateos, P. (2020). Evaluation of different solvents on flavonoids extraction efficiency from sweet oranges and ripe and immature Seville oranges. *Int. J. Food Sci. Technol.*, 55(9): 3123-3134.

[2] García, J. F., Olmo, M., García, J. M. (2016). Decay incidence and quality of different citrus varieties after postharvest heat treatment at laboratory and industrial scale. *Postharvest Biol. Technol.*, 118: 96-102.

[3] García-Martín, J. F., Olmo, M., & García, J. M. (2018). Effect of ozone treatment on postharvest disease and quality of different citrus varieties at laboratory and at industrial facility. *Postharvest Biol. Technol.*, 137: 77-85.

[4] Bordas, S. A. (2020). *Aceites esenciales: una gota de vida.* Accessed on 9 August, 2021. https://www.bordas-sa.com/aceites-esenciales-gota-vida/.

[5] Raffoul Orozco, A. K., Ávila González, A. E., Cancela Carral, J. M. (2018). Effects of naringin in combination with exercise on clinical outcomes: a Systematic Review. *Rev. Esp. Nutr. Hum. Diet.*, 22(1):21-30.

[6] Bordas, S. A. (2020). *Citrus bioflavonoids complex.* Accessed on 3 August, 2021. https://www.bordas-sa.com/citrus-bioflavonoid-complex/.

In: The Book of Flavonoids
Editors: Chao-Hui Feng et al.
ISBN: 978-1-68507-376-3

Chapter 12

FLAVONOIDS AS A STARTING POINT FOR THERAPEUTICS AGAINST COVID-19: CURRENT STATE-OF-THE ART RESEARCH ADVANCES

***Chao-Hui Feng*[1,2,*], *Chiko Otani*[2,3]**
and *Juan Francisco García Martín*[4]
[1]School of Regional Innovation and Social Design Engineering, Faculty of Engineering, Kitami Institute of Technology, Hokkaido, Japan
[2]RIKEN Centre for Advanced Photonics, Sendai, Japan
[3]Department of Physics, Tohoku University, Sendai, Miyagi, Japan
[4]Departamento de Ingeniería Química, Facultad de Química, Universidad de Sevilla, Seville, Spain

* Corresponding Author's E mail. feng.chaohui@mail.kitami-it.ac.jp.

Abstract

A large amount of biologically active compounds, including phenolic acids and flavonoids, are present in the citrus fruits' peel. If these compounds can be fully extracted from waste citrus peels and utilized, this could not only transform them into value-added by-products and cut the economic costs, but also reduce the environmental problem of their disposal. Flavonoids, which possess anti-cancer, antimicrobial, antioxidant, and anti-allergic properties, have been extensively applied in the pharmaceutical industry and skincare products. They have also been reported to potentially inhibit coronavirus. In this chapter, the properties of flavonoids will be first discussed. The bioavailability of flavonoids will also be covered. Following this, their potential capacity to inhibit COVID-19 will be illustrated. This chapter offers detailed information from recent studies focused on flavonoids and their extraction and widespread application in the food and pharmaceutical fields. As a starting point for therapeutics against COVID-19, the intake of certain amount of citrus fruits or their extracts could contribute to the prevention of COVID-19.

Keywords: citrus, COVID-19, flavonoids, hesperidin, viral receptor angiotensin-converting enzyme

Introduction

Flavonoids are widely presented in fruits, vegetables, cereals and crops and are one of the most important groups of dietary phenolics. They consist of two aromatic rings, linked via a 3-carbon bridge and usually in a heterocyclic ring form. Six distinct subclasses, i.e., flavonols, flavanols, flavones, flavanones, isoflavones and anthocyanidins are generated from the alteration in the heterocyclic ring's substitution pattern (Balasundram et al., 2006). Flavonoids have antioxidant, anticancer, anti-inflammatory, antimicrobial, and antiproliferative properties. (Feng et al., 2020) and at least 300 different flavonoid compounds are usually consumed via human diet (Bondonno et al., 2020). Hereto, a great attention has been paid due to their positive effects on health, especially to the potential therapeutics against Coronavirus disease 2019 (COVID-19).

FLAVONOIDS IN THE FOODSTUFFS

In most edible plants, flavonoids are present as *O*-glycosides with sugars such as glucose and/or rhamnose linked to the phenolic hydroxyl group. Flavonols, one subclass from flavonoids, are commonly present in whole organs of vegetables, fruits, grains and are abundant in green leaves (Terahara, 2015). Rutin, isoquercitrin, and quercitrin are the major flavonoids and are largely present in cranberry (*Vaccinium macrocarpon*), apple (*Malus sylvestris*), kale (*Brassica oleacea*), leaves of sweet potato (*Ipomoea batatas*), lettuce (*Lacuca sativa*), and red pepper (*Capsicum annuum*). Onion (*Allium cepa*) contains 4'-*O*-glucosides and 3,4'-di-*O*-glucosides of isorhamnetin and quercetin whilst spinach (*Spinacia oleracea*) contains rutin, spinacetin glycosides and patuletin glycosides (Terahara, 2015).

As for flavones, their main form in edible plants is glycosides of apigenin, luteolin, and diosmetin and can be mainly found in parsley (*Petroselinum crispum*), celery (*Apium graveolens*), and sorghum (*Sorghum bicolor*). The flavone apigenin has demonstrated a positive action against to the cancer cells *in vitro* (Shankar et al., 2017) and can enhance the inflammatory response according to studies with flavones on animals (Nicholas et al., 2007). Green pepper, paprika, lettuce, red pepper (*Capsicum annuum*), some herbs and even citrus fruits contain the glycosides of apigenin and luteolin. Vitexin, isovitexin, orientin, and isoorientin, as parts of C-glucosyl flavones, commonly occur in citrus fruits, peels, as well as cereals (Terahara, 2015).

It is known that 50% of flavonoid content of fruits and vegetables may decrease as a result of food preparation or processing such as water leaching or removal of the inedible parts but rich in flavonoids. The concentration of flavonoids is higher in colourful nutritional components.

It should be noted that the white pulp between the fruit and peel of citrus contains more flavonoids than the peel and red pulp parts.

FLAVONOID BIOAVAILABILITY

Bioavailability is defined as the part of the digested flavonoids that is uptaken and metabolized through regular pathways (Kamiloglu et al., 2020). It is an important parameter when the function of flavonoids in health of humans is investigated. The symptoms of flavonoids deficiency in the human being are easy bleeding, frequent bruising which takes long time to recovery, easy swelling after injured and so on. The deficiency of flavonoids may probably result in immunologic weakness, leading to susceptibility to cold or other infections. According to people's eating and cultural habits, the intake of flavonoids varies from 70 to 170 mg/day. Five to ten percent of the total intake of flavonoids can be absorbed in the small intestine with the structures of monomeric and dimeric. The residual part of flavonoids that reaches to colon is later metabolised to compounds with different physiological significance by the enzymatic action of the gut microbiota (Kamiloglu et al., 2020). There are two transformations (phases I & II) for ingested flavonoids in the human body. Phase I involves oxidation, reduction and hydrolysis whilst phase II is related to conjugation reactions where different metabolites (e.g., methyl, glucuronic and sulphate derivatives) are formed (Grootaert et al., 2015). Compared to phase I, phase II occurs more frequently and takes place in the liver and mostly in the intestine (Kamiloglu et al., 2020).

It is well described in the work of Kamiloglu et al. (2020) that both macronutrients (e.g., carbohydrates, lipid, proteins) and micronutrients (e.g., vitamins, minerals, etc.) can greatly improve the bio-accessibility and bioavailability of flavonoids in the gastrointestinal tract (Jakobek, 2015). For example, the bioavailability of flavan-3-ols is positively affected by the presence of sucrose in chocolate (Neilson et al., 2009).

EFFECTS OF FLAVONOID ON COVID-19

Coronaviruses are single-stranded RNA viruses (Lai et al., 1997) with large, enveloped and positive senses that can infect both animals and humans. Coronaviruses negatively influence the lifestyle of humans. Up to date, the human coronaviruses are regarded as one of the rapidly evolving viruses derived from their characteristic high genomic nucleotide replacement rates and recombination (de Wit et al., 2016). The common symptoms of COVID are coughing, high fever (over 38°C), chills, convulsions, headaches, dizziness, progressive radiographic changes of the chest, lymphopenia and so on (Jo et al., 2020).

There is a high demand to investigate new approaches or to exploit pharmaceutics to cure these diseases. Flavonoids in orange peel have antioxidant, anti-cancer, antiviral, anti-inflammatory, anti-allergic, antimicrobial and antiproliferative properties (Feng et al., 2020), and have been reported to potentially inhibit coronaviruses (Jo et al., 2020; Li et al., 2020). Figure 1 illustrates the main steps of Severe Acute Respiratory Syndrome Coronavirus (SARS-CoV)-2 virus invasion of the lung. By binding the spike glycoprotein of the virus with its receptor (ACE2) on the cell membranes, the viral particle is internalised in a vesicle, rendering the genomic RNA to be released into the cytoplasm (Figure 2). After replication, the sub-genomic RNAs produced via transcription are translated into structural protein. After being assembled with positive-sense RNA, new copies of the virus are generated and spread into environment, infecting other cells and organs in the body and forming a chain expansion. In order to halt the invasion of SARS-CoV-2, substances with a possible beneficial effect in coronavirus infection may act in various stages: 1) preventing the binding of the virus to the normal cells, 2) inhibiting viral replication by blocking, for instance, RNA polymerase, proteases or new particle assembly. It has been discovered that the molecule of hesperidin has a chemical physical structure that is suitable for binding to the key protein in the functioning of the SARS-CoV-2 virus (Bellavite & Donzelli, 2020). After testing 1066 natural substances with potential antiviral effect, hesperidin was reported to be the most suitable compound to bind to the "spike," resulting in the disruption of the interaction of ACE2 with receptor binding domain (RBD) complex (Wu et al., 2020). The most recent available literature also shows that flavonoids (e.g., hesperidin and rutin) possess a better binding affinity to the main protease of COVID-19 than nelfinavir and so they could be regarded as the starting point for therapeutics against COVID-19 (Adem et al., 2020).

The protease "3Clpro" (also called "Mpro") is the main enzyme that enables the processing of the first proteins that are transferred from the viral genome (pp1a and pp1ab) into functional proteins in the host cell. Thus, the protease "3Clpro" is regarded as a second theoretical site of low energy binding of SARS-CoV-2 with hesperidin (Wu et al., 2020). For this reason, this enzyme is currently used in the formulation of new chemical antiviral drugs. From the virtual screening among 7173 purchasable drugs, the active sites on chain A and chain B, derived from a catalytically active dimeric model, were screened by AutoDock Vina (an open-source program for doing molecular docking) independently. Hesperidin is reported to be the second most efficient for binding to chain A, with a free energy of -10.1 kcal/mol. The energy for binding to chain B is reported to be -8.3 kcal/mol (Chen et al., 2020). Besides, a further research study has proved that hesperidin can strongly bind to SARS-CoV-2 main protease as well as to the viral receptor angiotensin-converting enzyme 2 (ACE-2), compared with other natural molecules (Joshi et al., 2020). Jo et al., (2020) investigated the effects of flavonoids on inhibiting SARS-CoV 3CL protease and discovered that herbacetin, rhoifolin and pectolinarin efficiently blocked the enzymatic activity of SARS-CoV "3CLpro".

A number of researches regarding *in vitro* and *in vivo* studies have demonstrated the antioxidant activity of hesperidin, which not only limits its free radical scavenger activity but also induces enhanced cellular defence against oxidative stress and reduces inflammation makers via the ERK/Nrf2 signalling pathway (Roohbakhsh et al., 2015). Although the reasons why the severity COVID-19 disease varies from different ages are unknown, it may be probably explained by the high antioxidant capacity of children and the redox imbalance of elderly subjects with low antioxidant capacity. It is assumed that the intracellular redox environment changes the presentation of antigens and the expression or function of ACE2 (Delgado-Roche and Mesta, 2020; Keles, 2020).

Although flavonoids possess potential capacity to inhibit COVID-19 and provide good information for pharmaceutical industry for making medicine against COVID-19, the dose-response relationship between intake and disease still merits to be paid attention. Taking vitamin C for example, intravenous high doses of vitamin C has proved a complementary therapeutic effect on COVID-19 (Cheng, 2020) and clinical trials are currently undergoing (Liu et al., 2020). However, high doses of ascorbate may also be detrimental to human health (Lehene et al., 2020). Therefore, many preliminary studies and deep assays in actual conditions are still needed before an industrial production. A commercial therapeutic utilization of medicines containing flavonoids against COVID-19 needs to be developed.

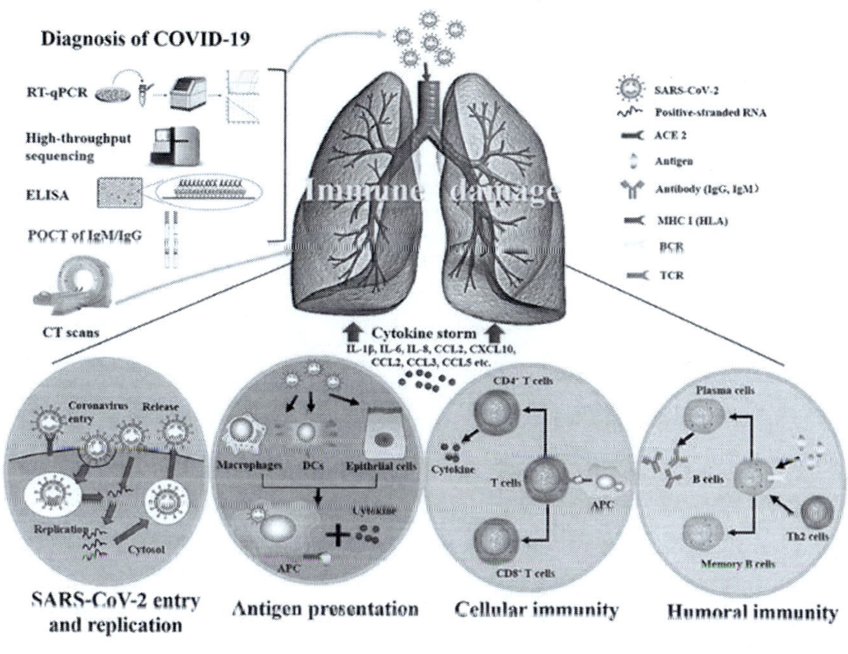

Figure 1. Invasion of SARS-CoV-2 virus into lung (from Li et al., 2020).

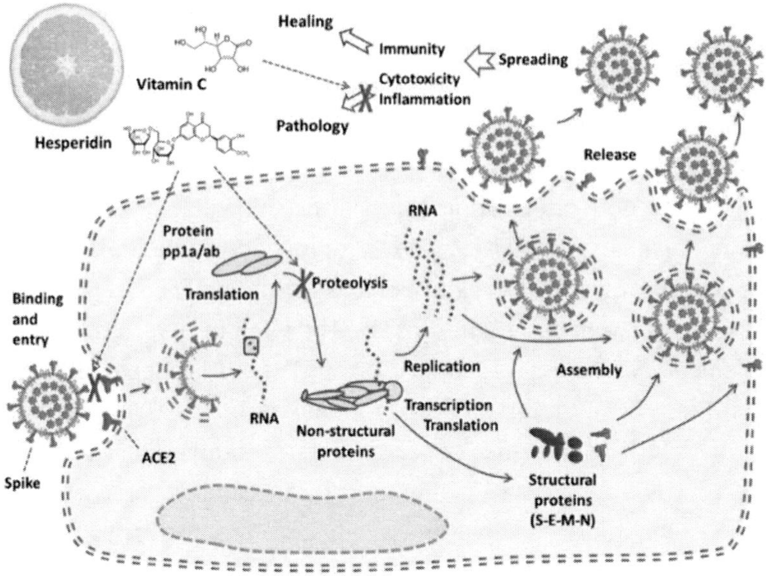

Figure 2. Cellular cycle of invasion of SARS-CoV-2 virus and the inhibition (indicated by "X") of virus-induced cellular and systemic pathology by hesperidin and vitamin C (from Bellavite & Donzelli, 2020).

Conclusion

This chapter overviews some recent discoveries and highlights the effects of flavonoids on COVID-19. With antioxidant, anticancer, anti-inflammatory and antimicrobial properties, flavonoids have been widely applied to the pharmaceutical industry. Recently, their effects on COVID-19 have aroused scientific interest: the suitable structure of hesperidin to bind to the key proteins in the functioning of SARS-CoV-2 virus. There is still a long way to confirm their effects on COVID-19, such as the effective dose, the flavonoids' bioavailability, *in vivo* metabolization and so on. Future studies regarding preclinical, epidemiological and clinical aspects are requested to corroborate the hypothesis that a certain intake of citrus fruits and their extracts could effectively contribute to the prevention of COVID-19.

ACKNOWLEDGMENTS

Chao-Hui Feng wishes to acknowledge the financial support of her research work under JSPS Grant-in-Aid for Early-Career Scientists (20K15477), Leading Initiative for Excellent Young Researchers (LEADER) from Ministry of Education, Culture, Sports, Science and Technology (MEXT) (2020L0277), FY 2021 President's Discretionary Grants, funded by Kitami Institute of Technology, Special Postdoctoral Researcher Program, and Exploratory "Collaboration Seed" fund at Riken.

REFERENCES

Adem, S., Eyupoglu, V., Sarfraz, I., Rasul, A. & Ali, M. (2020). Identification of Potent COVID-19 Main Protease (Mpro) Inhibitors from Natural Polyphenols: An in Silico Strategy Unveils a Hope against CORONA. *Preprints,* 2020030333.

Balasundram, N., Sundram, K., & Samman, S. (2006). Phenolic compounds in plants and agri-industrial by-products: Antioxidant activity, occurrence, and potential uses. *Food Chemistry,* 99, 191-203.

Bellavite, P. & Donzelli, A. (2020). Hesperidin and SARS-CoV-2: new light on the healthy function of citrus fruits. *Antioxidants,* 9, 742.

Bondonno, N. P., Lewis, J. R., Blekkenhorst, L. C., Bondonno, C. P., Shin, J. H. C., Croft, D. K., Woodman, R. J., Wong, G., Lim, W. H., Gopinath, B., Flood, V. M., Russell, J., Mitchell, P. & Hodgson, J. M. (2020). Association of flavonoids and flavonoid-rich foods with all cause mortality: the blue mountains eye study. *Clinical Nutrition,* 39, 141-150.

Chen, Y. W., Yiu, C. B., Wong, K. Y. (2020). Prediction of the SARS-CoV-2 (2019-nCoV) 3C-like protease (3CL (pro)) structure: Virtual screening reveals velpatasvir, ledipasvir, and other drug repurposing candidates. *F1000Research,* 9, 129.

Cheng, R. Z. (2020). Can early and high intravenous dose of vitamin C prevent and treat coronavirus disease 2019 (COVID-19)? *Medicine in Drug Discovery,* 5, 100028.

de Wit, E., van Doremalen, N., Falzarano, D. & Munster, V. J. (2016). SARS and MERS: recent insights into emerging coronaviruses. *Nature Review Microbiology,* 14, 523-534.

Delgado-Roche, L. & Mesta, F. (2020). Oxidative stress as key player in severe acute respiratory syndrome coronavirus (SARS-CoV) infection. *Archives of Medical Research,* 51, 384-387.

Feng, C. H., García Martín, J. F., Lavado, M. B., López-Barrera, M. C. & Álvarez-Mateos, P. (2020). Evaluation of different solvents on flavonoids extraction efficiency from sweet oranges and ripe and immature Seville oranges. *International Journal of Food Science and Technology*, 55, 3123-3134.

Grootaert, C., Kamiloglu, S., Capanoglu, E. & Van Camp, J. (2015). Cell systems to investigate the impact of polyphenols on cardiovascular health. *Nutrients,* 7, 9229-9255.

Jakobek, L. (2015). Interactions of polyphenols with carbohydrates, lipids and proteins. *Food Chemistry,* 175, 556-567.

Jo, S. Kim, S. Shin, D. H. & Kim, M. -S. (2020). Inhibition of SARS-CoV 3CL protease by flavonoids. *Journal of Enzyme Inhibition and Medicinal Chemistry,* 35, 145-151.

Joshi, R. S., Jagdale, S. S., Bansode, S. B., Shankar, S. S., Tellis, M. B., Pandya, V. K., Chugh, A., Giri, A. P. & Kulkarni, M. J. (2020). Discovery of potential multi-target-directed ligands by targeting host-specific SARS-CoV-2 structurally conserved main protease. *Journal of Biomolecular Structure and Dynamics,* 39, 3099-3114.

Kamiloglu, S., Tomas, M., Ozdal, T. & Capanoglu, E. (2020). Effect of food matrix on the content and bioavailability of flavonoids. *Trends in Food Science & Technology,* in press.

Keles, E. S. (2020). Mild SARS-CoV-2 infections in children might be based on evolutionary biology and linked with host reactive oxidative stress and antioxidant capabilities. *New Microbes and New Infections,* 36, 100723.

Lai, M. M. C. & Cavanagh, D. (1997). The molecular biology of coronaviruses. *Advances in Virus Research,* 48, 1-100.

Lehene, M., Fischer-Fodor, E., Scurtu, F., Hădade, N. D., Gal, E., Mot, A. C., Matei, A. & Silaghi-Dumitrescu, R. (2020). Excess Ascorbate is a Chemical Stress Agent against Proteins and Cells. *Pharmaceuticals*, 13, 107.

Li, X. W., Geng, M. M., Peng, Y. Z., Meng, L. S. & Lu, S. M (2020). Molecular immune pathogenesis and diagnosis of COVID-19. *Journal of Pharmaceutical Analysis,* 10, 102-108.

Liu, F., Zhu, Y., Zhang, J., Li, Y. & Peng, Z. (2020). Intravenous high-dose vitamin C for the treatment of severe COVID-19: Study protocol for a multicentre randomised controlled trial. *BMJ Open,* 10, e039519.

Neilson, A. P., George, J. C., Janle, E. M., Mattes, R. D., Rudolph, R., Matusheski, N. V. & Ferruzzi, M. G. (2009). Influence of chocolate matrix composition on cocoa flavan-3-ol bioaccessibility *in vitro* and bioavailability in humans. *Journal of Agricultural and Food Chemistry,* 57, 9418-9426.

Nicholas, C., Batra, S., Vargo, M. A., Voss, O. H., Gavrilin, M. A., Wewers, M. D., Guttridge, D. C., Grotewold, E. & Doseff, A. I. (2007). Apigenin blocks lipopolysaccharide-induced lethality *in vivo* and proinflammatory cytokines expression by inactivating NF-kappaB through the suppression of p65 phosphorylation. *The Journal of Immunology,* 179, 7121-7127.

Roohbakhsh, A., Parhiz, H., Soltani, F., Rezaee, R. & Iranshahi, M. (2015). Molecular mechanisms behind the biological effects of hesperidin and hesperetin for the prevention of cancer and cardiovascular diseases. *Life Sciences,* 124, 64-74.

Shankar, E., Goel, A. Gupta, K. & Gupta, S. (2017). Plant flavone apigenin: an emerging anticancer agent. *Current Pharmacology Reports,* 3, 423-446.

Terahara, N. (2015). Flavonoids in Foods: a review. *Natural Product Communications,* 10, 521-528.

Wu, C. R., Liu, Y., Yang, Y. Y., Zhang, P., Zhong, W., Wang, Y., Wang, Q. Q., Xu, Y., Li, M. X., Li, X. Z., Zheng, M. Z., Chen, L. X. & Li, H. (2020). Analysis of therapeutic targets for SARS-CoV-2 and discovery of potential drugs by computational methods. *Acta Pharmaceutica Sinica B,* 10, 766-788.

ABOUT THE EDITORS

Chao-Hui Feng has her expertise in estimating the meat quality by using hyperspectral imaging and terahertz spectroscopy, modifying natural hog casing by using surfactant solution and lactic acid, applying innovative cooling method (immersion vacuum cooling) to meat products. Up to now, she has 27 SCI papers accepted to be published in the peer-reviewed international journals (newest total impact factor: 190.812), has been invited as the peer-reviewer for several mainstream journals in food science field and has been invited as the Guest Associate Editor for Journal Frontiers in Plant Science (impact factor: 5.753), Sustainability for special issue "Innovative Production Technology to Meet Food and Fuel Needs" (impact factor: 3.251), the Editor for the upcoming Nova publication "The Book of Flavonoids", obtained 39 awards, 12 patents and secured 16 research fundings as a principal investigator. Recently, she was selected as a research representative from the early-career researchers and receive an invitation for contributing to a viewpoint article in Nature Reviews Physics (2-year Impact Factor: 31.068), it is a series of Nature.

Juan Francisco García Martín was born in Linares (Spain) in 1979. He earned a B.Sc. in Chemistry in 2001 and afterwards he obtained a PhD *summa cum laude* (European Doctorate Mention) in 2007, both from the University of Jaén (Spain). Shortly after, he was a postdoctoral researcher at University of Granada (Spain), University College Dublin (Ireland) and Instituto de la Grasa (Spanish National Research Council), and meanwhile did several research stays at Institute of Catalysis and Petrochemistry (Spanish National Research Council) and the Universities Paul Sabatier Toulouse III (France), Claude Bernard Lyon I (France) and Geisenheim (Germany). He was also a lecturer at the Universities of Jaén (Spain), Granada (Spain) and Málaga (Spain), and from 2016 he is Associate Professor at the Department of Chemical Engineering of the University of Seville (Spain).

INDEX

#

A

B

C

G

H

P

Q

R

S

T

U

V

W

X

Y

Z